Astronomy:
The Structure of the Universe

Astronomy: The Structure of the Universe

William J. Kaufmann, III

Astrophysics-Relativity Group
California Institute of Technology
and
Department of Astronomy
University of California at Los Angeles

Macmillan Publishing Co., Inc.
New York

Collier Macmillan Publishers
London

Cover Photo

THE EXPLODING GALAXY NGC 5128 (Centaurus A)

NGC 5128 is the visible galaxy associated with the strong radio source called Centaurus A. Centaurus A is one of the most powerful radio sources in the sky. It was first observed by Dr. J. G. Bolton in 1947 when radio astronomers first began looking for discrete sources of radio noise. Subsequent observations revealed that the radio structure of Centaurus A is extremely complex. In the early 1970s, it was discovered that this unusual galaxy is also emitting X rays. As an X-ray source, this object is called 3U1322-42.

NGC 5128 has a visual magnitude of 7 and is located at right ascension 13h 22.5m and declination $-42°$ 45' (1950 coordinates) in the constellation of Centaurus. It is believed that NGC 5128 is about 16 million light years away and is receding from us at a speed of approximately 300 miles per second. This remarkable color photograph was taken at the Cerro Tololo Inter-American Observatory in Chile.

© 1975 by the Association of Universities for Research in Astronomy, Inc.

Copyright © 1977, William J. Kaufmann, III, and Kaufmann Industries, Inc.

Printed in the United States of America

All rights reserved. No part of this book may be reproduced or transmitted in any form or by any means, electronic or mechanical, including photocopying, recording, or any information storage and retrieval system, without permission in writing from the Publisher.

Macmillan Publishing Co., Inc.
866 Third Avenue, New York, New York 10022

Collier Macmillan Canada, Ltd.

Library of Congress Cataloging in Publication Data

Kaufmann, William J
    Astronomy : the structure of the universe.

    Includes index.
    1. Astronomy.
QB45.K38   520   75-31505
ISBN 0-02-362130-3

Printing: 1 2 3 4 5 6 7 8   Year: 7 8 9 0 1 2 3

*To my son
William, IV,
with love*

Preface

Astronomy has the dual distinction of being both the oldest science and one of the most fascinating fields of modern research. In the mid-twentieth century, many scientists, both young and old, turned their attention to a variety of astronomical and astrophysical problems. The advent of the space program has made observations above the earth's atmosphere a reality. No longer are astronomers restricted to the optical and radio portions of the electromagnetic spectrum. Observations from astronomical satellites literally have given mankind new eyes with which to view the universe. On the theoretical side, the development of huge high-speed computers has allowed astrophysicists to perform intricate calculations that were unthinkable only a few decades ago. A clear understanding of many of the aspects of stellar evolution is one of the important results of these calculations. In addition, the revival of interest in general relativity, which started more than a decade ago, is beginning to bear fruit. It now appears that the general theory of relativity is extremely relevant to any in-depth understanding of physical reality.

One of the results of the increased activity of astronomers is that more than ever before, astronomy is a dynamic and rapidly changing field of science. What the astronomer does, the problems in which he is interested, and the subjects that fascinate him change almost weekly with every new issue of the *Astrophysical Journal*. Astronomy today is *very* different from what it was only ten or fifteen years ago. For example, the planets have absolutely nothing to do with modern astronomy. Astronomers before World War II spent many hours at their telescopes observing the planets, but this is no longer the case. The study of the objects that orbit our sun has been taken over by geologists, meteorologists, engineers, and other scientists who together have given rise to the new field of *planetary physics*. Similarly, those scientists whose primary interest is our sun also have broken away from the mainstream of modern astronomy.

Modern astronomy begins at distances greater than 1 light year from the earth. Modern astronomy deals with the nature and life cycles of stars, with the properties of galaxies, and with the mysteries of quasars. The modern astronomer finally has arrived at the point where he can ask the most fundamental questions concerning the over-all structure of the universe.

The purpose of this book is to present a clear, vivid, nonmathematical treatment of the frontiers of modern astronomy. As such, it is ideally suited for a one-semester or one-quarter general astronomy course. The book begins with a rather "traditional" approach. The student becomes conversant and familiar with basic concepts. About one third of the way through the book, attention is focused on the properties and evolution of stars. The student hears of the latest developments with regard to pulsars and learns why many astronomers are convinced that black holes must exist. In view of the rising interest in black holes, Chapters 10 and 11 are devoted entirely to relativity theory. These chapters set the tone and mood for the rest of the book. The student realizes that as he looks out into space he is also looking backward in time. From that point, he is aware that he must always think of time as a fourth dimension. The last quarter of the book deals with cosmology. Special emphasis is placed on quasars and the controversies surrounding the redshifts.

Beginning in the fall of 1974, I went on sabbatical leave from the directorship of the Griffith Observatory. One of the primary purposes of this sabbatical was to write a text to meet the need for a quality astronomy book for the general astronomy course. During my sabbatical, I received joint appointments to the faculty of the Department of Physics at the California Institute of Technology and to the Department of Astronomy at UCLA. I am especially grateful to Dr. Kip S. Thorne, who found room for me in the Astrophysics-Relativity Group at Caltech. Working alongside people like Dr. Stephen Hawking and Sir Fred Hoyle at Caltech made 1975 one of the most productive, rewarding, and exciting years of my life. I am also indebted to Dr. George O. Abell, chairman of the Department of Astronomy at UCLA. During the spring quarter, I was pleased to be able to teach the general astronomy course at UCLA, thereby allowing me to use this text in a classroom setting long before it went to the publisher.

With regard to the actual production of this book, it is a pleasure to acknowledge the very fine work of Louise Nelson. Ms. Nelson spent many long evenings typing page after page of illegible handwriting. Her knowledge of astronomy proved to be a valuable asset in catching errors in the first draft. The first draft of this book was then given to Pamela Gentry. Ms. Gentry patiently and carefully scrutinized every word of the manuscript for style and literary content. The readability of the final product is entirely due to Ms. Gentry's gallant efforts. Pamela and Louise made writing this book a totally pleasurable experience.

Finally, no list of acknowledgments would be complete without mentioning Patricia Whitt. Without her dedicated efforts and hard work, along with the assistance she received from her father and his associates, this book *never* would have been written.

William J. Kaufmann, III

Contents

1. Introducing the Universe — 1

   1.1 Prologue — 1
   1.2 The Course of Astronomy — 5
   1.3 Stars and Constellations — 10
   1.4 The Celestial Sphere — 17

2. The Birth of Modern Astronomy — 24

   2.1 Ancient Astronomy and Ptolemy — 24
   2.2 The Copernican Revolution — 31
   2.3 Kepler's Cosmology — 36
   2.4 Newton and Gravitation — 44

3. The Earth in Astronomy — 54

   3.1 The Telling of Time — 54
   3.2 The Seasons — 59
   3.3 Tides and Precession — 63
   3.4 Eclipses — 69

4. The Solar System — 78

   4.1 Inventory of the Solar System — 78
   4.2 Our Moon — 85
   4.3 Mariner and Mars — 92
   4.4 Venus and Mercury — 102
   4.5 The Jovian Planets — 108
   4.6 Interplanetary Matter — 113

5. The Physics of Light — 123

   5.1 Electromagnetic Radiation — 123
   5.2 Atoms and Spectra — 129
   5.3 Radiation Laws — 138
   5.4 Information from Spectra — 143

6. The Astronomer's Tools — 151

- 6.1 Refracting Telescopes — 151
- 6.2 Reflecting Telescopes — 157
- 6.3 Spectroscopy and Photometry — 164
- 6.4 Radio Telescopes — 168
- 6.5 IR, UV, and X-Ray Astronomy — 174

7. The Nature of Stars — 186

- 7.1 Distances and Magnitudes — 186
- 7.2 Stellar Spectra — 190
- 7.3 The Hertzsprung–Russell Diagram — 195
- 7.4 Our Sun—A Typical Star — 204
- 7.5 Binary Stars — 216
- 7.6 Some Unusual Stars — 221

8. Stellar Evolution — 230

- 8.1 The Birth of Stars — 230
- 8.2 Maturity and Old Age — 235
- 8.3 Star Clusters — 239
- 8.4 How Stars Die — 246

9. White Dwarfs and Neutron Stars — 251

- 9.1 White Dwarfs As Dead Stars — 251
- 9.2 The Discovery of Pulsars — 257
- 9.3 Pulsars and Neutron Stars — 265
- 9.4 Black Holes in Space? — 273

10. The Special Theory of Relativity — 280

- 10.1 The Crisis in Classical Physics — 280
- 10.2 The Birth of Special Relativity — 286
- 10.3 Traveling Near the Speed of Light — 298
- 10.4 Do Tachyons Exist? — 305

11. The General Theory of Relativity — 314

- 11.1 Gravitation and Warped Space-Time — 314
- 11.2 Classical Tests — 319
- 11.3 The Schwarzschild Black Hole — 325
- 11.4 The Kerr Solution — 331
- 11.5 Discovering Black Holes — 337

12. The Realm of the Galaxies — 345

    12.1 Our Galaxy — 345
    12.2 The Mystery of the Nebulae — 353
    12.3 Hubble and Galactic Distances — 357
    12.4 Hubble Classification — 365
    12.5 Clusters of Galaxies — 368
    12.6 The Redshift — 373

13. Quasars and Exotic Galaxies — 379

    13.1 The Discovery of Quasars — 379
    13.2 The Redshifts of Quasars — 384
    13.3 Peculiar Galaxies — 394
    13.4 Intrinsic Redshifts and New Ideas — 405

14. Cosmology and the Universe — 418

    14.1 The Expanding Universe — 418
    14.2 The Big-Bang, Steady-State, and "Other" Cosmologies — 423
    14.3 The Young Universe — 431
    14.4 General Relativity and Cosmology — 437
    14.5 The Fate of the Universe — 444

Glossary — 452

Appendix — 471

    A.1 The Planets (Physical Data) — 471
    A.2 The Planets (Orbital Data) — 471
    A.3 Satellites of Planets — 472
    A.4 The Brightest Stars — 473
    A.5 The Nearest Stars — 474
    A.6 The Brightest Galaxies — 475
    A.7 The Nearest Galaxies — 476
    A.8 Messier Catalogue — 476

Index — 479

Monthly Star Charts

Astronomy:
The Structure of the Universe

Introducing the Universe

1.1 Prologue

For thousands of years men have looked up into the star-filled night sky, seen what we see, and asked many of the questions we ask. All of us have experienced being out on the desert or high in the mountains at night and, gazing into space, have wondered about the meaning and order of the universe. These are among the most fundamental questions man can ask, questions that certainly were posed by man before the dawn of recorded history. The attempt to formulate answers—to understand the nature, properties, and origin of what we see in the sky—constitutes the basis of the field of science known as astronomy.

The rising and setting of the sun, the moon going through its phases, the planets wandering among the constellations of the zodiac, and the drama of an eclipse were phenomena that first inspired our ancestors to study astronomy. But few human beings are satisfied with an endless collection of observations. It is not enough to make long lists of what we see in the sky. Rather, we would like to know why things are the way they are. We would like to know how and why the universe works the way it does. If we see an eclipse of the moon on one evening, we are naturally prone to ask when the next eclipse will occur. In short, we would like to have a *cosmology*.

In the broadest sense, a cosmology is a collection of ideas and hypotheses that forms a theoretical framework from which we can understand our universe. Within the framework of a cosmology we are given the feeling that the universe does indeed make sense. Every civilization and every religion ever to come forth on the earth has had a cosmology at the core of its teachings. Such cosmologies, which often require an obscure act of faith, tell us why there is a sun and a moon, how they were created, and what their nature is.

Although it may seem surprising, modern cosmology and the science of astronomy also require an act of faith. In earlier times this faith was primarily of a religious nature, but the modern scientist begins by believing that there is order in the universe. The regular rising and setting of the sun, or the moon going through the same phases every $29\frac{1}{2}$ days, clearly implies order rather than chaos to the rational human mind. Secondly, the modern scientist believes that there are fundamental laws of nature—rules followed by the physical uni-

Figure 1-1. The Planet Mars. By studying the motions of the planets around the sun, Newton was able to formulate some of the most fundamental laws of physics. (*Lick Observatory*)

verse—that can be discovered through careful observation or experimentation. And, finally, the modern scientist contends that these physical laws, once discovered and formulated, can be applied to the problem of understanding the universe as a whole. Of course, over the years new discoveries are made and old ideas, once thought to be the final word, are revised or discarded for better ones. Nevertheless, the basic picture remains unchanged: there are fundamental laws at work in the physical world that can be used to understand the complex and varied phenomena we observe in the sky.

It may seem that discoveries often are made in someone's laboratory that the astronomer then applies to celestial phenomena. But more often than not, the reverse is actually the case. For example, Sir Isaac Newton formulated some of the most basic laws of mechanics as a result of his efforts to explain the motions of the planets. This was possible because in the motions of the planets these laws of mechanics are revealed in their simplest and purest form, unhampered by the wind resistance or friction encountered in laboratory experiments. One of the important motivations for the field of optics was the demand of the astronomer for better observing equipment. The element helium, furthermore, was discovered to be a constituent of the solar atmosphere even before this familiar gas was found here on the earth.

This pattern of astronomical discoveries as the precursors or motivating factors in the formulation of fundamental physical laws was dramatically illustrated at the turn of the twentieth century. At that time, geologists were presenting strong evidence that the earth was extremely old, perhaps several billion years old. Concurrently, astronomers and physicists realized that by any known physical process, the sun could not shine for more than a few hundred million years. If the geologists were correct, then the sun should have burned out long ago. The great mysteries then were "Why is the sun still shining? What is the mechanism of the sun's energy production?"

The answers to these dilemmas required the development of quantum me-

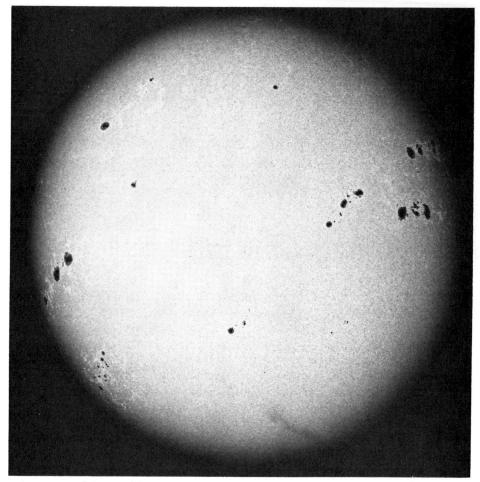

Figure 1–2. The Sun. Up until only seventy years ago we still had no true understanding of why the sun shines. The solution of this mystery resulted in major advances in our knowledge of the physical world. (*Hale Observatories*)

chanics, nuclear physics, and the special theory of relativity. Today we know that at the sun's center hydrogen is converted into helium with the accompanying release of enormous amounts of energy according to Einstein's famous equation $E = mc^2$. Thus, the answer to the simple question "Why does the sun shine?" led mankind to an understanding of thermonuclear reactions. It is now a matter of conscious choice as to whether mankind ultimately will use this understanding for destructive or humanitarian purposes.

There are, today, dilemmas facing modern astronomers that are painfully reminiscent of the problem of the sun's energy. For example, for almost two decades we have been aware of peculiar starlike objects in the sky known as *quasars*. With the traditional understanding and theories now in our possession,

Figure 1–3. A Quasar. Quasars have been known to exist since the 1960s, yet their nature and properties are still a complete mystery. Perhaps understanding how and why quasars shine so brightly will result in the discovery of new fundamental laws at work in the physical world. (*Hale Observatories*)

astronomers find that they are completely unable to explain the incredible energy output of these mysterious sources. Perhaps the answer will lead to the discovery of new fundamental laws at work in the physical world.

At first glance it may seem that many of the problems and mysteries at the forefront of modern astronomy are irrelevant to us and to civilization as a whole. After all, who really cares whether or not we manage to figure out why and how quasars shine? But to anyone with even the most rudimentary understanding of the course of history over the last several centuries, the critical role of basic science in shaping our world is blatantly obvious. In our society such factors as the standard of living are directly related to the health of the economy. The economies of virtually all Western countries are intimately associated with industry and, therefore, technology. And, finally, advances in technology are possible only as a result of advances in pure science, which is founded on basic research. Thus, basic scientific research ultimately plays an important role in our everyday lives. In this regard, when someone demands a justification for thus-and-such a research project, we are reminded of the words of Michael Faraday who, following his discovery of electromagnetic induction in the nineteenth century, responded to similar questions by asking about the value of a newborn baby. Parenthetically we note that electromagnetic induction is the fundamental principle by which electric motors operate.

In conclusion, the advances in our understanding of the universe can have a profound effect on the course of history and our lives. We would not know how to build an automobile without Newtonian mechanics, and the H-bomb—a major factor in international politics—could not have been invented if we were still ignorant of the thermonuclear processes at the sun's center. Unfortunately, the general public traditionally has taken little interest in what is going on at the frontiers of science. This situation was perhaps acceptable in the distant past when there was a substantial lag time between a scientific discovery and its practical application. Today, however, virtually every major company and industry employ scientists to do basic research and this lag time can be terrifyingly short. It is, therefore, the moral responsibility of the modern scientist to bring to the general public an appreciation of what is going on at the forefront of human knowledge, lest in the face of an aggressive technology we find ourselves impotently trying to cope with dubious "benefits" and ill-conceived applications of basic research.

1.2 The Course of Astronomy

To our ancestors the earth must have seemed vast and immobile. It was the stage of all human experience. The stars, sun, moon, and planets appeared to revolve around this vast stage, paying daily homage to man's unique abode at the center of the universe. No wonder that up until a few hundred years ago astronomy and astrology were intimately related. After all, since the very heavens appeared to revolve around the earth, it seemed quite reasonable to suppose that celestial phenomena played an important role in human events. Of course, there were very practical reasons for ancient man to study the stars. As a result he developed a workable calendar to tell when crops should be planted and when the Nile would flood. Nevertheless, the mystical, or metaphysical, motivation seems to have been stronger than the practical applications. The planets were thought to be the physical embodiments of the gods themselves, and specific qualities were attributed to the constellations of the zodiac through which they moved. It was only natural for our ancestors to infer the validity of astrology. Every respected astronomer had to be familiar with astrology, and his duties included predicting famines or plagues, the birth or death of kings, and auspicious times to build monuments or go to war.

Ancient man's preoccupation with the heavens is truly remarkable. The pyramids in Egypt were aligned with the stars and Stonehenge, which also is about 5,000 years old, clearly demonstrates sophisticated knowledge of astronomy. Actually Stonehenge is one of many hundreds of Neolithic astronomical sites scattered through the British Isles. In all cases the alignments of stones, posts, mounds, or holes in the ground unmistakably point to key directions on the horizon, thereby preserving for posterity information such as the rising and setting locations of the sun and moon at important times throughout the year. Some researchers in this field of astroarchaeology even argue that monuments such as Stonehenge could have been used to predict eclipses. This is especially impressive when we realize that the builders of Stonehenge were literally in the Stone Age and, as far as we know, had not yet invented either the written word or the wheel.

Faced with the astounding astronomical accomplishments of our ancestors,

Figure 1-4. Stonehenge. Five thousand years ago the Neolithic inhabitants of Great Britain erected this huge astronomical monument. Clearly they were gifted, careful, and patient observers of the heavens. (*Griffith Observatory*)

certain authors of crank literature in the 1970s have taken the viewpoint that the astronomical monuments of the past were actually constructed with the help of visitors from outer space. The assumption that ancient man was quite stupid, that his brain was as primitive as his technology, is a necessary prerequisite of such a hypothesis. Yet, when we examine other endeavors of our ancestors such as poetry, sculpture, or literature, they appear to be in no way inferior to those of modern man. Did little green men from flying saucers also help Homer write the *Iliad* and the *Odyssey?* Did they also assist the anonymous sculptors of Ramses's temples at Abu Simbel? If so, why in Britain did these creatures, with an advanced technology capable of interstellar space flight, do nothing better than set a number of large stones in a circle?

It indeed seems far more reasonable to give credit to ancient man for the ingenuity he obviously possessed. Not only was the astronomer expected to be ingenious, but precision and accuracy were also required of him. Legend has it that more than 4,000 years ago the Chinese astronomers Hi and Ho were executed for having failed to predict an eclipse. Some historians still argue about the alleged crime, but whether it was errors in eclipse prediction or in the preparation of the official calendar, the sentence was unquestionably harsh.

Perhaps what is most difficult for us to accept or appreciate is the dedication and preoccupation of our ancestors with astronomy. For example, virtually the entire Mayan city of Chichen Itza in Mexico was laid out with an astronomical orientation. The rulers, priests, architects, and citizenry of eons past were far more motivated by astronomical phenomena than modern man living in the so-called Space Age. With this in mind, our own efforts on the moon or at Mount Palomar Observatory take on a new and more humble perspective.

Although the earliest astronomical endeavors of man were related to the motions of the sun and moon, as time went on he turned his attention to the planets. Compared to the motions of the sun and moon, those of the planets are extremely complex. In this regard, we note that ancient man labored under a *geocentric cosmology*, or the assumption that the planets went around the earth. In spite of these difficulties, the Greeks were able to devise a system, the details of which will be discussed later in this book, whereby they could calculate the paths of the planets against the background of stars. As presented in its final form by Claudius Ptolemy in the second century A.D., this geocentric cosmology survived intact for more than a thousand years. It was not until the work of Copernicus in the sixteenth century that new ideas made their way into our thinking about the universe. Copernicus proposed the revolutionary concept that the sun and not the earth is at the center of the universe. Indeed he was successful in showing that it is much easier to explain the motions of the planets using a *heliocentric cosmology* than by trying to make things fit into the geocentric picture.

At this point in history, astronomy and astrology parted company forever. Copernicus, Kepler, and Newton demonstrated that the earth was just one of several planets all revolving about the sun. No longer did the earth have a special place in the universe; no longer was it possible simply to infer the validity of astrology from our ideas about how the universe works. In fact, exactly the opposite was the case. As mankind moved into the seventeenth century it was becoming clear that the validity of science was demonstrable by experimentation and observation, whereas the astrologer had to rely increasingly on the belief and faith of his followers.

With a basic understanding of the laws of physics also came the ability to build telescopes. Man was then able to look out into space with the aid of instruments capable of magnifying what he saw. During the eighteenth and nineteenth centuries, telescopic observations were devoted primarily to the planets and other objects in our solar system. The stars, however, remained incredibly remote and no one knew what to make of the strange, fuzzy objects called nebulae.

A little more than one century ago we finally had developed the necessary observational and mathematical tools needed to measure the distances to the stars. In addition, the invention of the spectroscope and the formulation of atomic theory made it possible for us to analyze and understand the light from the stars. Therefore, by the turn of the twentieth century, astronomers were focusing their attention mainly on the stars and questions of stellar structure. It was out of this understanding of the properties of certain stars that Edwin Hubble in the mid-1920s was able to measure the distances to some of these nebulae, which are now called *galaxies*. Their incredible remoteness opened a whole new chapter in the history of astronomy. With the work of Hubble it became clear that matter in the universe is contained in galaxies that are themselves huge, swirling aggregates of billions of stars.

Another fundamental advance in our understanding of the universe occurred in around 1930, again as the result of the work of Hubble. Having measured the distances and speed of a number of galaxies, Hubble was able to show that

Figure 1-5. A Galaxy. Seen through a telescope, a typical galaxy looks like a small fuzzy patch of light. With a time exposure, however, a galaxy is found to consist of billions of stars. (*Lick Observatory*)

almost all galaxies are moving away from us and that their speeds are directly related to their distances from the earth. This does not mean, as we will learn in the final chapter, that we are at the center of the universe. Rather the relationship of the speeds and distances of the galaxies is incontrovertible evidence that the universe itself is in the process of expanding! One inference from this observational evidence is that billions of years ago all the matter and energy in the universe were confined to a very small volume. Then, for some unknown reason, the entire universe experienced a stupendous primordial explosion, commonly called the big bang, the remnants of which we still observe today in the recession of the galaxies.

It is truly remarkable to realize that in spite of thousands of years of man's preoccupation with the heavens, up until only a century ago we really did not know where the stars were. And only a few decades ago we had little understanding of the nature of the galaxies, the largest physical objects in the universe. Our ideas about the nature of the universe have changed remarkably over the last hundred years, and the revolution is still going on.

That our concepts of the universe are in a state of flux becomes obvious when we consider what astronomers are doing today. Space flight enables astronomers to place telescopes in orbit above the earth's atmosphere. This is of primary importance because the earth's atmosphere is opaque to most of the radiation coming from outer space. X rays and ultraviolet radiation, for example, simply cannot get through the layer of air surrounding our planet. This is, of course, fortunate because many of these radiations are lethal. But while this protective atmosphere has shielded us from these deadly rays, it has kept us in ignorance. Up until very recently we simply could not see what the universe looked like in X rays, gamma rays, and ultraviolet or infrared radiation. With the construction of orbiting astronomical observatories that contain exotic telescopes, however, mankind has literally been given new eyes with which to view the universe. We

Figure 1–6. *An Astronomical Satellite.* Satellites such as this are placed in orbit above the earth by astronomers. The telescopes on board detect radiations coming from objects in deep space. Such observations cannot be made from the earth's surface. (*NASA*)

can predict confidently that these new observations will have a profound effect on our understanding of the cosmos, that virtually no idea and no concept will be left unexamined by what man is currently discovering about the universe in which he lives.

Our exploration of the universe is one of the greatest intellectual adventures of the twentieth century. The achievements of astronomers and astrophysicists during the last few decades form the basis of a revolution in our concepts of the universe unparalleled in the whole of recorded human history. In this book we will place special emphasis on the frontiers of human knowledge, hopefully to convey the true nature of this remarkable scientific revolution.

1.3 Stars and Constellations

For those of us who live in a city or in the suburbs, it is easy to forget what the night sky really looks like. Only on rare occasions when we get far away from the smog and the city lights can we appreciate the splendor of an unobscured view of the heavens. Lying on your back looking up at the dark, clear sky, you might think that you can see millions of stars. Actually there are only about six thousand stars over the entire sky visible to the unaided naked eye. At any one time during the night you can see only about half of this number.

When ancient man looked up at the night sky, certain groupings of stars suggested mythical creatures or the gods and heroes of his culture. Perhaps for this reason, as well as man's desire to see order in the random scattering of the stars, our ancestors invented the *constellations*. A constellation is simply an apparent grouping of stars in the sky. In reality the stars are trillions of miles apart, but from the earth they appear to be close together. The apparent groupings called constellations often still retain their ancient names, such as Leo (the lion), Orion (the hunter), and Taurus (the bull).

Civilization developed in the Northern Hemisphere, and as a result our ancestors could not see many of the southern stars, which are normally visible only from places like Australia and Antarctica. This in part explains why only forty-eight constellations were identified in antiquity. The modern astronomer, familiar with the entire sky, recognizes eighty-eight constellations.

The modern astronomer still finds it useful to talk about the constellations. Of course, we do not imagine that we see mythical animals or heroes. Rather, we find it convenient to use the constellations to denote certain areas in the sky. Thus, the astronomer talks about "the galaxy NGC 224 in Andromeda" or "the quasar 3C 273 in Virgo."

The eighty-eight modern constellations cover the entire sky, but the boundaries of these constellations were in dispute for many years. Everyone who made a chart of the heavens would draw the constellations covering slightly different areas. In addition, the total number of constellations would vary from one star

Figure 1-7. Photograph of Orion. The region of the sky around Orion contains many of the bright stars which dominate the winter nights. Since this is a time-exposure photograph, many faint stars normally invisible to the human eye are recorded on the film. (*Hale Observatories*)

Figure 1-8. Drawing of Orion. Many of the old star charts contain beautiful drawings of the constellation figures. This drawing of Orion is from a star chart published in the early nineteenth century. (*Griffith Observatory*)

chart to the next. These difficulties were resolved in 1928, when astronomers from around the world finally agreed on the constellations and their boundaries. The constellations they set down will be in use for many years to come.

It is perhaps instructive to examine one particular constellation. In Figure 1-7 we see a photograph of the sky in the area of Orion, which is located high above the southern horizon during winter nights. This particular photograph is a time exposure and, therefore, many very faint stars, normally invisible to the naked eye, appear on the photograph. In Figure 1-8 we again see the same area of the sky, this time from a beautifully illustrated nineteenth-century star chart. The figure of Orion, the hunter, is shown with upraised club and shield and the familiar belt and sword. Finally in Figure 1-9 we see a modern star chart showing the accepted boundaries in use today. All such boundaries always run either north-south or east-west. This third example is the kind of chart to which the modern astronomer often refers.

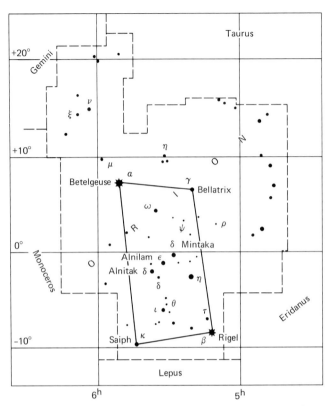

Figure 1-9. The Modern Orion. Star charts used by modern astronomers all recognize the official I.A.U. constellation boundaries established in 1928. As shown in this chart of Orion, these boundaries always run either north-south or east-west.

If one takes the time to familiarize himself with the constellations in the sky, it becomes immediately apparent that the stars rise in the east and set in the west, just as the sun and moon rise and set. This is due to the earth's rotation. But in addition to rotating, the earth also goes around the sun once each year. This means that at different times during the year, different constellations will be seen in the night sky. This effect is demonstrated in Figure 1–10. As the earth moves along in its orbit, the dark side of the earth faces a slightly different direction from day to day. As a result, the stars and constellations visible in the night sky also gradually change.

The combined effects of the rotation of the earth about its axis and the revolution of the earth around the sun result in a given star or constellation rising four minutes earlier each day. Although this is hardly noticeable from one day to the next, the changing appearance of the night sky is very apparent over a period of weeks or months. Following the appendixes of this book there is a series of twelve star charts showing the appearance of the sky, as seen from the

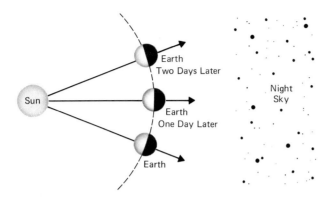

Figure 1-10. The Changing Night Sky. The dark side of the earth is always the nighttime side of the earth. As the earth goes around the sun, the dark side of the earth gradually faces different directions from night to night as shown here.

United States, at about 9:00 P.M. during the middle of each of the 12 months. Careful examination of these star charts shows the changes in the night sky in the course of a year.

In addition to the names of the familiar constellations, the names of many of the brighter stars in the sky also come down to us from antiquity. Although some stars still carry their Greek or Roman names, most of the star names used today are of Arabic origin. This is because during the Dark Ages, the study of astronomy was taken over by Islamic cultures. For example, the two bright stars marking the shoulders of Orion are called Betelgeuse and Bellatrix and the nearby bright reddish star in Taurus is Aldebaran. Many of these ancient names sound interesting, if not romantic, but it should be clear that learning the names of the stars could be a monumental task. Many of these names are almost impossible to pronounce, let alone spell or remember. To alleviate this difficulty, Johann Bayer, in 1603, devised a method for naming stars that is largely still in use. Each star in a given constellation receives a letter from the Greek alphabet. Only lower-case Greek letters are used, and the usage is approximately in order of decreasing brightness. The full star name consists of a Greek letter followed by the Latin genitive form of the constellation name. Thus, the bright star Sirius in the constellation Canis Major, the large dog, is also called α *Canis Majoris*. And Denebola, the second brightest star in Leo, is β *Leonis*. Sometimes this naming of stars is not exactly in the order of decreasing brightness. In Orion, for example, Rigel is slightly brighter than Betelgeuse. Yet, Rigel is β *Orionis* and Betelgeuse is α *Orionis*.

Since there are only twenty-four letters in the Greek alphabet, Bayer's system applies only to the brightest stars in a given constellation. However, especially during the nineteenth century, astronomers who were making extensive star catalogues with the aid of telescopes needed a system for the naming of thousands of stars. In the mid-1800s, F. W. Argelander at the Bonn Observatory in Germany compiled a massive catalogue containing one third of a million stars. This catalogue is called the *Bonner Durchmusterung* and the stars are designated by their *BD numbers*. For example, BD+5°1668 is a star in the constellation of Monoceros, the unicorn. Similarly, at the Harvard College Observatory, a team of astronomers compiled the so-called *Henry Draper Catalogue* in which stars are denoted by their *HD numbers,* such as HD 247572. Modern astronomers frequently use the BD and HD numbers when referring to faint stars.

As we look up at the nighttime sky, it is immediately obvious that stars vary widely in apparent brightness. Some stars are so bright that they can be seen shortly after sunset, whereas others are so dim that they are visible only under the clearest conditions. As with constellation and star names, a system of stellar brightnesses or magnitudes also comes to us from the distant past. In the second century B.C., the Greek astronomer Hipparchus invented a method of denoting a star's brightness. The brightest appearing stars in the sky are of the *first magnitude* and the faintest of the *sixth magnitude.* Stars of intermediate brightness receive an intermediate designation. Of course, Hipparchus' system was based on naked-eye estimates. Nevertheless, the basic concepts of this system are still in use.

It should be noted that in this discussion we are referring only to *apparent visual magnitudes,* as opposed to other types of magnitudes (for example, absolute or photographic), which will be discussed later in the book. For convenience, apparent visual magnitude is often simply called apparent magnitude.

In 1856, N. R. Pogson refined Hipparchus' system by proposing a method of assigning magnitudes to objects in the sky. In observing stars, Pogson noticed that a first magnitude star seemed to shine one hundred times brighter than a fifth magnitude star. In other words, it would take one hundred fifth magnitude stars piled on top of each other to appear as bright as a first magnitude star. This is the starting point of all magnitude scales—namely, that by definition a difference of five magnitudes shall correspond exactly to a difference of one hundred times in luminosity or light output. This, in turn, means that in going from one magnitude to the next, there is a difference of 2.512 in luminosity since

$$2.512 \times 2.512 \times 2.512 \times 2.512 \times 2.512 = 100.$$

Thus, for example, it would take about $2\frac{1}{2}$ third magnitude stars to appear as bright as a second magnitude star. Or, for example, it would take $6\frac{1}{3}$ ($= 2.512 \times 2.512$) sixth magnitude stars to appear as bright as a fourth magnitude star.

Now that the relationship between differences in magnitudes and differences in luminosity was established, Pogson needed to tie down his magnitude scale to some reference stars. For this purpose he chose Aldebaran in the winter sky and Altair in the summer sky. Both Aldebaran and Altair are first magnitude stars; their magnitudes are exactly 1.0. Since the scale was tied down to these two reference stars, it was then possible to assign magnitude values to objects we see in the sky. It is further possible to extend the magnitude scale to brighter or fainter objects. For example, the brightest appearing star in the sky is Sirius. From observation we conclude that it emits about ten times more light than Aldebaran; it would take ten Aldebarans to appear as bright as one Sirius. A factor of 10 in luminosity corresponds to a difference of 2.5 in magnitude.

Therefore, the apparent magnitude of Sirius is 1.0 − 2.5 = −1.5. In concluding that the magnitude of Sirius is −1.5, we see that we have had to go to negative numbers. This is the natural extension of the magnitude scale for brighter objects. To the human eye here on earth, the sun seems to shine ten billion times brighter than Sirius. This corresponds to a difference of 25 magnitudes. Therefore, the apparent magnitude of the sun is −26.5.

Similarly we can extend the magnitude scale to fainter objects. The dimmest star visible to the naked eye under the best observing conditions has a magnitude of 6.5. With the aid of optical instruments, we can see many more even fainter stars. In this regard, the astronomer speaks of the *limiting magnitude* of a telescope, which is simply the magnitude of the dimmest star that can be seen. A typical homemade telescope for use by an amateur has a limiting magnitude of

Figure 1–11. *The World's Largest Telescope.* This dome on Mount Palomar in California houses the world's largest telescope. Looking through this telescope you can see stars as faint as the 20th magnitude. (*Hale Observatories*)

about 13. The largest telescope in the world, the 200-inch telescope at Mount Palomar in California, has a limiting magnitude of 20.

Even though we have spoken here only of visual limiting magnitudes, it should be clear that with the aid of time-exposure photography, far dimmer stars can be reached. The astronomer, therefore, refers to the *photographic limiting magnitude* of a telescope as the magnitude of the dimmest star that can be photographed by it. The photographic limiting magnitude of the giant Mount Palomar telescope is about 23.5.

By way of summary, we have listed here a variety of magnitudes ranging from that of the sun to the photographic limit of the world's largest telescope. This range corresponds to a factor of 1 quintillion in luminosity.

Object	Apparent Magnitude
Sun	$-26\frac{1}{2}$
Full moon	$-12\frac{1}{2}$
Venus (at brightest)	-4
Jupiter; Mars (at brightest)	-2
Sirius	$-1\frac{1}{2}$
Aldebaran; Altair	1
Naked-eye limit	$6\frac{1}{2}$
Binocular limit	10
Telescope limit (amateur)	13
200-inch (visible) limit	20
200-inch photographic limit	$23\frac{1}{2}$

There is one important refinement on Pogson's system that we should not fail to mention. Pogson tied his magnitude scale down to only two stars, Aldebaran and Altair. Since that time, astronomers have realized that it is far more convenient to have a large number of "standard stars" scattered throughout the sky. Therefore, Aldebaran and Altair have been discarded as the standards, and the modern astronomer keys his measurements of apparent magnitudes to a large number of new standard stars scattered across the sky. In this new system, the magnitudes of Aldebaran and Altair are not quite exactly equal to 1.0.

In this section we have come to appreciate the fact that the names of the stars and the constellations as well as the basis for the modern system of stellar magnitudes all come to us from the distant past. This is a small part of the debt we owe to antiquity, to those men who first looked up into the star-filled sky and wondered at what they saw.

Chapter 1 Introducing the Universe   17

1.4 The Celestial Sphere

Standing out underneath the nighttime sky, looking up at the stars, we can imagine ourselves at the center of a huge hollow sphere. This impression of the spherical shape of the sky is so strong that in ancient times Greek astronomers actually thought that we were indeed at the center of a hollow crystalline sphere on which the stars were embedded like jewels. Presumably this sphere rotated once each day carrying with it the sun, moon, planets, and stars, as shown schematically in Figure 1–12. The diameter of this sphere, the Greeks reasoned, must have been very large because otherwise they would have noticed a slight shifting of the stars as they traveled from one place to another on the earth.

Today, of course, we realize that the stars in the sky are not embedded in a huge sphere, but rather are scattered through space. Nevertheless, it is often very useful for the modern astronomer to speak of the stars *as if* they were on such a sphere with the earth at the center. This imaginary sphere is called the *celestial sphere*.

One way in which the concept of the celestial sphere is extremely useful is that it provides a convenient mechanism for expressing the directions of the stars and other objects in the sky. For example, you might look up into the night and say that "there is a bright star 20° above the southeastern horizon." In actuality you are referring to the direction of the star on the celestial sphere. In particular, this kind of a reference to location is closely related to the so-called *horizon system*.

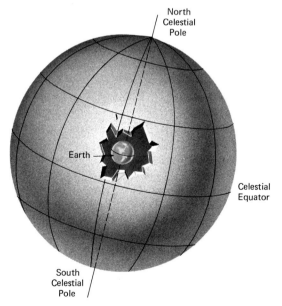

Figure 1–12. The Celestial Sphere. Ancient astronomers believed that the earth was at the center of a huge sphere on which the stars were embedded like jewels. Even today modern astronomers sometimes find it useful to refer to the celestial sphere.

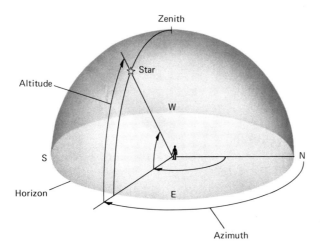

Figure 1-13. The Horizon System. In this coordinate system the position of a star is specified by the altitude and the azimuth.

The horizon system is the simplest of all coordinate systems. As shown in Figure 1–13, the position of a star in this system is given by two angles called the *altitude* and the *azimuth*. To establish these two angles, we imagine a circle drawn through the star in question and the *zenith*, or overhead point, in such a way that the circle is perpendicular to the horizon. The altitude of the star is then simply the angular distance, measured in degrees, along this circle from the horizon to the star. The azimuth is the angle, also measured in degrees, from the north around to where the imaginary circle intersects the horizon, as shown in the diagram.

Unfortunately, the horizon system has a severe limitation. Different observers on the earth have different zeniths and horizons. As a result, the altitude and azimuth of an object in the sky vary from observer to observer. It would clearly be far more useful if we could invent a coordinate system that would be the same for *all* observers on the earth.

To see how we might be able to construct a coordinate system that would be the same for everyone, consider the earth. Any location on the earth's surface can be expressed by giving the *longitude* and *latitude* of that location. Longitude is measured in degrees either east or west of Greenwich, England, and latitude is measured in degrees either north or south of the equator. Thus, for example, the location of Moscow in the Soviet Union is 37°37′E and 55°45′N, which means that Moscow lies 37°37′ east of the meridian passing through Greenwich and 55°45′ north of the equator. The earth's equator is a natural reference circle from which we measure latitude. The fact that we use the meridian passing through Greenwich as the reference circle for measuring longitude is due to a historical accident.

If we think of the sun, moon, planets, and stars as being on the celestial sphere, it is then clear that we should be able to devise a coordinate system similar to longitude and latitude for the purpose of denoting the location of astronomical objects. We first must begin, however, by establishing reference circles and points from which angles can be measured.

Chapter 1   Introducing the Universe

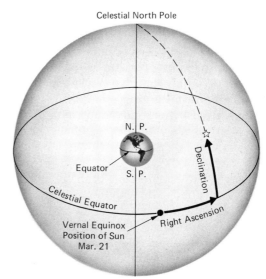

Figure 1-14. *The Equitorial System.* In this coordinate system the position of a star is specified by the right ascension and the declination. This system is used more than any other by astronomers today.

Imagine the earth at the center of the celestial sphere, as shown in Figure 1–14. On the earth there are several geographical reference points and circles that can help us in devising a coordinate system on the celestial sphere. For example, the earth's rotation axis passes directly through the North and South Poles of the earth. We can imagine extending the earth's axis of rotation out into space until it intersects the celestial sphere. This intersection will occur at two points, one directly above the earth's North Pole and one directly above the earth's South Pole. These two points are known as the *celestial north pole* and the *celestial south pole,* respectively.

Similarly we could imagine extending the plane of the earth's equator out into space to the celestial sphere. This projection of the earth's equator onto the celestial sphere gives us a very important reference circle called the *celestial equator.*

While the celestial poles and celestial equator are direct analogies of the poles and equator of the earth, we cannot project the Greenwich meridian onto the celestial sphere and hope to get something useful. The reason is that the earth rotates once in about 24 hours and, therefore, the stars overhead at Greenwich are constantly changing. Rather, we must select some relatively fixed point on the celestial sphere just as the city of Greenwich is a fixed point on the earth. By mutual agreement, the point chosen on the celestial sphere for this purpose is the location of the sun on the first day of spring, usually March 21. On the first day of spring, the sun is directly on the celestial equator, as shown in Figure 1–14, at a point we call the *vernal equinox.* We then can imagine drawing a circle on the celestial sphere passing through the vernal equinox and the celestial poles. This circle among the stars is our second reference circle; it plays a role similar to the meridian passing through Greenwich.

We are now in a position to set up a coordinate system for the purpose of denoting the positions of astronomical objects. Just as latitude is measured in degrees north or south of the earth's equator, the astronomer defines *declination* as the angular distance (degrees, minutes, seconds) of a star north or south of the celestial equator. If a star is north of the celestial equator its declination is preceded with a plus sign, whereas a minus sign is used in connection with declinations in the southern half of the sky. The Greek letter delta (δ) is commonly used by astronomers as an abbreviation for the word declination.

As an analogy to longitude, the astronomer defines *right ascension* as a measure of angular distance eastward from the vernal equinox. Unlike longitude, however, right ascension is *not* measured in degrees. As we mentioned earlier in this chapter, one of the practical motivations for ancient man to study astronomy had to do with the telling of time. Due to this historical motivation, as well as to the fact that the earth rotates about its axis once every 24 hours, right ascension is measured in units of time (hours, minutes, seconds) eastward from the vernal equinox. The Greek letter alpha (α) is commonly used by astronomers as an abbreviation for the term right ascension.

Since declination is measured from the celestial equator to the celestial poles, the values of δ can range from 0° to +90° in the northern half of the sky, or from 0° to −90° in the southern half. Right ascension is measured in units of time and α has a range of 0^h to 24^h.

Using right ascension and declination we can describe the location of any astronomical object in the sky. This coordinate system is used extensively in books such as *The American Ephemeris and Nautical Almanac* (sometimes simply called *The Ephemeris*), published annually by the United States Naval Observatory. It contains such data as the locations of the sun, moon, and planets over the year. By way of illustration, on page 225 of the 1974 edition we find that the location of Saturn on August 27 of that year was

$$\alpha = 7^h 05^m 59^s$$
$$\delta = +22°03'00''$$

This places the planet Saturn in Gemini, the twins, approximately between the two stars ζ Geminorum and δ Geminorum.

This so-called *equatorial system*, using right ascension and declination, is by far the most commonly accepted system in astronomy today. Whenever an astronomer refers to the location of an object on the celestial sphere, he is almost invariably speaking of right ascension and declination.

Whether we look up into the sky or examine the contents of *The Ephemeris*, we notice that the stars remain relatively stationary with respect to each other, whereas the sun, moon, and planets gradually change their position from day to day. Indeed the word *planet* comes from the Greek word meaning "wanderer." This meandering of the sun, moon, and planets is not random but rather is confined to a specific region of the sky. The sun, for example, moves eastward by about 1° each day. During the course of a year, the sun makes one complete journey around the celestial sphere. This apparent path of the sun against the background of stars on the celestial sphere is called the *ecliptic*. When we draw

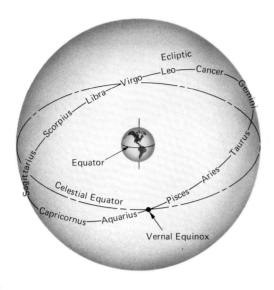

Figure 1–15. *The Ecliptic and Zodiac.* The apparent path of the sun among the stars, the ecliptic, is inclined to the celestial equator. The constellations of the zodiac are centered approximately on the ecliptic.

the ecliptic on the celestial sphere, as shown in Figure 1–15, we notice that it is inclined at an angle to the celestial equator. As we will see in Chapter 3, this inclination of the ecliptic to the celestial equator causes the seasons on earth.

Ancient astronomers noticed that the moon and planets are never very far from the ecliptic. Although their individual motions over a period of time appear to be complex, they never stray more than a few degrees north or south of the sun's apparent path. It was, therefore, believed that the constellations along the ecliptic had some special significance. There are twelve such constellations and they comprise the so-called *zodiac*. Today we realize that the motions of the sun, moon, and planets are confined to the zodiac because the orbits of the moon and planets are very nearly in the same plane. The twelve constellations of the zodiac along with their approximate locations in the sky are listed here.

Name	Position α	δ
Pices (fishes)	1^h	$+10°$
Aries (ram)	3^h	$+20°$
Taurus (bull)	4^h	$+15°$
Gemini (twins)	7^h	$+20°$
Cancer (crab)	9^h	$+20°$
Leo (lion)	11^h	$+15°$
Virgo (virgin)	13^h	$0°$
Libra (scales)	15^h	$-15°$
Scorpius (scorpion)	17^h	$-35°$
Sagittarius (archer)	19^h	$-25°$
Capricornus (sea goat)	21^h	$-20°$
Aquarius (water carrier)	23^h	$-15°$

Some of these constellations contain bright stars and are easy to recognize—for example, Taurus, Leo, and Scorpius. Some are very difficult to find in the sky because they contain only faint stars—for example, Cancer, Libra, and Aquarius.

We should note in passing that the constellations of the zodiac used in astronomy have virtually nothing to do with the "signs" of the zodiac used in astrology. When Ptolemy set up the rules of astrology 2,000 years ago, the signs and constellations of the zodiac coincided. However, due to an effect called precession, which we will discuss in Chapter 3, the position of the vernal equinox moves very slowly westward among the stars. As a result, the signs of the zodiac used by most astrologers are now substantially displaced from their corresponding constellations. Thus, if you were born on December 27, the astrologer will tell you that the sun was in the sign of Capricorn. In reality on that date the sun is actually in the constellation of Sagittarius.

In this chapter we have reviewed the purpose, scope, and history of astronomy. We have seen how the sky is divided up into constellations and how the stars were named. We have learned how the astronomer expresses the magnitudes and position of the stars. We are ready now to discuss the planets and will find that by understanding the workings of the solar system we will discover the force of gravity: the force that dominates the interactions of all material objects in the astronomical universe.

Questions and Exercises

1. Where is Stonehenge?
2. Who, according to legend, were Hi and Ho?
3. Where is Chichen Itza?
4. What is meant by a geocentric cosmology?
5. When did Ptolemy live?
6. In what century did Copernicus propose his heliocentric cosmology?
7. In what century were the distances to the galaxies first successfully measured? By whom?
8. Why is it necessary for an astronomer to make observations from *above* the earth's atmosphere if he wants to detect X rays coming from objects in outer space?
9. If you look up into the night sky under the clearest of conditions, about how many stars can you see with your naked eyes?
10. What is a constellation?
11. How many constellations are recognized by astronomers today?
12. In what year were the boundaries of the modern constellations finally agreed on?
13. Describe Bayer's system for naming the stars.

Chapter 1 Introducing the Universe

14. What do the letters BD and HD stand for when used in connection with star names?
15. Name the Greek astronomer who devised a system of stellar magnitudes similar to that used today.
16. Who was Pogson?
17. How many sixth magnitude stars would it take to have a combined total apparent brightness equal to a second magnitude star?
18. What is the name and apparent magnitude of the brightest star in the night sky?
19. What is the apparent magnitude of the sun?
20. What is meant by limiting magnitude?
21. What is the apparent magnitude of the faintest object ever seen by man?
22. Where is the world's largest optical telescope?
23. What is meant by the celestial sphere?
24. In the horizon system, what angles are used to denote the position of a star or planet?
25. Over what well-known geographical feature on the earth are the north and south celestial poles located?
26. How is the celestial equator related to the earth's equator?
27. What is the vernal equinox?
28. What is meant by declination?
29. What is meant by right ascension?
30. What is the ecliptic?
31. How many constellations are in the zodiac?
32. From memory, name half of the constellations in the zodiac.

The Birth of Modern Astronomy

2.1 Ancient Astronomy and Ptolemy

Any creature endowed with intelligence, with the capacity of self-knowledge, with the ability to examine the process of thinking, and with the awesome awareness of the inevitability of death, would naturally be prone to ask fundamental questions about the surrounding world. There is a compulsion to feel that existence has a purpose and that the universe makes sense. These motivating factors are the true source of a variety of human activities that range from religion to astronomy. It is, therefore, not surprising that every civilization to come forth on this planet has had a cosmology at the core of its teachings.

If you had lived in ancient Egypt, you would have learned that the heavens were actually the starry arched body of the goddess Nut across which the sun, Ra, made his daily journey in a boat. Or, if you were born thousands of years ago in India, you would have been taught that the earth is resting on the backs of three huge elephants. In ancient Babylonia you would have believed that the sky was shaped like the inside of a jar, and if you were an Arab in the Middle Ages you would have regarded the heavens as an enormous tent.

One common characteristic of most of these early cosmologies is that they were rooted in mythology or religion. The gods themselves supposedly revealed the nature of the universe to the high priests and occult initiates. It was not until the time of ancient Greece, slightly more than 2,000 years ago, that we find the first major and significant departure from this approach to cosmology. At this point in history, man realized that knowledge of the universe could come from observing natural phenomena and by using reason to try to understand what was seen.

As a result of this innovative approach, the Greeks made several important advances in astronomy. As early as the fifth century B.C., Anaxagoras correctly explained why the moon goes through phases. He argued that the moon must be spherical in shape and that it shines by reflected light from the sun. Comparing the moon to a ball, Anaxagoras reasoned that if a ball is illuminated by a single source of light, such as the sun, then one side of the ball is lighted while the other side is dark. As the moon goes around the earth, as shown in Figure 2–1, the

Chapter 2   The Birth of Modern Astronomy   25

phase of the moon is simply determined by how much of the lighted side we can see. When the sun and the moon are approximately in a straight line on the same side of the earth, we can see only the dark side of the moon, resulting in a phase known as *new moon*. However, when the moon is opposite the sun, we see the fully illuminated side, which gives us a *full moon*. In between new moon and full moon are *first quarter* and *last quarter* phases, at which times we see half of the lighted side and half of the dark side.

The phases of the moon can be subdivided further. We say that there is a *crescent moon* when less than half of the illuminated side is visible, or a *gibbous moon* when more than half can be seen. In going from new moon to full moon, we progressively see more and more of the illuminated side from night to night. During this half of the lunar cycle we, therefore, speak of a *waxing crescent* or *waxing gibbous* (waxing means getting bigger). And, during the second half of the lunar cycle, we speak of *waning crescent* and *waning gibbous*, since we observe less and less of the lighted side of the moon from one night to the next (waning means getting smaller).

It takes $29\frac{1}{2}$ days for the moon to go through all its phases. Thus, for example, the time between first quarter and full moon is approximately one week. Furthermore, our moon keeps the same physical side (which may be dark or illuminated) facing toward the earth at all times. As a result, the moon as seen during its monthly cycle will appear as shown in Figure 2–2.

Figure 2–1. **Phases of the Moon.** The phases of the moon are determined by how much of the illuminated side of the moon can be seen from earth.

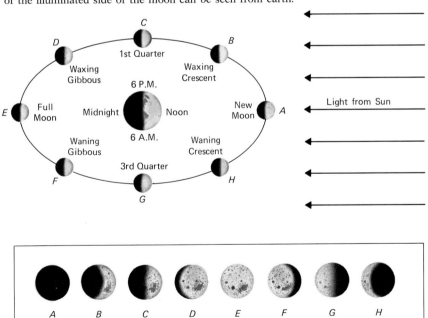

Figure 2–2. *The Appearance of the Moon.* This series of photographs shows the moon at various times during the lunar month. (*Lick Observatory*)

It is interesting to note that the phases of the moon are related in an approximate fashion to the times of moonrise and moonset. For example, if there is a full moon on a particular night, then moonrise will occur at the time of sunset, around 6:00 P.M., because the moon and sun are on opposite sides of the earth. The moon must come up at the same time the sun does down. Similarly, by realizing the relative positions of the sun, moon, and earth, the first quarter moon rises at noontime, whereas a last quarter moon rises at approximately midnight.

Figure 2–3. *Geometry of a Solar Eclipse.* When the moon passes directly between the earth and the sun, the moon casts its shadow onto the earth. People standing inside the moon's shadow will see an eclipse of the sun.

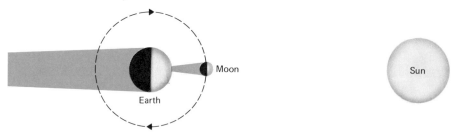

Figure 2–4. *Geometry of a Lunar Eclipse.* When the moon passes through the earth's shadow, an eclipse of the moon occurs. A lunar eclipse is visible to anyone standing on the nighttime side of the earth.

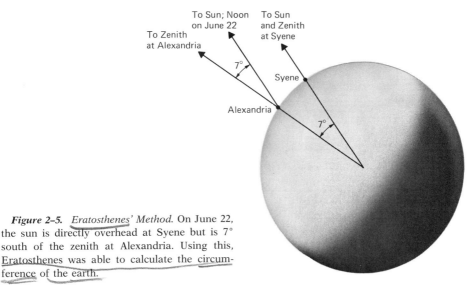

Figure 2-5. *Eratosthenes' Method.* On June 22, the sun is directly overhead at Syene but is 7° south of the zenith at Alexandria. Using this, Eratosthenes was able to calculate the circumference of the earth.

In addition to understanding why the moon goes through phases, the Greeks also knew why eclipses occurred. On rare occasions at the time of new moon, the alignment of the sun, moon, and earth is so precise that the moon casts its shadow directly onto the earth. If you happen to be lucky enough to be standing in that shadow, you see a *solar eclipse* as the moon passes in front of the sun, as shown in Figure 2-3. Similarly, sometimes during a full moon, the moon passes through the earth's shadow, as shown in Figure 2-4, resulting in a *lunar eclipse.*

From observing eclipses of the moon, the Greeks were able to argue that the earth is spherical. During a lunar eclipse the earth's shadow is projected onto the moon. The edge of this shadow is always circular, and it seems entirely reasonable that only a sphere could always cast such a shadow. Indeed, during the third century B.C., Eratosthenes actually measured the size of the earth with surprising accuracy. It turns out that on June 22 at noon, the sun would always be directly overhead as seen from the ancient city of Syene, not far from the modern Aswan Dam. Eratosthenes, however, lived in Alexandria where the sun is never directly at the zenith. In fact, in Alexandria on June 22 at high noon, the sun is 7° south of the zenith. Therefore, as shown in Figure 2-5, the angle

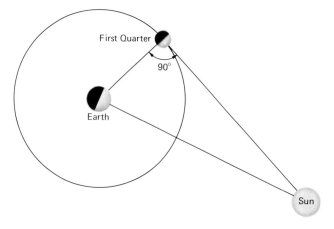

Figure 2-6. *The Geometry of First Quarter.* At first quarter, the sun, moon, and earth form a right triangle. From this fact, Aristarchus attempted to measure the relative distances to the sun and the moon.

between Syene and Alexandria as measured from the center of the earth must also be 7°. Since there are 360° in a circle and since 7° is about one-fiftieth of 360°, the distance from Syene to Alexandria must be one-fiftieth of the circumference of the earth. Eratosthenes knew how far it was from Syene to Alexandria, so he simply had to multiply this distance by 50 to get the earth's circumference.

With much less success, Aristarchus attempted to measure the relative distances to the sun and moon. He reasoned that at first quarter, when we see exactly half of the moon illuminated and half in darkness, the angle between the directions to the moon and the sun as seen from the earth must be slightly less than 90°. On the other hand, as seen from the moon, the angle between the

Figure 2-7. *The Path of Mars.* During August of 1971, Mars moved through the constellation of Capricornus. From mid-July to early September Mars was moving retrograde.

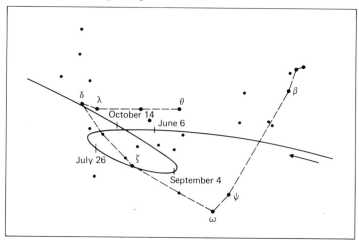

directions to the earth and the sun must be exactly 90°. As a result, the configuration of the sun, moon, and earth will form a right triangle, as shown in Figure 2–6. The Greeks had an excellent understanding of the geometrical properties of right triangles, and by measuring the angular separation of the sun and moon in the sky at the exact moment of first quarter, Aristarchus hoped to establish the relative distances to the sun and moon. The method he used actually involved comparing the time from first quarter to last quarter with the time from last quarter back to first quarter. Although this observation should work in principle, in practice it is very difficult to tell exactly when first or last quarter really occurs. Nevertheless, Aristarchus' measurements during the third century B.C., did prove that the sun was many times farther from the earth than the moon.

Even though the Greeks had a clear understanding of a variety of phenomena related to the sun and moon, one of the main thrusts of ancient astronomy was to explain the complex motions of the planets. If we observe a planet in the sky night after night, we notice that it gradually changes its position relative to the fixed stars. Planets usually move eastward among the stars. However, at certain intervals a planet will appear to stop its eastward motion and move westward for a period of time before resuming its eastward course. We refer to the eastward motion of a planet as *direct*, whereas the westward or backward motion of a planet is called *retrograde*. As a result, over a period of time, a planet will trace out a loop against the background of stars, as shown in the case of Mars in Figure 2–7.

As we mentioned in the previous chapter, all cosmologies at this time in history were *geocentric*, meaning that the earth was assumed to be at the center of the universe. The Greeks were, therefore, faced with the task of devising a system for explaining planetary motions assuming that the orbits were centered on the earth. Although several schemes were proposed, the most successful theory was based on a system first invented by Hipparchus during the second century B.C. Although Hipparchus' work dealt primarily with the motions of the sun and moon, it was found that his ideas could be applied readily to the planets. This cosmology assumed that each of the planets moves along a small circle, called an *epicycle*, which in turn moves along a larger circle, called a *deferent*, which is centered on the earth. The orbit of a planet, therefore, involves two circles, as shown in Figure 2–8. It is further assumed that the speed of the planet around the epicycle is greater than the speed of the epicycle around the deferent. Using these assumptions, the motion of the planet around the epicycle will add to the motion of the epicycle around the deferent, producing an apparent eastward movement of the planets among the stars. However, at certain intervals the motion of the planet along the epicycle will be in the opposite direction of the motion of the epicycle along the deferent. At such times, the greater speed of the planet along the epicycle will result in an apparent westward, or retrograde, motion of the planet among the stars.

Although the groundwork for this cosmological system was laid in the second century B.C., it was not until the time of Claudius Ptolemy, in around A.D. 140 that the details were worked out. First of all, Ptolemy found that Hipparchus' system

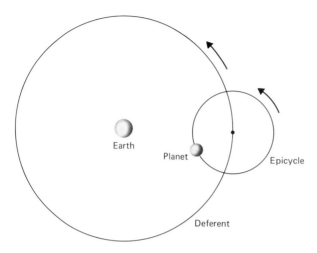

Figure 2–8. The Ptolemaic System. Ptolemy's system accounts for the retrograde motion of planets by assuming that the planets go around epicycles which in turn revolve around a deferent centered approximately on the earth.

had to be modified slightly by assuming that the earth was not exactly at the center of every deferent. Secondly, he had to assume that the uniform angular motion of the epicycle around the deferent had to be measured, not from the location of the earth, nor from the center of the deferent, but rather from a third point called the *equant point*. Using these assumptions as well as the astronomical data that had accumulated since the time of Hipparchus, Ptolemy was able to figure out the relative sizes and orientations of the deferents and epicycles for each planet, along with the speeds of the planets on the epicycles and the epicycles on the deferents. All of the details of this cosmological system were then compiled in a thirteen-volume work called the *Almagest*.

This Ptolemaic system was a triumph for ancient astronomy. For the first time the positions of the planets could be predicted with surprising accuracy. Nevertheless, Ptolemy's system is both contrived and artificial. The brief treatment presented in the previous paragraph really does not do justice to the incredible complexity of his system. The planets, the sun, and the moon are all moving around little circles at various speeds and inclinations that, in turn, all revolve about larger circles set off slightly from the earth and with angular velocities measured from a variety of equant points in space. In view of this complexity, in the *Almagest* Ptolemy makes no claim that his system is a true description of reality. It was simply the only known theory by which the planetary position could be calculated.

During the Middle Ages, the Ptolemaic system became accepted as the absolute authority in astronomy. It is indeed a great tribute to Ptolemy that his work was so precise and thorough that it survived for more than 1,000 years as the correct description of the workings of the universe. It was not until the sixteenth century that man began to reexamine some of the basic assumptions of this geocentric cosmology and found, as we will see in the next section, that everything would be a lot simpler if the sun were at the center of the universe.

2.2 The Copernican Revolution

It was clear to ancient man that our vast and immobile earth must be at the center of the universe. The daily movement of the stars, planets, sun, and moon across the sky seemed to support this geocentric hypothesis, which no intelligent man would dare question. Yet, any system capable of explaining the intricate motions of the planets would have to be extremely complex. Ptolemy's geocentric cosmology with all its epicycles, deferents, and equants is an excellent example.

During the third century B.C., and in spite of public opinion, Aristarchus came up with a better idea. The central problem with the planets is that although they usually move eastward among the stars, sometimes they stop and go backward (or westward) for a period of time. It was this apparent variable speed of the planets with direct and retrograde motion that necessitated a complex and artificial geocentric cosmology.

Imagine that you are driving down a street at 50 miles per hour. If you pass a car going in the same direction at 45 miles per hour, it will appear to you that this slower moving vehicle is going backward relative to you. It was from this type of an analogy (probably using chariots rather than cars) that Aristarchus proposed the first heliocentric cosmology. He realized that if the earth and the planets were all moving about the sun, as the earth passes by a slower-moving planet, that planet will appear to go backward for a period of time.

Perhaps the best way of illustrating Aristarchus' explanation of retrograde motion is with the aid of a diagram such as Figure 2-9. In this diagram, the earth catches up with and overtakes a slowly moving planet such as Mars. While the earth and the planet are far apart, the planet appears to be moving eastward among the stars. But when the earth is overtaking this outer planet, retrograde

Figure 2-9. *Retrograde Motion.* As the earth overtakes and passes a slower outer planet, that planet will appear to stop its usual eastward motion and back up for a period of time.

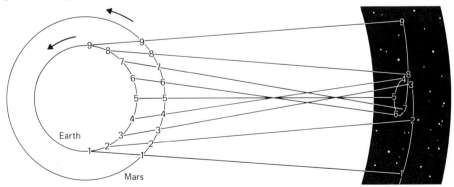

motion occurs because the earth is moving at a higher speed. Finally, when the separation between the earth and the planet is again large, the planet resumes its usual eastward course.

Although this scheme obviously would explain retrograde motion, it is necessary to assume the then unpopular idea that the earth is moving. In addition, Aristarchus evidently did not have the data or motivation to prove in detail that this heliocentric cosmology could, in fact, be used to predict the courses of the planets. As a result his ideas were dismissed as clever fantasy. After all, if the earth is moving, why don't we fall off?

As we mentioned in the previous section, the Ptolemaic system worked so well and seemed so reasonable that it completely dominated all of astronomy for more than 1,000 years. But as the centuries passed, it became clear that there were discrepancies between the observed positions of the planets and their predicted position from the *Almagest*. To correct this, astronomers added more epicycles. Planets then were assumed to travel around epicycles that, in turn, moved along additional epicycles. By the early 1500s the geocentric picture of the universe contained no less than seventy-nine separate circles. It should have been clear that this system was so artificial and had so many circles that it was in imminent danger of collapsing under the sheer weight of its own complexity. The time was at hand for a basic change in thinking.

In the early part of the sixteenth century, a young monk visited the Vatican. The Vatican Library contained copies of Aristarchus' works, but it is unknown whether or not the young man ever read about Aristarchus' heliocentric ideas. Nevertheless, this humble monk, who was also a gifted artist, a distinguished physician, and a renowned economist, was destined to give to the world a new way of thinking about the nature of the universe. His name was Nicholas Copernicus.

Figure 2–10. Nicholas Copernicus, 1473–1543. (*Yerkes Observatory*)

Copernicus' main contribution is that he was the first astronomer to work out all the details of a heliocentric cosmology. Unlike Aristarchus, who merely showed that it was plausible to assume that the sun was at the center of the universe, Copernicus used mathematics to prove that the positions of the planets could be predicted with accuracy in such a system. His heliocentric cosmology was published in a book called *De Revolutionibus Orbium Celestium* in 1543, the year of his death.

It should be emphasized that Copernicus' cosmology was entirely empirical. In other words, he set up his theory in such a way that he would get the right answers for the positions of the planets. For example, Copernicus explained retrograde motion by assuming that the order of the planets, starting nearest the sun, must be Mercury, Venus, Earth, Mars, Jupiter, and Saturn. With any other ordering, he would not be able to account for their motions. Also, he assumed that the nearer a planet is to the sun, the greater is its orbital speed. Thus, Mars goes around the sun faster than Jupiter. Using these two assumptions, he was able to make his heliocentric system fit the observations.

From this ordering of the solar system, we are now able to distinguish between so-called *inferior planets* and *superior planets*. An inferior planet is one that has an orbit inside the earth's orbit. There are only two such planets, Mercury and Venus. Both Mercury and Venus always appear near the sun in the sky. Under favorable conditions they are seen either in the west shortly after sunset, in which case they are sometimes called evening stars, or in the east just before sunrise, when they are called morning stars. Due to the brilliance of the sun, the best time to observe these planets is when they are farthest from the sun. As shown in Figure 2–11, this occurs when the inferior planet is at either *greatest eastern elongation* (in the evening sky) or *greatest western elongation* (in the morning sky).

From time to time, an inferior planet passes between the earth and the sun. We refer to such an event as *inferior conjunction*. On the other hand, when Mercury or Venus passes behind the sun, we speak of *superior conjunction*. Although it might seem that Mercury and Venus should be seen as little black dots against the solar disc at every inferior conjunction, this rarely occurs. Actually the orbits of Venus and Mercury are inclined to the earth's orbit and, therefore, the inferior planets usually pass to the north or south of the sun. Those rare occasions when they actually cross directly in front of the sun are called *transits*. The most recent transit of Venus occurred in the nineteenth century; the next transit is not due until June 8, 2004. Transits of Mercury are far more common, with about thirteen occurring each century.

Superior planets have orbits that are larger than the earth's orbit. Obviously a superior planet such as Mars, Jupiter, or Saturn can never appear at inferior conjunction. But they do, however, pass behind the sun. As shown in Figure 2–12, when the sun lies between the earth and a superior planet, we say simply that the planet is in *conjunction*. Unlike Venus and Mercury, the superior planets can appear directly opposite the sun at what we call *opposition*. When a superior planet is in opposition, the earth lies between the planet and the sun. At such times, the planet is seen high in the nighttime sky at midnight.

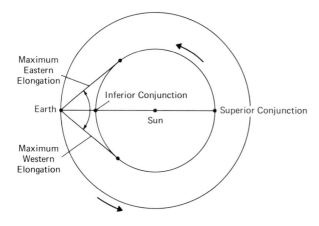

Figure 2–11. Aspects of an Inferior Planet. Various configurations of an inferior planet are defined as shown here. There are only two inferior planets: Mercury and Venus.

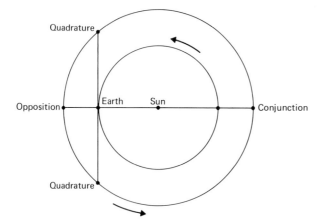

Figure 2–12. Aspects of a Superior Planet. Various configurations of a superior planet are defined as shown here. Any planet whose orbit is larger than the earth's orbit is called a "superior" planet.

In setting up his heliocentric theory, Copernicus realized that he had to distinguish between the *sidereal period* and the *synodic period* of a planet. The sidereal period of a planet is its real orbital period. It is simply how long it takes for a planet to go once all the way around the sun. For example, the sidereal period of Mars is 687 days and, thus, you might say that a "year" on Mars lasts for 687 earth days.

Unfortunately, we can never directly measure the sidereal period of a planet from the earth because the earth itself is moving. What can be measured is the synodic period, which is the time it takes the planet to go from one configuration back to the same configuration. For example, the synodic period of Mars is 780 days, meaning that slightly more than 2 years pass between one opposition of Mars and the next. Copernicus was able to prove that the sidereal and synodic periods of the planets are related in a very simple mathematical fashion. Thus, if we measure the synodic period of a planet, we can always calculate its sidereal period. A table giving the sidereal and synodic periods of the planets known to Copernicus is shown here:

Chapter 2 The Birth of Modern Astronomy

Planet	Sidereal Period	Synodic Period (in days)
Mercury	88 days	116
Venus	225 days	584
Earth	365 days	
Mars	687 days	780
Jupiter	11.9 years	390
Saturn	29.5 years	378

Knowing the various periods of the planets, and measuring their positions in the sky, Copernicus was able to figure out the relative sizes of their orbits. Of course, he had no way of coming up with answers in miles or kilometers, so he expressed his results in terms of the size of the earth's orbit. These answers agree remarkably well with modern results as shown here.

Planet	Distance of Planet from Sun	
	Copernicus	Modern
Mercury	0.38	0.39
Venus	0.72	0.72
Earth	1.00	1.00
Mars	1.52	1.52
Jupiter	5.22	5.20
Saturn	9.17	9.54

The work of Nicholas Copernicus marks an important turning point in the course of human thinking. He started a revolution that spread through Europe during the century following the publication of his book and that culminated with the brilliant accomplishments of Sir Isaac Newton. Yet, in the interim, the Copernican cosmology met stiff opposition, primarily from the Church. *De Revolutionibus* was banned and burned, and had Copernicus lived in Rome, he most certainly would have been executed for heresy.

In spite of the successes of his heliocentric cosmology, Copernicus' system still had a few problems. Following the example of many astronomers before him, Copernicus chose to express the planetary orbits using only circles. As explained in the next section, the true orbits of planets are not precisely circular, and therefore he resorted to epicycles. Although Copernicus was able to dismiss the idea of the earth's central location in the universe, he still clung in his theory to circles and circular motion. Two generations would pass before Johannes Kepler would prove that ellipses, not circles, describe the true orbits of planets. Kepler banished forever the troublesome epicycle from the field of astronomy.

2.3 Kepler's Cosmology

How could Copernicus possibly be right? How could something as heavy and sluggish as the earth be in motion about the sun? To make matters worse, it was virtually impossible to reconcile a heliocentric cosmology with the Bible or with the great writings of Aristotle and Plato.

It occurred to some of Copernicus' contemporaries that observations might prove or disprove his theory. After all, if you move from one place on the earth to another, the scenery changes; the relative position of distant mountains, trees, and houses appears to change as you walk down the street. If the earth is really moving, shouldn't we observe a similar effect with the stars? It seemed reasonable to suppose that if Copernicus were right, then the relative positions of the stars should move slightly as the earth allegedly goes around the sun. Clearly, to decide this issue, extremely accurate measurements of the positions of the stars were needed, far more precise than anything that had ever been done before.

Three years after the death of Copernicus, Tycho Brahe was born in Denmark. Extravagant, arrogant, and often obnoxious in his behavior, Tycho Brahe met the challenge of the times by establishing a superb observatory on the Danish island of Hveen. For 20 years, this colorful astronomer and his assistants carried out a program of making the most complete and precise astronomical observations of the locations of the sun, moon, stars, and planets ever produced.

Tycho Brahe assumed that the stars were only seven thousand times farther away from the earth than the sun. He, therefore, expected to detect slight changes in the positions of the stars over the period of a year. No such changes were observed, and he therefore concluded that Copernicus must have been wrong. Actually, the real distances to the stars is more like 300,000 times the distance from the earth to the sun. Thus, although slight changes in stellar positions do indeed occur, they are so small that Tycho Brahe could not have detected them with his sixteenth-century instruments.

In spite of his rejection of a heliocentric cosmology, Tycho Brahe's observations proved invaluable to astronomy in the seventeenth century. After his death in 1601, most of Brahe's notes and records were given to his gifted young assistant, Johannes Kepler. The precision of the measurements in these records allowed Kepler to proceed in establishing the groundwork for all of modern celestial mechanics.

During his youth as a theology student in Germany, Kepler became an early convert to the Copernican system. It was soon found, however, that Copernicus' theory did not accurately predict the positions of the planets as measured by Tycho Brahe. So Kepler began working out the details of planetary motions. He chose to concentrate his attention on the particularly troublesome planet Mars. For nearly 10 years Kepler tried all kinds of epicycles, deferents, and equants in a heliocentric system in an effort to reproduce Brahe's observations. Every attempt met with failure. In desperation, it occurred to Kepler that perhaps the

Figure 2-13. Johannes Kepler, 1571–1630. (*Yerkes Observatory*)

cause of his difficulties was the use of circles. For thousands of years, all astronomers had used circles to describe the motions of the planets. Perhaps the circle was simply the *wrong* curve. After several attempts with ovals, Kepler discovered that the orbit of Mars conformed extremely well with a curve called an *ellipse*.

An ellipse is a basic geometrical curve that can be drawn with the aid of two thumbtacks and a loop of string, as shown in Figure 2-14. To draw an ellipse, first stick two thumbtacks into a piece of paper on a drawing board. Then place the loop of string over the thumbtacks and use a pencil to hold the string taut. As you move the pencil around the taut loop of string, an ellipse will be drawn.

Each thumbtack in this construction is located at a *focus* of the ellipse. An ellipse has two foci and the distance between the foci plays an important role in

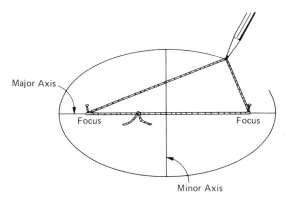

Figure 2-14. The Ellipse. An ellipse may be drawn with the aid of two thumb tacks and a loop of string. Each thumb tack is then located at a focus of the resulting ellipse.

how flat or round the ellipse will be. If the foci are far apart in relation to the size of the loop of string, then the ellipse will be narrow. If the foci are close together, then the ellipse will have a circular appearance. In fact, you can think of a circle as an ellipse in which one focus is on top of the other. In addition, just as we speak of the diameter of a circle, the size of an ellipse can be specified by giving the lengths of its *major axis* and *minor axis.* As shown in the diagram, the major axis is simply a straight line passing through both foci. It is the maximum diameter of the ellipse, and half of this length is called the *semimajor axis.* The minor axis is the perpendicular bisector of the major axis, and the *semiminor axis* is equal to one half of the minor axis.

This new curve provided Kepler with the tools for the first major breakthrough in astronomy since the publication of *De Revolutionibus.* Not only did Kepler succeed in showing that the orbit of Mars is an ellipse, he found that the orbits of *all* the planets are ellipses—provided that you assume the sun to be located at one focus of each elliptical orbit. This discovery was published in 1609 in *The New Astronomy* and has come to be known as *Kepler's first law.* It can be stated concisely as follows:

> *Each planet moves about the sun in an orbit that is an ellipse, with the sun at one focus of the ellipse.*

In addition to establishing the true shapes of the orbits of the planets, Kepler also investigated the speeds of the planets along their orbits. He found that planets move faster when they are near the sun, and slower when they are far from it. After considerable research, he discovered that there was a fairly simple way of expressing how fast a planet moves along its orbit at different times. This discovery is called the law of equal areas, or more commonly Kepler's second law. It also was published in *The New Astronomy* and can be stated as follows:

> *A straight line joining the sun and a planet sweeps out equal areas in space in equal intervals of time.*

To demonstrate the meaning of Kepler's second law, consider the orbit of a hypothetical planet, as shown in Figure 2–15. Of course we now realize that this orbit is an ellipse with the sun at one focus; the other focus is empty. Suppose it takes one month for the planet to go from point *A* to point *B*. In doing so, a line joining the planet and the sun sweeps out a triangular area in space. Kepler's second law tells us that during any other month, a line joining the sun and the planet must sweep out an area equal in size to the original triangular segment. Thus, if it takes the planet one month to go from point *C* to point *D*, the two shaded triangular segments in the diagram must have equal areas.

Although this example illustrates Kepler's second law, it perhaps erroneously suggests that the orbits of the planets are flattened ellipses. Actually, all of the planetary orbits are very nearly circular. If you draw the orbit of Mars to scale on a piece of paper, it will look almost circular. It is, therefore, incredible that

Chapter 2 The Birth of Modern Astronomy

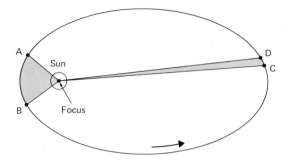

Figure 2-15. The Law of Equal Areas. If it takes a planet one month to go from "A" to "B", it will also take the planet one month to go from "C" to "D" because the shaded areas are equal.

Kepler was able to discover that the orbits of planets are ellipses at all.

The final major contribution Kepler made in building the foundations of modern astronomy is found in his third law. In general terms, this law tells us the size of the orbit of each of the planets. Just like Copernicus, Kepler could not figure out the sizes of the orbits in miles or kilometers. Instead, he chose to express the scale of the solar system relative to the size of the earth's orbit. For this purpose it is extremely useful to define the *astronomical unit* (AU) as the length of the semimajor axis of the earth's orbit. Since the earth's orbit is very nearly circular, we can say that the average distance from the earth to the sun is

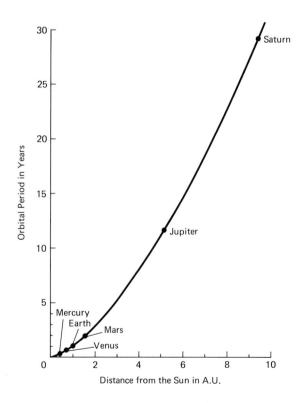

Figure 2-16. Kepler's Third Law. The sidereal periods of the planets are plotted against the sizes of their orbits. Since a smooth curve fits through all the points, there is a simple relationship between these two qualities.

Figure 2–17. Galileo Galilei, 1564–1642. (*Yerkes Observatory*)

1 AU. It was not until the twentieth century that astronomers succeeded in accurately measuring the astronomical unit, arriving at the value of 92,956,000 miles for the average earth-sun distance.

The easiest way to express Kepler's third law is with the aid of a graph, as shown in Figure 2–16. Here the orbital periods (for example, sidereal periods) of the planets are plotted against the lengths of the semimajor axes of their orbits (expressed in AUs). The fact that a smooth curve fits through all the points on the graph proves that there must be a simple mathematical relationship between the sidereal period of a planet and the length of its semimajor axis. This relationship, published in *The Harmony of the Worlds* in 1619, is called Kepler's third law and can be stated as follows:

> *The squares of the sidereal periods of the planets are in direct proportion to the cubes of the semimajor axes of their orbits.*

This may sound complicated, so we will consider Mars as an example. The sidereal period of Mars is 687 days, or 1.88 years. This is how long it takes Mars to go once around the sun. Thus, the "square of the sidereal period" is simply $1.88 \times 1.88 = 3.54$. We then realize that the length of the semimajor axis of Mars' orbit must be 1.52 AU since "the cube of the semimajor axis" $1.52 \times 1.52 \times 1.52$

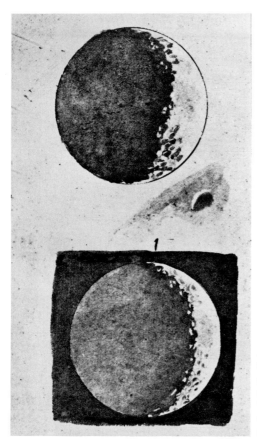

Figure 2–18. Galileo's Drawings of the Moon. When Galileo trained his telescope on the moon, he saw mountains, valley, craters, and plains. This was very troubling to his contemporaries who thought that the moon (up in "heaven") should be perfectly smooth since it was made by a perfect God. (*Yerkes Observatory*)

also equals 3.54. Parenthetically note that given the modern value of nearly 93 million miles for the AU, the average Mars-sun distance must be 330 million miles.

While Kepler was doing primarily theoretical work in northern Europe, important new observations were being made in Italy by Galileo Galilei. In 1609, Galileo heard about the invention of a remarkable device called a telescope. After some experimentation with lenses, he succeeded in building several telescopes, the best having a 30-power magnification. Galileo's major contribution to astronomy is that he was the first person to point the telescope toward the stars. He promptly discovered that the moon was covered with craters and mountains and that the sun had "sunspots." This, of course, was considered heresy at that time since heavenly bodies created by God must be perfect, uncorruptible, and without blemish. So Galileo found himself in big trouble with the Church. Nevertheless, he persevered and discovered that Venus goes through phases, just like our moon. Furthermore, after a couple of years he found that when Venus presents a gibbous phase, it appears small in size. However, when Venus shows a crescent phase, it appears large through the telescope. As shown in Figure 2–19, the relationship between the phases and sizes of Venus constitutes conclusive proof that Venus goes around the sun.

Galileo also noticed that there were four little "stars" near Jupiter. From night

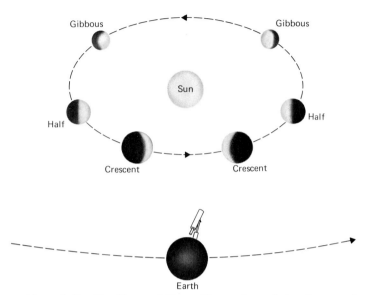

Figure 2-19. *The Phases of Venus.* Observations of the phases and sizes of Venus demonstrate that Venus must be in orbit about the sun, *not* about the earth.

to night these stars seemed to move back and forth with respect to the planet. After only a few nights of observation, it was clear to Galileo that these stars were not stars at all, but rather must be four moons in orbit about Jupiter. News of the discovery of four new moons in the universe spread quickly. Upon learning the details of Galileo's observations, Kepler immediately proved that his third law worked for the moons of Jupiter by demonstrating that the cubes of the distances of the Jovian satellites from Jupiter were proportional to the squares of their orbital periods around the planet.

With the brilliant work of Johannes Kepler and the important observations of Galileo Galilei, it was now totally impossible to go back to a geocentric picture of the universe. On the theoretical side, Kepler had succeeded in formulating the laws of planetary motion about the sun. Galileo had made a number of observations and discoveries that only make sense in a heliocentric cosmology. In spite of the efforts of the Church, the geocentric universe of Ptolemy that had dominated astronomy for twelve centuries was dead and would never be resurrected.

Finally, it should be noted that, similar to that of Copernicus, all of Kepler's work was empirical. In other words, Kepler tried all kinds of different curves for the orbit of Mars until he finally came up with one (the ellipse) that would fit the observations of Tycho Brahe. But it would be much more elegant if we could make some fundamental assumptions about the nature of the physical world and then, using mathematics, actually prove that the orbits of the planets around the sun *must* be ellipses. Accomplishing such a feat would be a monumental achievement of the human intellect.

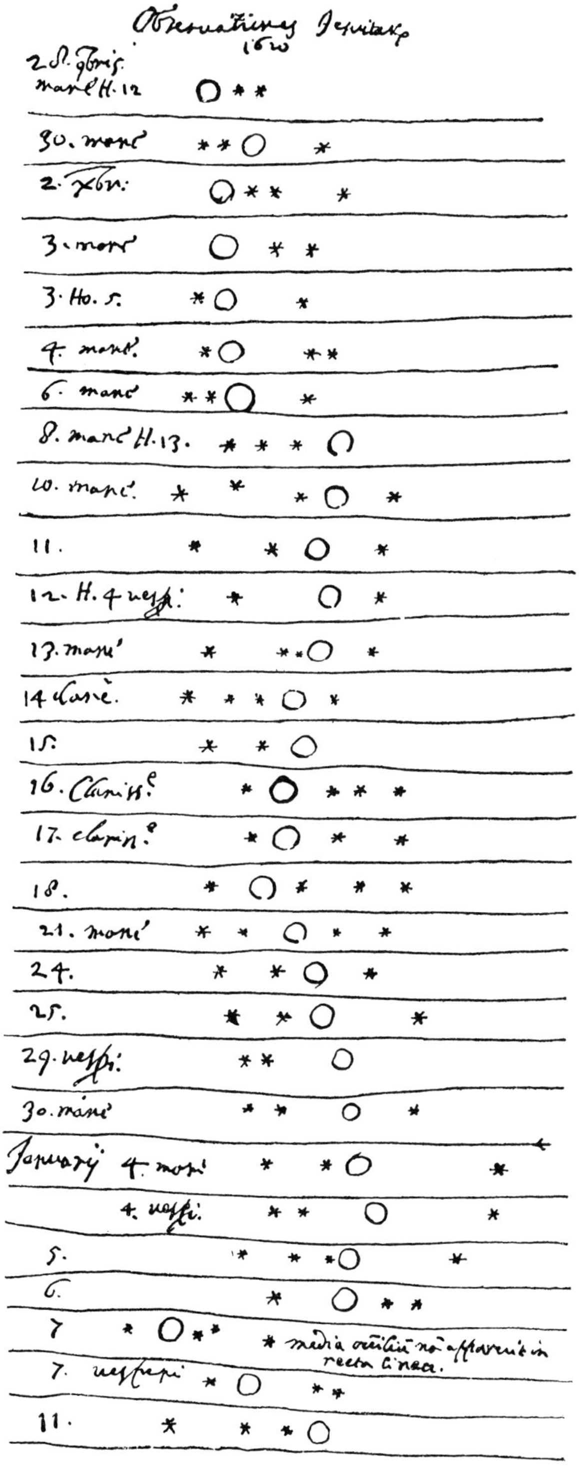

Chapter 2 The Birth of Modern Astronomy

Figure 2–20. *Galileo's Drawings of Jupiter and Its Satellites.* Galileo noticed that there were little "stars" near Jupiter. From night to night, these "stars" seemed to move back and forth from one side of Jupiter to the other. Galileo concluded that Jupiter must have four moons which orbit that planet. (*Yerkes Observatory*)

2.4 Newton and Gravitation

One of the greatest figures in the entire history of science, Sir Isaac Newton, was born in England in 1642, the same year in which Galileo died. From the time of his youth, Newton had been preoccupied with trying to understand the physical reasons behind the heliocentric cosmology of Copernicus and Kepler. What was the physical force that kept the planets going around the sun? It was apparent to Newton that a truly first-rate approach to these and related questions could be achieved only by making some fundamental assumptions about the nature of the universe. Using mathematics he hoped to be able to prove the validity of Kepler's laws.

To begin with, Newton made three basic assumptions about the nature of the physical world, now known as *Newton's laws of motion*. The groundwork for his first law actually had been laid by Galileo several decades earlier. This first law simply states that

> *In the absence of outside forces, a body at rest will remain at rest, or a body in motion will remain in motion at a constant speed in a straight line.*

At first glance, this law of motion might seem unreasonable. For example, if you shove a chair and let it slide across the floor, the chair will come to rest rather than continue "in motion at a constant speed in a straight line." But you realize

Figure 2-21. Sir Isaac Newton, 1643–1727. (*Yerkes Observatory*)

that there were, in fact, "outside forces" acting on the chair. It was the force of friction between the legs of the chair and the floor that caused the chair to slow down. If the floor had been perfectly smooth and frictionless, the chair would indeed have continued unhampered until it hit the wall or some other object.

Using the first law, Newton immediately realized that some kind of force must be acting on the planets. As demonstrated by Kepler, the planets move about the sun in ellipses with the sun at one focus. The planets do not move in straight lines at constant speeds. Therefore, there must be a force constantly acting on the planets causing them to move along elliptical orbits and preventing them from flying straight off into space. Using mathematics, Newton was able to prove that this force on the planets is always pointed directly at the sun, as shown schematically in Figure 2–22. It is as if the sun exerts a force on the planets directly across empty space, thereby keeping them in their orbits.

With further study, Newton realized that he had to make a clear distinction between the two commonly confused concepts of *mass* and *weight*. Every material object has mass. Mass is an inherent property of matter; it is that property of matter that resists a change in its state of motion. If you push on an object, the object will begin to move or to accelerate. If the mass of the object is small, the object will give only a little resistance to the force with which you push it. The object will accelerate very easily. If, however, the mass is large, the object will put up much more resistance to the force with which you push, and therefore the object will accelerate much more slowly. Thus, if the car you are driving runs out of gas and it is a Volkswagen, you will find it much easier to push the car to a gas station than if it is a Cadillac. The mass of a Cadillac is greater than the mass of a Volkswagen. This relationship between mass, force,

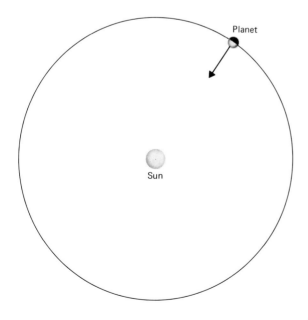

Figure 2–22. An Orbit About the Sun. Since the planets do not travel in straight lines, a force must be acting on them. Newton proved that this force is always pointed toward the sun.

and acceleration is stated concisely in Newton's second law:

> *If a force acts on a body, it produces an acceleration which is proportional to the force and inversely proportional to the mass of the body.*

Although the details of Newton's second law are not important to the basic theme of this text, we nevertheless realize that mass is a fundamental property of matter. An object has a certain mass regardless of where that object is or what it is doing. Thus, it is always more difficult to push a Cadillac than a Volkswagen, regardless of whether these cars are on the earth, on Jupiter, or floating in free space.

Weight, however, is a different subject. The weight of an object is a measure of how hard something pushes downward under the influence of gravity. Thus, a Cadillac weighs more than a Volkswagen because the Cadillac pushes down harder on the pavement of a street than a Volkswagen does. But unlike mass, the weight of an object does depend on where that object is. On Jupiter or on the moon, these cars would weigh differently than here on earth. In fact, floating in outer space, they would have no weight at all; they would be *weightless.*

As a final example of the difference between mass and weight, consider a man who weighs 150 pounds on the earth. It turns out that this same man would weigh only 25 pounds on the moon, whereas he would tip the scales at 380 pounds on Jupiter. Floating in space, he would weigh 0 pounds. Yet, in all these examples, he has exactly the same mass. There are the same number of atoms in his body and his body will give the same resistance to your pushing and shoving regardless of where he is.

In discussing the distinction between mass and weight, we have jumped the gun by introducing the concept of *gravity.* The earth obviously exerts a force on everything around us. This force is called gravity and, without it, tables, chairs, and people would go floating off into space. In a rudimentary fashion, this fact was apparent to any intelligent person in the seventeenth century. Yet, Newton realized that while gravity exerts a force on tables, chairs, and people, these objects exert an equal and opposite force back on the earth. This idea is contained in very general terms in his third law of motion:

> *For every action there is an equal and opposite reaction.*

Thus, for example, if you weigh 150 pounds, you are pushing down on the floor with a force of 150 pounds. But the floor is pushing back up against your feet with a force also equal to 150 pounds. Similarly, if the sun is exerting a force on the planets that keeps them in elliptical orbits, each planet must be exerting an equal and opposite force back on the sun.

In thinking about the concept of gravity, Newton wondered whether this force of gravity could be the same force that causes the planets to stay in orbit about the sun. Perhaps the same force that pulls an apple toward the earth could also

Chapter 2 The Birth of Modern Astronomy

be the force that pulls Jupiter toward the sun. But in order to proceed, he realized that he needed a better understanding of how gravity works.

One reason why mass and weight are so easily confused is that there is an intimate relationship between mass, gravity, and weight. If one object has twice the mass of another object, under the influence of gravity it will also have twice the weight. Similarly, if a Volkswagen weighs 2,000 pounds and a Cadillac weighs 4,000 pounds, by pushing the two cars we find that the mass of the Cadillac is twice the mass of the Volkswagen. From a clear understanding of this relationship between mass, gravity, and weight, Newton was able to conclude that the gravitational force between objects must be proportional to their masses. The bigger the mass, the bigger the force, and conversely. Yet, how does gravity vary with distances between objects? Is the gravitational force between the sun and Mercury larger or smaller than the gravitational force between the sun and Mars? To answer this, Newton realized that Kepler's laws worked. He, therefore, could reduce the question to "What must I *assume* the relationship between gravitational force and distance to be in order to come out with the planets traveling along elliptical orbits according to Kepler's laws?" After some considerable mathematical work, he found a simple answer, which he stated concisely in his *Universal Law of Gravitation:*

> Two bodies attract each other with a force that is proportional to the product of their masses and inversely proportional to the square of the distance between them.

$$F = \frac{M_1 M_2}{R^2}$$

Before we proceed to the remarkable implications of this powerful law, perhaps we should examine a few examples of what it means. Imagine two objects in space. Newton's law says that they will attract each other; they will exert a force on each other that tries to pull them together. The strength of this force first of all depends on the masses of the objects. If you double the mass of one of the objects, the force will be twice as strong. Triple the mass, and the force will go up by a factor of 3. Secondly, Newton's law states that the strength of the force depends "inversely on the square of the distance." To see what this means, suppose these two objects are 1 foot apart and exert a force of 1 pound on each other. If you were to separate them by 2 feet, the force would be reduced to $\frac{1}{4}$-pound because $2 \times 2 = 4$. If the separation were increased to 3 feet, the force would go down to $\frac{1}{9}$-pound since $3 \times 3 = 9$. This behavior of the gravitational force with distance is shown graphically in Figure 2–23.

By way of another example, consider a man weighing 150 pounds standing on the earth. If he overeats substantially, so as to double the number of atoms in his body, his mass will have doubled and he will weigh 300 pounds. On the other hand, if he had gone on a starvation diet, so as to lose half of the atoms in his body, his mass would then be one half of its original value and he would weigh only 75 pounds.

To see the dependence of gravitation on distance, let us assume that our hypothetical man neither diets nor overeats; the number of atoms in his body

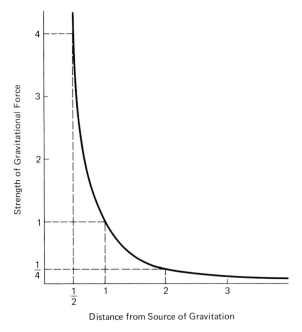

Figure 2-23. Newton's Law of Gravitation. This graph shows how the strength of the gravitational force of a body varies with distance from that body. Move twice as far away, and the force is only one-quarter as great.

remains constant. In standing on the earth, this 150 pound man is 4,000 miles from the center of the earth. If he climbs to the top of a 4,000-mile-long ladder, he is now 8,000 miles from the center of the earth. Since his distance is doubled, a bathroom scale at the top of the ladder will show that he weighs only one-fourth as much as he did originally, or $37\frac{1}{2}$ pounds. It is important to note that this situation is identical to doubling the size of the earth. If we simply were to blow up the earth to twice its original size without adding any new matter (for example, if we double the distances among all the atoms inside the earth), our man would again find himself 8,000 miles from the earth's center and he would again weigh only $37\frac{1}{2}$ pounds. Conversely, if we were to squeeze the earth down to half of its original size, being sure not to gain or lose any matter in the process, our man would be only 2,000 miles from the center of this compressed earth. As a result, his weight would be four times its original value, or 600 pounds. If we were to keep squeezing the earth down to smaller and smaller sizes, even though we did not add any new material, the man's weight would get greater and greater. In fact, if we could squeeze the earth down to zero volume, his weight would be infinite. Although this might sound absurd, we will later find that, under certain conditions, stars can collapse to zero volume, thereby producing incredibly intense gravitational fields and giving rise to *black holes*.

Armed with his laws of motion and the law of gravitation, Newton proceeded to tackle the problem of the solar system. The careful reader will notice that Newton has cheated a little bit. He used Kepler's laws to discover the nature of gravity. Rigorously speaking, however, we now take the position of assuming the validity of the laws of motion and the law of gravitation. From these assump-

Chapter 2   The Birth of Modern Astronomy   49

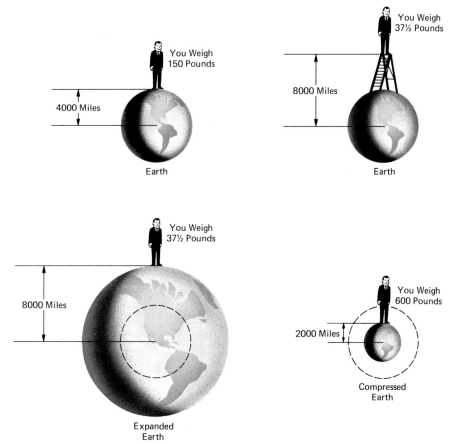

Figure 2–24. *Man and Earth.* This series of examples illustrates Newton's law of gravitation. (See text for discussion.)

tions, Newton found that he could easily prove Kepler's laws as well as predict a great deal more. Specifically, in calculating the orbits of planets around the sun, Newton found that allowable orbits could be any *conic section* and not just ellipses. A conic section is any curve obtained by cutting a cone with a plane, as shown in Figure 2–25. There are four types of curves that result: circle, ellipse, parabola, and hyperbola. Thus, in addition to proving Kepler's first law, Newton showed that parabolic and hyperbolic orbits also were possible. Parabolic and hyperbolic orbits are "open," whereas circular and elliptical orbits are "closed." If an object travels along one of these open curves, it will pass by the sun only once and return to the reaches of interstellar space from whence it came. Comets sometimes travel along such orbits.

The precise orbit that an object follows is determined by how much energy it has. To see how this is so, consider a satellite in a circular orbit about the sun. Suppose this satellite has a rocket attached to it and at some point we turn on the

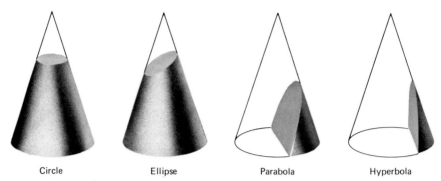

Figure 2–25. *Conic Sections.* A conic section is any one of a family of curves which can be obtained by cutting a cone with a plane. By slicing the cone at different angles you obtain one of four curves: circle, ellipse, parabola, and hyperbola.

rocket engines for a short time. This boost of energy will enable the satellite to get a little farther from the sun, provided the thrust from the rocket is parallel to the direction in which the satellite is moving. As a result of this added speed, the satellite will go into an elliptical orbit. The longer the rocket burns, the bigger the resulting elliptical orbit. In fact, if the rockets are very powerful, the satellite could achieve a speed high enough to escape from the sun's gravitational pull altogether. Such a speed is called the *escape velocity.* An object traveling at the escape velocity moves along a parabolic orbit. And, finally, if the satellite has a speed greater than the escape velocity, it will rapidly fly off into interstellar space along a hyperbolic orbit. These three kinds of orbits are contrasted in Figure 2–26.

The power of Newton's work, or *Newtonian mechanics* as it is sometimes called, lies in the fact that in addition to explaining everything that had been

Figure 2–26. *Orbits About the Sun.* Newton proved that orbits about the sun can be any conic section (not just ellipses).

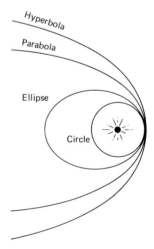

known earlier, his methods could be used to predict new phenomena. For example, Newton's good friend Edmond Halley had a great interest in comets. In particular, he noticed that the bright comets seen in 1531, 1607, and 1682 seemed to have almost identical orbits. After some study, he concluded that these three comets were actually the *same* comet returning to the sun on three separate occasions. Using Newton's ideas about orbits, Halley predicted the return of this comet in 1758. The comet was sighted on Christmas night of that year, thereby providing a dramatic verification of Newtonian mechanics. Since that time this comet has been known as *Halley's comet*. It has been observed since 240 B.C. and passes near the sun approximately every 76 years.

It is perhaps instructive to mention a few more details about Halley's comet. Actually, the orbit is *not* a perfect ellipse with the sun at one focus. This does not mean that Kepler's first law is wrong, but rather that the orbit is affected by the planets, especially Jupiter and Saturn. When the comet passes near a planet, the gravitational force of the planet pulls on the comet. This causes the comet to deviate slightly from its elliptical orbit. Such minor variations are called *perturbations*. In calculating the orbit of any object about the sun, astronomers must be sure to include perturbations; they can be important.

One of the most dramatic confirmations of Newtonian theory occurred in the mid-nineteenth century. In March 1781, the great astronomer Sir William Herschel was making a routine survey of the sky with his telescope. While examining a region near Gemini, he noticed a faint fuzzy object that he initially thought to be a comet. After several weeks of observation, however, it became apparent that its orbit was almost a perfect circle outside the orbit of Saturn. This object unquestionably was a new planet and was named Uranus.

By the early 1800s a precise orbit for Uranus had been calculated. Its sidereal

Figure 2–27. *Halley's Comet.* Using Newton's ideas about orbits, Edmond Halley correctly concluded that the bright comets seen in 1531, 1607, and 1682 were the *same* comet. This comet is due to return in the spring of 1986. (*Lick Observatory*)

period is 84 years and the average Uranus-sun distance came out to be 19AU. However, by 1840 it was clear that Uranus was *not* following its predicted orbit. For some mysterious reason, this new planet was deviating slightly from its predicted path as though something was pulling on it. Several years after this peculiar behavior of Uranus had been noticed, two astronomers independently hypothesized that there must be another planet beyond Uranus. They found that they could account for the mysterious motions of Uranus in terms of the perturbations caused by this new, more distant planet. Their calculations predicted the positions of this unknown planet. On September 23, 1846, the predicted positions were received at the Berlin Observatory. Astronomers turned their telescopes to this location in the sky and on that very same evening the planet Neptune was discovered.

This discovery of the eighth planet from the sun was a great triumph for Newtonian mechanics. In a very real sense, the discovery was made with pencil and paper. Newton's laws and the mathematical tools he developed were so universal and powerful that astronomers could theoretically predict the existence of a new planet. As a result of these and similar triumphs, Newtonian theory has become a major cornerstone in all of physical science. Even today scientists essentially use Newton's same ideas to calculate the trajectory for the Apollo astronauts to the moon or the orbits of interplanetary spacecrafts to Jupiter and beyond.

Although Newtonian theory is extremely successful, we will see that in the early 1900s Albert Einstein proposed an even better theory of gravitation. In this new theory, the *general theory of relativity*, gravity is not talked about as a force at all. Rather, Einstein preferred to speak in terms of the geometry of warped space-time. And, just as Newton's theory predicted the existence of a new planet, Einstein's theory predicts the existence of bizarre objects called black holes.

Questions and Exercises

1. Who was Anaxagoras?
2. Explain why we see phases of the moon.
3. What is the difference between waning gibbous and waxing gibbous phases of the moon?
4. Approximately how long does it take the moon to go once through all its phases?
5. Approximately at what time of the day does a new moon rise? When does it set?
6. What time of the day does the first quarter moon rise? When does it set?
7. Explain what happens to cause an eclipse of the moon.

Chapter 2 The Birth of Modern Astronomy

8. Who was Eratosthenes?
9. Who as Aristarchus?
10. What is the difference between retrograde and direct motion of a planet?
11. Briefly describe Ptolemy's geocentric cosmology. How did he account for retrograde motion?
12. What is the *Almagest?*
13. How did Copernicus account for the retrograde motion of the planets? How did his explanation differ from that of Ptolemy?
14. What is *De Revolutionibus?*
15. What is the order of the planets from the sun?
16. How many inferior planets are there? What are their names?
17. Why can't Jupiter appear at inferior conjunction?
18. What is meant by greatest eastern elongation and greatest western elongation? To which planets do such terms apply?
19. What is the difference between the sidereal period of a planet and the synodic period of a planet?
20. Who was Tycho Brahe?
21. How would you draw an ellipse?
22. How are ellipses related to the orbits of planets about the sun? Who discovered such a relationship?
23. Explain how Kepler's second law tells us that planets move faster when they are near the sun and slower when they are far away.
24. What is an AU?
25. Who wrote *The New Astronomy* and *The Harmony of the Worlds?*
26. What major contributions did Galileo make to astronomy?
27. Explain why the phases and sizes of Venus clearly demonstrate that Venus goes around the sun and *not* around the earth.
28. Explain the difference between mass and weight.
29. How did Newton conclude that a force must be acting on the planets?
30. Briefly describe the sequence of events or reasons leading up to Newton's formulation of his universal law of gravitation.
31. What is a conic section?
32. What is meant by escape velocity?
33. Who was Edmond Halley?
34. Describe the sequence of events leading up to the discovery of the planet Neptune? Discuss some of the implications of this discovery.

The Earth in Astronomy

3.1 The Telling of Time

Ever since men first looked up into the sky, one of the primary tasks of astronomers has been to establish ways of telling time. The need for accurate time telling is universal. The Pharaoh wanted to know when the Nile would flood. The timing of religious events was extremely important to the Hebrews and Moslems. Even today, if we were to lose track of April 15 we would find ourselves in deep trouble with the Internal Revenue Service.

For ancient man there were three natural clocks in the sky. The rising and setting, or *diurnal motion*, of the sun and stars provided a way of measuring day and night. The phases of the moon defined the length of a month. And the apparent motion of the sun along the ecliptic gave rise to the concept of a year.

Unfortunately, however, nature has not been very kind to us. For example, there is no integer number of days in the year; the year lasts for approximately $365\frac{1}{4}$ days. There is no integer number of days in the lunar month; it takes the moon about $29\frac{1}{2}$ days to go through all its phases. As a result, there is no integer number of lunar months in a year. Since $365\frac{1}{4}$ divided by $29\frac{1}{2}$ equals $12\frac{2}{5}$, the moon goes through slightly more than 12 cycles of its phases as the earth goes once around the sun.

These and similar complications meant that better ways of measuring time had to be devised. Since the earth rotates, it is quite natural that we would want to measure time in such a way that it is related to the earth's rotation. This was accomplished by defining the *meridian*, an imaginary line in the sky passing through the north and south points of the horizon and the zenith overhead. With the aid of the meridian, a means of measuring the diurnal motion of objects such as the sun and stars is established. Indeed we can now invent different kinds of "days."

The simplest and most obvious kind of day is the *apparent solar day*. A solar day is the length of time it takes for the sun to go from one passage or *transit* across the meridian to the next. It is, therefore, the time from one high noon to the next. A clock based on the apparent solar day would measure *apparent solar time*. Sundials measure apparent solar time.

Astronomers find it useful to define a different kind of day relative to succes-

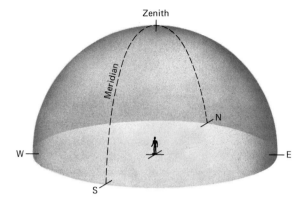

Figure 3–1. *The Meridian.* The meridian is an imaginary line running from north to south and passing through the zenith. The transits of celestial objects can be used to measure time.

sive transits of some object on the celestial sphere other than the sun. In other words, for the purpose of his observations, the astronomer prefers to measure time according to the stars rather than the sun. To accomplish this, recall from Chapter 1 that right ascension is measured in units of time ($0^h00^m00^s$ to $24^h00^m00^s$) eastward from the vernal equinox. We can, therefore, define the *sidereal day* as the time between successive transits of the vernal equinox. *Sidereal time*, which is based on the sidereal day, is very useful to the astronomer. There is a simple relationship between the right ascension of an object and the sidereal time at which the astronomer wants to observe that object. This allows the astronomer to point his telescope accurately at whatever he wants to study.

To a high degree of accuracy, we may think of the solar day as measured relative to the sun and the sidereal day as measured relative to the stars. They are not equal. The solar day is about 4 minutes longer than the sidereal. The reason for this is shown in Figure 3–2. As the earth rotates about its axis, it also revolves about the sun. During the course of one sidereal day, the earth rotates once with

Figure 3–2. *The Sidereal and Solar Days.* The sidereal day is measured in reference to the stars, while the solar day is measured in reference to the sun. Because the earth is moving along in its orbit, the solar day is a little longer than the sidereal day.

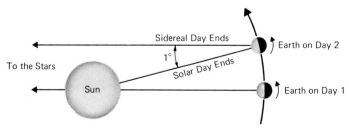

respect to the stars. But during a solar day, the sun appears to move slightly (approximately 1° per day, eastward) along the ecliptic. Therefore, to complete 1 solar day, the earth has to rotate a little bit more to catch up with the sun. During a sidereal day, the earth rotates through 360° with respect to the stars. But during a solar day the earth rotates through 361° with respect to these same stars. This extra 1° of rotation translates into 4 minutes of time.

At this point you might think that we have come to the end of the story. You might be inclined to say "Let's leave sidereal time to the astronomers. I will measure my day by the sun, since when I am awake or asleep depends on the sun, not the stars. I will define the apparent solar day to be exactly 24 hours long and be done with it." No such luck. Nature does not oblige us by taking such a simple approach. The earth's orbit around the sun is not a perfect circle; rather, the orbit is an ellipse with the sun at one focus. Kepler's second law says that the earth is moving faster at *perihelion* (when it is nearest the sun) than at *aphelion* (when it is farthest from the sun). This variable speed of the earth along its orbit means that the speed of the sun along the ecliptic will not be constant throughout the year. In other words, the exact length of a solar day changes slightly over the course of a year. If you had said that there will be 24 hours in an apparent solar day, then you would find that the duration of an "hour" or a "minute" would vary slightly from one day to the next in order to agree with the sun. This would be terribly inconvenient.

So the sun is really not a good timekeeper. Timekeeping would be a lot simpler if the sun moved at a constant speed in the sky. To circumvent this difficulty, an imaginary "average" sun was devised. This imaginary sun, which is called the *mean sun*, is sometimes a little ahead of or behind the real sun in the sky. The real sun and the mean sun are displaced by just enough so that the mean sun does in fact move along the celestial equator at a constant speed. The *mean solar day* is, therefore, the time between successive transits of the mean sun. By doing things in this fashion, we have guaranteed that the mean solar day is equal to the average length of an apparent solar day. On any given date, the apparent solar day will be a little longer or a little shorter than this average value, but the mean solar day is constant throughout the entire year. Twenty-four hours on your wrist watch or clock is exactly one mean solar day. *Mean solar time* is based on the mean solar day and passes at a uniform rate throughout the year.

From what we have said, it might seem that your wrist watch measures mean solar time. This is not true, and the reason does not have anything to do with astronomy. Recall that time is measured by transits of the meridian. For example, 12:00 noon in mean solar time is defined as the instant that the center of the imaginary mean sun crosses the meridian. But there are different meridians for different locations on the earth. Someone in San Francisco has a different zenith and meridian than someone in Los Angeles. Therefore, strictly speaking, there is a small difference in mean solar time at these cities. If you traveled from San Francisco to Los Angeles and you wanted your wrist watch to read mean solar time, you would constantly be having to reset your clock during your journey. This would be extremely inconvenient.

Chapter 3   The Earth in Astronomy   57

Figure 3–3.   Time Zones. The earth is divided up into time zones so that people living in various geographic areas have synchronized clocks. The standard time in a zone may be thought of as an *average* mean solar time across the zone.

   To cope with this difficulty, it is advantageous for everyone in nearby cities to agree on the same time. For this purpose, *time zones* were invented. By mutual agreement, all the clocks of everyone living in a particular time zone are synchronized. Thus, all the clocks from Seattle to San Diego read the same time, Pacific standard time.
   By dividing the world into time zones we say that we have standardized mean solar time for particular regions. There are four such time zones in the continental United States, and clocks across the country read one of four types of time: eastern standard time (EST), central standard time (CST), mountain standard time (MST), and Pacific standard time (PST). Each time zone differs by one hour from each adjacent time zone so that, for example, 10:00 A.M. EST in New York is the same as 7:00 A.M. PST in Los Angeles. These time zones are selected so that the standard time in a zone is approximately equal to the average mean solar time for all cities in that time zone. Thus, central standard time on a clock in Chicago is near, but not exactly equal to, mean solar time for that precise location on earth.
   During the summer months, the sun is in the sky longer than during the winter months. The sun rises earlier and sets later in June than it does in December. To take advantage of these extra daylight hours, *daylight savings time* was invented. From March through October clocks are set ahead by one hour. The reason for doing this is that it seems to be more convenient for the average person. For example, on a particular day during the summer the sun would rise at 4:00 A.M. and set at 8:00 P.M. if clocks measured standard time. But with daylight savings time, the sun on that day would rise at 5:00 A.M. and set at 9:00 P.M., which more closely reflects when the typical person is awake or asleep.
   At this point, our understanding of time can be summarized as follows:

   1. Apparent solar time is based on the sun. The length of an apparent solar

day varies throughout the year from about 23 hours, $59\frac{1}{2}$ minutes to 24 hours, $\frac{1}{2}$ minute.

2. Sidereal time is based on the stars. The length of a sidereal day is 23 hours, 56 minutes, 4 seconds.

3. Mean solar time is based on an imaginary mean sun. The length of a mean solar day is exactly 24 hours and is constant throughout the year.

4. Apparent solar time, sidereal time, and mean solar time are measured with reference to the local meridian of an observer. They all depend on the location of the observer on the earth.

5. Standard time was invented so that all the clocks in a time zone would be synchronized. In a time zone, standard time may be thought of as the *average* mean solar time.

6. Daylight savings time was invented to take advantage of the extra daylight hours during the summer months. To get daylight savings time, add one hour to standard time.

Therefore, the time on a wrist watch is based partly on astronomy and what is seen in the skies, and partly on what is convenient and useful. The same is true with the calendar and the year. The problem with the year is that the earth rotates $365\frac{1}{4}$ times on its axis in one complete revolution about the sun. More precisely, the *tropical year* is defined as the time it takes the sun to go once around the ecliptic with respect to the vernal equinox. The tropical year turns out to be equal to 365.242199 mean solar days. It is this fraction of a day that causes some complications.

First of all, suppose the year were exactly 365 days long. This means that we would be in error by about 1 day every 4 years. As a result, the date of the first day of spring would gradually change. After a few decades, the first day of spring would actually occur in April, rather than on or near March 21. And after a few centuries, summer would occur during December and January. It was generally agreed that this would be an undesirable situation, and so a method was invented of setting up the calendar to assure that this would not happen.

About 2,000 years ago, it was apparent that the method of telling time was a complete mess. Every town and city in the Roman Empire had its own calendar, few of which were the same. Thus, a traveler would find himself going from year to year as he went from one town to the next. To straighten things out, Julius Caesar instituted calendar reform in 46 B.C. According to the best information available, the length of the year was $365\frac{1}{4}$ days. In order to account for this fraction of a day, Caesar decreed that every 4 years an extra day would be added. Those years with 366 days are called *leap years*, and the method is still used. Once every 4 years we add February 29. Every year that is evenly divisible by 4 is a leap year. Thus, for example, 1976, 1980, and 1984 are leap years.

This system would work just fine if the length of a year were 365.250000 mean solar days instead of 365.242199. But this difference amounts to the loss of 1 day every 128 years. Thus, by the time of the Renaissance, the calendar was again in need of revision. Indeed, during the reign of Pope Gregory XIII, the first day of spring was occurring on March 11.

Chapter 3 The Earth in Astronomy

In 1582, Pope Gregory instituted a further calendar reform. In order to get closer to the true value of the tropical year, it was decreed that in addition to Caesar's method for establishing leap years, centuries not divisible by 400 would not be leap years. Thus, 1700, 1800, and 1900, which are evenly divisible by 4 but not by 400, are common years in Pope Gregory's system.

In the Gregorian calendar, the length of the year comes out to be 365.2425 mean solar days. The difference between this value and the true value amounts to an error of only 1 day in 3,300 years.

The calendar we use today is the Gregorian calendar, with one small modification. The years 4000, 8000, 12000, and so on, which would have been leap years according to Pope Gregory, will now be common years. The calendar is, therefore, accurate to 1 day in 20,000 years. As a result of all these rules, reforms, and regulations, we can all rest comfortably, secure in the knowledge that for many years to come, it will snow on Christmas and be hot and sticky on the Fourth of July, not vice versa.

3.2 The Seasons

The existence of the four seasons is apparent to anyone who has taken the time to observe nature. Long, cool winter nights and warm summer days have a direct effect on the world around us and on our behavior. Even the clothes we wear are determined by the seasons. It is also obvious that the seasons have something to do with astronomy, with the relationship between the earth and the sun.

A common misconception is that the seasons are a result of the distance between the earth and the sun. It is erroneously assumed the earth is near the sun during the summer and farther away during the winter. In the Northern Hemisphere, this is an incorrect and backward assumption. The earth's orbit is indeed elliptical and the earth-sun distance varies from about $91\frac{1}{2}$ million miles at perihelion to about $94\frac{1}{2}$ million miles at aphelion. But perihelion occurs during the first week in January, whereas aphelion occurs during the first week of July. Thus, it is winter in the United States when the earth is nearest the sun and summer when the earth is farthest from the sun. The small variation in the distance between the earth and the sun has virtually no effect on the seasons.

The real cause of the seasons has to do with the fact that the earth's axis of rotation is not exactly perpendicular to the earth's orbit around the sun. As shown in Figure 3-4, the earth's axis is inclined at an angle of $23\frac{1}{2}°$ with respect to the perpendicular. This tilt of the earth's axis is also reflected in the fact that the ecliptic and the celestial equator are inclined at an angle of $23\frac{1}{2}°$, as was mentioned in Chapter 1.

To see why the inclination of the earth's axis is responsible for the seasons, consider someone living in the continental United States. During the summer months, the North Pole of the earth is tilted toward the sun. This means that the

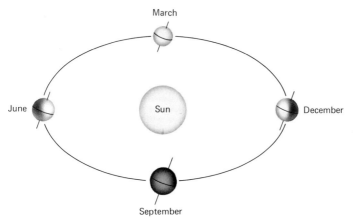

Figure 3–4. *The Inclination of the Earth's Axis.* The earth's axis of rotation is *not* perpendicular to the earth's orbit. The axis is tilted at an angle of $23\frac{1}{2}°$ and therefore the sun appears south of the celestial equator for six months of the year and north of the celestial equator for the remaining six months.

sun is north of the celestial equator and, therefore, the sun appears high in the sky for most of midday. But during the winter months, the North Pole of the earth is tilted away from the sun. As a result, the sun is south of the celestial equator, and even at noontime the sun will appear low in the sky, near the southern horizon.

As a result of the varying positions of the sun in the sky over the course of a year, the number of daylight hours also changes from month to month. During the summertime, the sun rises in the northeast, passes high overhead, and sets in the northwest. But during the wintertime, the sun rises in the southeast, passes low in the southern sky, and sets in the southwest. As shown in Figure 3–5, the path of the sun in June across the sky is much longer than the path of the sun

Figure 3–5. *The Path of the Sun Across the Sky.* As seen by a typical observer in the United States, the path of the sun across the sky varies during the year. In the summer months, the sun passes high overhead, while during the winter it is low in the southern sky.

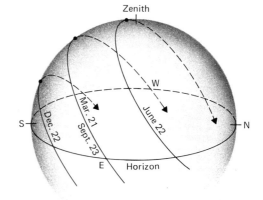

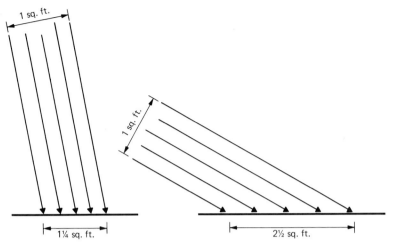

Figure 3-6. *The Heating Effect of the Sun's Altitude.* When the sun is high in the sky, one square foot of sunlight is concentrated onto a smaller portion of the earth's surface than when the sun is low in the sky.

across the sky in December. Therefore, in order to make its long journey across the sky in the summer, the sun must spend more hours above the horizon than in the winter. Thus, a typical person in the United States might have 16 hours of daylight in June as opposed to only 8 hours of daylight in December. On days when the sun is in the sky for a long period of time, the air and the ground will be heated to a higher temperature than on days when the duration of daylight is short. This is one of the primary reasons why the summers are hot and the winters are cold.

Although the duration of daylight plays an important role in the seasons, there is a second important effect. When the sun is high in the sky, a patch of the earth's surface receives more concentrated sunlight than when the sun is low, nearer the horizon. As shown in Figure 3-6, at high noon in summer, 1 square foot of sunlight illuminates only $1\frac{1}{4}$ square feet of the earth's surface. But at high noon in the winter, since the sun is low in the sky, this same square foot of sunlight is spread over nearly $2\frac{1}{2}$ square feet of ground. Thus, the sun heats up the ground much more efficiently in July than in January. Therefore, the seasons are caused by the combined effect of the duration of daylight and the elevation of the sun above the horizon.

As the sun moves along the ecliptic, it reaches its most northerly point on June 22. On that date, the sun is $23\frac{1}{2}°$ north of the celestial equator; this point on the ecliptic is called the *summer solstice*. Therefore, June 22 is called the *date of the summer solstice*; it is the first day of summer. Since the sun is at its most northern declination at the time of the summer solstice, the sun is above the horizon longer than at any other time during the year. Roughly speaking, June 22 is the "longest day of the year."

Similarly, we notice that the sun is at its southernmost declination, $23\frac{1}{2}°$ south of the celestial equator, on December 22. The location of the sun on the ecliptic on this day is called the *winter solstice*. On this *date of the winter solstice*, the sun spends the least amount of time above the horizon. December 22 is the "shortest day of the year"; it is the first day of winter.

In between the summer and winter solstices there are two times when the sun is exactly on the celestial equator. As we saw in Chapter 1, one of these points is called the *vernal equinox*. The sun is at the vernal equinox on March 21, the first day of spring. Similarly, on the first day of autumn, September 23, the sun is again on the celestial equator at a point called the *autumnal equinox*. On the first day of spring and the first day of fall, the sun rises exactly in the east and sets exactly in the west. Daylight and nighttime each last for 12 hours.

For the purposes of illustration, it is useful to restrict our discussion of the seasons to the Northern Hemisphere. In Australia, however, it is hot in December and cool in June. The first day of spring for someone in Canada is the first day of autumn for someone in New Zealand. The seasons in the Southern Hemisphere are simply reversed. When the North Pole is tilted away from the sun, giving rise to winter at the northern latitudes, the South Pole is tilted toward the sun, thus producing summer for someone living "down under."

In thinking about the seasons and the inclination of the ecliptic to the celestial equator, there are some places on the earth where strange things happen. At the North Pole and at the South Pole, daylight and nighttime each last for 6 months. At the North Pole, the sun rises on the first day of spring and stays above the horizon until the first day of autumn, half a year later. The maximum altitude of the sun above the horizon is $23\frac{1}{2}°$, which occurs on the date of the summer solstice. The situation is reversed for the South Pole.

Actually, there is a region of the earth surrounding the North Pole where the sun stays above the horizon for a full 24 hours on at least 1 day during the year. On such days, the midnight sun can be seen. The southernmost limit of this region is called the *Arctic Circle*, $23\frac{1}{2}°$ south of the pole (at a latitude of $66\frac{1}{2}°$ north). On the Arctic Circle, the midnight sun can be seen only on one day, the date of the summer solstice.

A similar situation exists for a region surrounding the South Pole. The northernmost limit of the midnight sun is called the *Antarctic Circle*, $23\frac{1}{2}°$ north of the pole (at a latitude of $66\frac{1}{2}°$ south). In between the arctic and antarctic circles, the midnight sun can never be seen.

On the equator, the sun is directly overhead at high noon on the first day of spring and the first day of autumn. During the summer months, the sun always passes north of the zenith, whereas during the winter months, the sun is always south of the zenith at high noon. There must be a region of the earth surrounding the equator where the sun appears directly overhead on at least one day during the year. The northernmost limit of this region is called the *Tropic of Cancer* at a latitude of $23\frac{1}{2}°$ north. On the Tropic of Cancer, the sun is at the zenith at high noon only on the date of the summer solstice. Similarly, the southernmost limit of this region is called the *Tropic of Capricorn* at $23\frac{1}{2}°$ south.

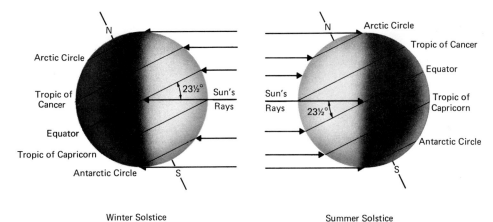

Figure 3–7. *The Earth at the Solstices.* At the time of a solstice, the sun is at its greatest declination, either north or south of the celestial equator. At this time of the year, one of the poles is experiencing 24 hours of continuous daylight, while the other is in constant darkness.

At this tropic, the sun is directly overhead only at noon on the date of the winter solstice. These various regions of the earth are illustrated in Figure 3–7, for the dates December 22 and June 22.

From this brief discussion of the seasons note that some of the most fundamental properties of our environment are related directly to astronomical considerations. Whether a certain region on the earth will support a tropical jungle or will be eternally covered with ice and snow is entirely a result of the relationship between the sun and our planet.

3.3 Tides and Precession

In the previous discussion of the seasons, we found that the relationship between the earth and the sun has a profound effect on the environment. Yet, this discussion dealt only with the relative orientations of the earth, its axis of rotation, and the sun. To understand the cause of the seasons, it was not necessary to talk about the gravitational forces between the earth and the sun. Of course, all the bodies in the solar system exert gravitational forces on the earth. Since the sun is tens of thousands of times more massive than anything else in the solar system, the orbits of the planets are completely dominated by the enormous gravitational field of the sun. But to a much lesser extent, all the other bodies in the solar system are pulling on each other. Thus, as we learned at the end of Chapter 2, the orbits of the planets are not exactly perfect ellipses. The earth's orbit, for example, deviates slightly from an ideal ellipse due to the

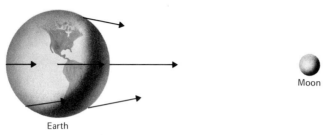

Figure 3-8. *The Moon's Pull on the Earth.* The gravitational force of the moon varies from one side of the earth to the other. The force is strongest nearest the moon, and weakest on the opposite side of the earth.

gravitational perturbations of the moon and other planets. But there are other ways in which the gravitational fields of the sun and the moon profoundly affect the earth.

Consider the earth and the moon, as shown in Figure 3-8. There are some parts of the earth that are nearer to the moon than others. Since the strength of gravity varies with distance, those parts of the earth nearest the moon experience a greater attraction toward it than those parts that are farther away. In other words, there will be a difference in the gravitational pull of the moon from one side of the earth to the other. This variation across the earth is called a *differential gravitational force*, or *tidal force*; it tries to deform the earth by stretching it slightly in the direction of the moon. Of course the earth does not stretch too easily, but its surface is covered with water. Since water flows more easily than rocks, the oceans of the earth are deformed by this differential gravitational force. The final result is that we observe *tides*.

Due to the fact that the moon is much closer to the earth than any other body in the solar system, the differential gravitational force of the moon dominates and controls the tides. Obviously a high tide occurs on that part of the earth closest to the moon since, at that region, the moon's pull is greatest; the water in the oceans is drawn to that area. As a result, the time of high tide at a certain location will be near the time at which the moon is on the meridian. But, in

Figure 3-9. *The Shape of the Oceans.* The tidal force of the moon deforms the shape of the oceans. High tides occur at locations nearest and farthest from the moon.

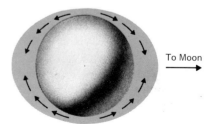

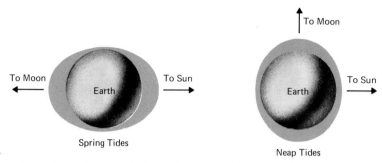

Figure 3–10. *Spring and Neap Tides.* Spring tides occur when the sun, moon, and earth are lined up. Neap tides occur when the moon and sun are at right angles, at first and last quarter.

addition, a second high tide will occur 12 hours later. This is because the oceans are deformed into the shape of an "oblate spheroid," which looks like a flattened ball, as shown in Figure 3–9. Although the pull of the moon is strongest on that part of the earth closest to the moon, the pull is weakest on the opposite side of the earth, weaker than at the earth's center. Since the earth receives an intermediate pull from the moon, the earth takes an intermediate position between the oceans on either side. Therefore, the net effect produces a high tide nearest the moon and a high tide farthest from the moon. Thus, if one high tide occurs at 10:00 A.M., a second high tide will occur around 10:00 P.M., with low tides in between at approximately 4:00 A.M. and 4:00 P.M.

Although the tides are dominated by the moon, our sun does have a noticeable, if smaller, effect. As you might expect, when the sun and moon are lined up, the combined tidal forces of both bodies tend to exaggerate the tides. Thus, at new moon and at full moon, the high tides are very high and the low tides are very low. This situation is referred to as *spring tides,* shown in Figure 3–10. But when the sun and moon are at right angles, at first and last quarter, the tidal forces of these two bodies act against each other. The variation in the heights of the tides is least pronounced at these times, resulting in the so-called *neap tides.*

Ancient man was well aware of the relationship between the tides and the moon, thousands of years before Newton and gravity. Any sailor knew, for example, that high tide occurred when the moon was high in the sky. Low tides were known to occur at approximately the time of moonrise and moonset. Yet, these ancestors were also aware of a far more obscure effect of the moon and the sun on the earth. Through centuries of observations it had become apparent that the vernal equinox was moving slowly among the fixed stars.

The tides are an obvious, easily observable phenomenon related to the gravitational pull of the moon and sun, but the gravity of these two bodies has another important effect on the earth. It is well known from geography that the earth is not a perfect sphere but rather has an *equatorial bulge.* Our earth is fatter at the equator than at the poles; the diameter of the earth at the equator is about

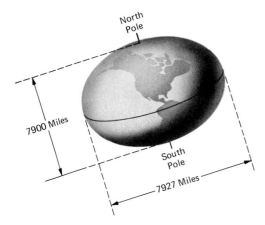

Figure 3–11. *The Equatorial Bulge.* The earth is fatter at the equator than at the poles. This bulge is tilted at an angle of 23½° with respect to the ecliptic.

27 miles longer than the distance from pole to pole. But it is also known that the earth's axis of rotation is tilted by 23½° away from the perpendicular direction to the earth's orbit. Therefore, this equatorial bulge is tilted by 23½° out of the plane of the ecliptic. Since the moon is always very near the ecliptic, both the sun and the moon are usually either above or below this equatorial bulge, as shown in Figure 3–11. Because the equatorial bulge of the earth is inclined to the ecliptic, the gravitational forces of the sun and moon try to "straighten up" the earth; they try to tip the earth up so that the equatorial bulge lies in the plane of the ecliptic. If the earth did not rotate, billions of years ago the sun and moon would have tipped the earth up so that today the earth's axis would be exactly perpendicular to its orbit. But since our planet still has seasons, it is obvious that this has not happened. To see why this is so, we will examine the behavior of a child's spinning top.

Imagine a child playing with a toy top on a sidewalk. With the aid of a string, he starts the top spinning and sets it down on the ground. Of course, gravity is acting on the top and we might expect to see the top fall over promptly. This does not happen. Rather, the spinning top begins to wobble around in a circle even though the top is inclined at a precarious angle. The axis of rotation of the top slowly traces out a circle, as shown in Figure 3–12. This motion is called *precession*. The reason for this unique motion is, in essence, that the axis of rotation of the top tries to remain fixed in space. But, simultaneously, the earth's gravity tries to pull the top over. The combination of these two actions produces precession. Precession may be thought of as a compromise between the top's desire to keep its axis of rotation pointed in a constant direction and gravity's desire to pull the top over.

This situation is analogous to the earth. The child's top is not spherical; it is fatter in one direction than the other. The earth is not spherical; it has an equatorial bulge. Just as gravity does not succeed in immediately knocking over the rotating top, the sun and moon do not succeed in straightening up the

Chapter 3   The Earth in Astronomy

rotating earth. And just as the top slowly precesses, our earth also slowly precesses.

The rate of precession in the case of the earth is extremely slow. Gradually, from one year to the next, the North Pole of the earth traces out a circle among the stars. Right now, the north celestial pole is near the star Polaris at the end of the handle of the Little Dipper. When the pyramids were built in ancient Egypt, the star Thuban, in the constellation of Draco, was the North Star. In 12,000 years, the bright star Vega will be very near the celestial pole. It takes 26,000 years for the earth's pole to complete one circle among the stars, as shown in Figure 3–13.

There is one very important side effect that results from the precession of the earth. Obviously, if the north celestial pole moves among the stars, the south celestial pole must also be moving among the stars, tracing out a similar circle in the southern skies. But if both poles are moving, then the celestial equator must be gradually shifting its orientation in space. As the celestial equator changes its position, the vernal equinox, which is the point where the ecliptic and the celestial equator cross, also must be moving gradually. In other words, over the centuries, the location of the sun on the first day of spring moves slowly among the constellations of the zodiac. The vernal equinox is now in the constellation of Pisces and will soon enter the constellation of Aquarius. This is what astrologers and other misguided people mean by the "dawning of the Age of Aquarius."

More importantly, recall that we tied our coordinate system of right ascension and declination to the location of the vernal equinox. Right ascension is meas-

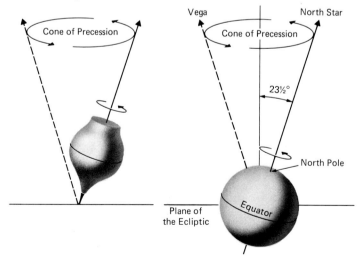

Figure 3–12. Precession. Due to the action of gravity on a spinning top, the top precesses. Similarly, due to the action of the gravitational forces of the moon and sun on the earth's equatorial bulge, the earth's axis of rotation also precesses.

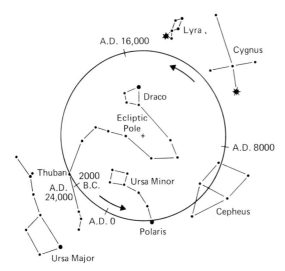

Figure 3–13. *The North Celestial Pole.* Due to precession, the north celestial pole is gradually moving among the stars. At this time, Polaris is the "pole star."

ured eastward from the vernal equinox, which we now realize is moving. As a result of this gradual motion of the vernal equinox, the coordinates of the stars listed in a star catalogue will change by small amounts from one year to the next. For example, the right ascension and declination of the bright star Regulus, in Leo, the lion, are listed as

$$\alpha = 10^h 3^m 3^s$$
$$\delta = +12°27'21''$$

in a star catalogue prepared for 1900. But, in 1950, the position of this same star was given as

$$\alpha = 10^h 5^m 43^s$$
$$\delta = +12°12'44''.$$

The star has not moved; the coordinate system has moved. This does *not* mean that we have done things wrong. We are located here on the earth and it is very convenient and natural to use a coordinate system tied to the earth in order to express the locations of the stars. Unfortunately, there is a minor complication. The earth precesses. This is dealt with by simply noting in all our star catalogues the year, or so-called *epoch*, for which the catalogue was prepared. Most star charts and catalogues in use today are for epoch 1950, which means that the positions given in the books are precise for the first day of the year 1950. Of course, such positions are incorrect today, but the error is very small and a professional astronomer who needed precise positions would know how to make the necessary minute corrections.

This phenomenon of precession was first discovered more than two thousand years ago by Hipparchus, who correctly deduced that the north celestial pole is gradually changing its position in the sky. Both Ptolemy and Hipparchus measured the rate at which the earth precesses and obtained values that were fairly accurate. This rate of precession can be expressed by saying that the vernal equinox moves westward along the ecliptic at a speed of approximately $1\frac{1}{2}°$ per century. It, therefore, takes about 26,000 years for the vernal equinox to go once all the way around the zodiac.

3.4 Eclipses

For ancient man, eclipses were certainly among the most terrifying of all natural phenomena. Without warning, the sun begins to disappear from the sky, as though it is being swallowed by some vast demonic creature and "day turns into night." As darkness descends, the stars come out one by one, and all of nature responds as though it were late in the evening. Birds return to their nests, crickets begin to sing, chickens go to roost, and the cows start their slow journey back to the barn. Only man seems to realize that something very strange is happening. In only a few minutes, the sun, which supplies us with light and warmth, has mysteriously vanished. In many primitive cultures it is believed that these events can be remedied by making a lot of noise to scare off the demon who is swallowing the sun. Of course, the method always succeeds.

The ancient Greeks seems to have been the first really to understand what goes on during an eclipse. During a *solar eclipse*, the moon passes directly between the earth and the sun, and people standing in the path of the moon's shadow see an eclipse of the sun. During a *lunar eclipse*, the moon moves through the earth's shadow, and people on the nighttime side of the earth see an eclipse of the moon. Obviously, solar eclipses can occur only at the time of new moon, and lunar eclipses can only occur at full moon.

It is well known that eclipses are rare events. Most people go through their entire lives without ever seeing a solar eclipse, and few have ever experienced an eclipse of the moon. Yet, there is a full moon or a new moon every 2 weeks. At first glance, therefore, we might wonder why we do not see an eclipse every 14 days. The reason has to do with the orientations of the orbits of the moon and earth.

The moon, of course, is in orbit about the earth. However, the moon's orbit is tilted slightly with respect to the earth's orbit, as shown in Figure 3-14. The angle between the plane of the earth's orbit (that is, the plane of the ecliptic) and the plane of the moon's orbit is about 5°. These two planes intersect along a very important line called the *line of nodes*. As a result of this inclination of the moon's orbit, the moon is usually above or below the ecliptic. Most of the time, new moon and full moon phases occur under conditions when the sun, moon,

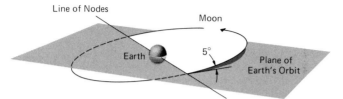

Figure 3–14. The Line of Nodes. The moon's orbit is tilted slightly with respect to the earth's orbit. The plane of the moon's orbit intersects the plane of the earth's orbit along the line of nodes.

and earth are *not* in perfect alignment. A perfect alignment between the sun, moon, and earth is *only* possible when new moon and full moon occur along the line of nodes, as shown in Figure 3–15. The times when perfect alignments are possible are called *eclipse seasons*, and only at such times can eclipses occur. An eclipse of the sun is seen only when the new moon is very near the line of nodes, because only then will the moon's shadow fall directly on the earth. Similarly, lunar eclipses are seen only when the full moon is very near the line of nodes because only then will the moon pass directly through the earth's shadow.

During a solar eclipse, only the very tip of the darkest part of the moon's

Figure 3–15. When Eclipses Occur. Eclipses can only happen when the full moon or new moon occurs on or very near the line of nodes. At all other times, a perfect alignment between the sun, moon, and earth is not possible.

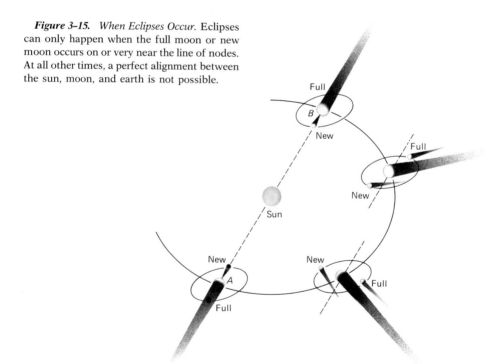

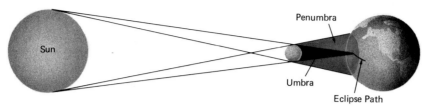

Figure 3-16. *The Geometry of a Total Solar Eclipse.* During a total solar eclipse, the moon's umbra traces an eclipse path across the earth. Totality is seen only inside the eclipse path; a partial eclipse is seen in those regions covered by the moon's penumbra.

shadow touches the earth, as shown in Figure 3-16. As the earth rotates and as the moon moves along in its orbit, the moon's shadow moves across the earth's surface at a speed slightly greater than 1,000 miles per hour. The path of the moon's shadow on the earth is called the *eclipse path*, and only those people located in the eclipse path will be treated to a total eclipse of the sun. The eclipse path is usually only a few miles wide, and due to the speed of the moon's shadow across the earth, the *duration of totality* is only a few minutes. Theoretically, the eclipse path is never more than 167 miles wide, and the maximum possible duration of totality is $7\frac{1}{2}$ minutes.

Because the sun is a sphere and not a "point source" of light, there are two parts to the moon's shadow. The darkest part of the moon's shadow is called the *umbra*. No sunlight at all falls inside the umbra; the sun is completely covered by the moon for anyone located in the umbra. By contrast, for a person standing in the *penumbra*, only part of the sun is covered. Such a person sees only part of the sun obscured by the moon; he is observing a *partial solar eclipse*. As shown in Figure 3-16, the penumbra covers a large region of the earth's surface during a solar eclipse. People living in a zone 2,000 miles on either side of the eclipse path will see a partial eclipse of the sun.

Since the orbit of the moon about the earth is known, astronomers can calculate when and where eclipses occur. The results of these calculations are published in books such as the *Ephemeris*. Figure 3-17 shows a typical eclipse map taken from the *Ephemeris*. From such a map, you can learn immediately where you must go on the earth's surface to observe a total eclipse of the sun.

An interesting complication with regard to solar eclipses becomes apparent with a little more thinking about the details of the moon's orbit. Neglecting perturbations, the moon's orbit is an ellipse with the earth at one focus. This means that sometimes the moon is nearer the earth than at other times. Indeed, the earth-moon distance over the course of a lunar month varies by about 10 per cent from the average value of 238,900 miles. The point in the orbit nearest the earth is called *perigee*, and the point farthest from the earth is called *apogee*. This means that the apparent size of the moon, as seen from the earth, varies over the course of a lunar month. At perigee, the moon actually looks a little bigger in the sky than at apogee.

TOTAL SOLAR ECLIPSE OF 1973 JUNE 30

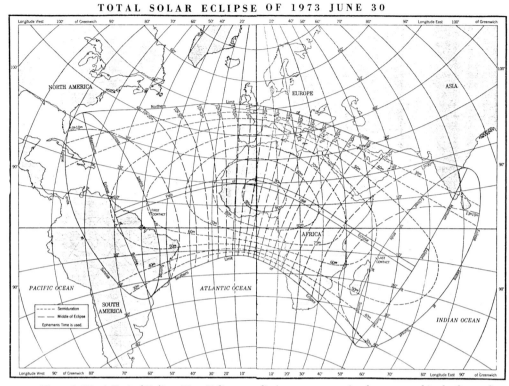

Figure 3–17. A Typical Eclipse Map. Eclipse predictions are conveniently expressed in the form of a map which shows the eclipse path and the region of the earth covered by the penumbra along with related information.

The average apparent angular diameter of *both* the sun and moon as seen from the earth is very nearly $\frac{1}{2}°$. This coincidence is responsible for the fact that, during a total solar eclipse, the moon appears to "fit" exactly over the sun. However, if a solar eclipse happens to occur when the moon is near apogee, then the moon has an apparent angular diameter slightly less than $\frac{1}{2}°$. This means that the moon, as seen in the sky, is not big enough to cover the entire disc of the sun completely. As shown in Figure 3–18, under these circumstances, the umbra does *not* reach all the way down to the earth's surface. Therefore, no one sees a completely total eclipse of the sun. Rather, at mideclipse we can still see the outer edge of the sun around the moon. Such an eclipse is called an *annular eclipse* because a ring, or annulus, of sunlight is seen around the moon at mideclipse.

Even though very few people have ever seen a solar eclipse, the average person may observe several lunar eclipses during his life. This does not mean that lunar eclipses occur more frequently than solar eclipses. Quite to the contrary, there are at least two (but never more than five) solar eclipses some-

Chapter 3 The Earth in Astronomy

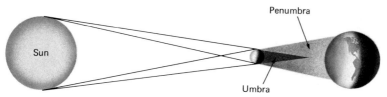

Figure 3–18. *The Geometry of an Annular Eclipse.* Annular eclipses are seen when the new moon is on the line of nodes *and* at apogee. Under these conditions, the moon's umbra does not reach all the way down to the earth.

where on the earth each year. But the number of lunar eclipses during a 12-month period ranges from zero to three. In order to see a total eclipse of the sun, you must be standing in the path of totality. But when a total eclipse of the moon occurs, *anyone* living on the nighttime side of the earth can see it. This is apparent from the geometry of a lunar eclipse shown in Figure 3–19.

A total lunar eclipse begins when the moon moves into the earth's penumbra. If you were standing on the moon, you would see the earth begin to cover up the sun, yet some sunlight would still be falling on you. Since the moon at this point is still in partial sunlight, the full moon appears just a little dimmer than usual. It is very difficult to notice this slight dimming when the moon is in the penumbra. However, when the moon moves into the umbra and totality begins, all direct sunlight is cut off. The moon seems to disappear almost completely from the sky. Totality ends when the moon leaves the umbra, and the entire eclipse is over

Figure 3–19. *The Moon and the Earth's Shadow.* If the full moon passes completely through the umbra, we see a total eclipse. If only part of the moon passes through the umbra, there is a partial eclipse. Penumbral eclipses occur when the moon passes only through the penumbra.

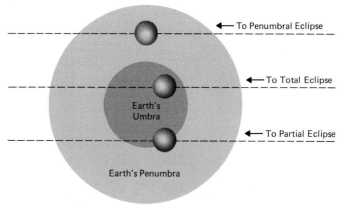

Figure 3-20. Partial Lunar Eclipse. The photograph was taken during a partial eclipse of the moon in March 1961. The illuminated portion of the moon is inside the earth's penumbra. The obscured portion is in the earth's umbra. (*Griffith Observatory*)

when the moon exits the penumbra. To the unaided eye, only the umbral phase of the eclipse is easily seen.

A *partial lunar eclipse* occurs when the moon does not move completely into the earth's umbra. Some part of the lunar surface always stays in the penumbra, and some piece of the moon always remains easily seen. A *penumbral eclipse* occurs when the moon moves only through the penumbra and not at all through

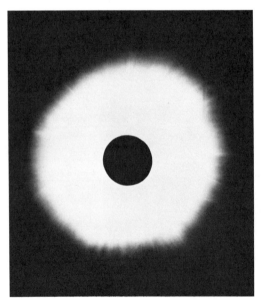

Figure 3-21. The Solar Corona. The shape of the solar corona varies from one eclipse to the next. Astronomers have learned that these variations are due to changes in the structure of the sun's magnetic field. (*Yerkes Observatory*)

the umbra. Penumbral eclipses go virtually unnoticed. Even at mideclipse, a trained observer would see only a slight over-all dimming in the appearance of the full moon.

Lunar eclipses are of virtually no interest to the modern astronomer. Of course, it is always nice to see an eclipse, but there is nothing new that we can learn from observing an eclipse of the moon. On the other hand, solar eclipses are extremely important. The sun is the only star in the sky that does not look like a pinpoint of light. It is the only star that we can study up close. Most of the time, the solar astronomer is concerned with examining the complex phenomena on the surface of the sun. (We will discuss many of these phenomena later.) However, during a total solar eclipse, the moon blocks out the blinding solar surface. Only under such conditions can we see and study the tenuous structure of the sun's upper atmosphere. During the precious few minutes of totality, the sun's *corona* flashes into view. The corona consists of extremely thin, hot gases extending hundreds of thousands of miles out from the sun's surface. From examining the nature of the corona, a great deal can be learned about the nature

Figure 3-22. Prominences. During a solar eclipse, it is possible to see huge jets of gas at the sun's edge surging up many thousands of miles into space. These jets of gas erupting from the sun are called "prominences." (*Lick Observatory*)

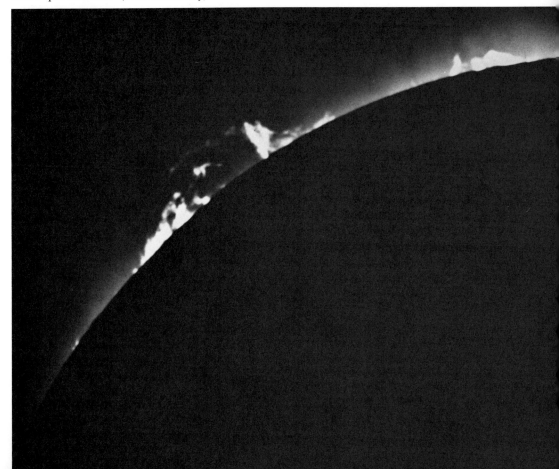

of our star. For example, the shape of the corona varies remarkably from one eclipse to the next, as shown in Figure 3–21. Such changes reflect corresponding changes in the structure of the sun's magnetic field.

Since studying the entire corona is only possible during a total solar eclipse, astronomers go to great lengths to insure good observations. Jet airplanes are loaded with telescopes and related equipment, and the solar astronomer flies along the path of totality during mideclipse. By chasing the moon's shadow across the earth, he can make totality last a little longer than it does on the ground. And, of course, by being above the clouds it is guaranteed that he will not be rained out.

Man has been observing eclipses for thousands of generations, and during those centuries, these dramatic natural phenomena have been a source of awe and wonder. Yet, only recently have we developed the insight and technology to use eclipses to probe the secrets of the universe.

Questions and Exercises

1. What is the meridian?
2. What is the difference between a solar day and a sidereal day?
3. Why is the solar day not constant throughout the year?
4. What is mean solar time? Why is the mean sun more useful for measuring time than the real sun?
5. Why do we find it useful to have time zones?
6. Why do we have leap years?
7. At what time of the year is the earth at perihelion?
8. Describe the two basic causes for the seasons.
9. What are the solstices and the equinoxes?
10. What is meant by the Arctic and Antarctic Circles?
11. What is the cause of the tides in the oceans?
12. What is the difference between spring tides and neap tides?
13. Describe the phenomenon of precession.
14. What are some of the effects of precession?
15. How long does it take for the vernal equinox to travel once around the ecliptic due to precession?
16. Who first discovered the phenomenon of precession?
17. What is the line of nodes?
18. What role does the line of nodes play with regard to eclipses?
19. What is meant by the eclipse season?
20. What is the maximum possible duration of totality in a solar eclipse?
21. What is the difference between the umbra and penumbra?

Chapter 3   The Earth in Astronomy   77

22. Under what conditions do annular eclipses occur?
23. Why can't an eclipse occur at first quarter phase of the moon?
24. What is a penumbral eclipse of the moon?
25. Give two reasons why it is advantageous for modern astronomers to observe solar eclipses from airplanes.
26. What is the maximum number of solar eclipses that can possibly occur in one year?

The Solar System

4.1 Inventory of the Solar System

For thousands of years, our ancestors regarded the vast and immobile Earth as being at rest at the center of the universe. The sun, moon, and planets performed an intricate celestial ballet among the constellations of the zodiac as the heavens themselves revolved about our Earth. It was only 400 years ago that man gradually began to accept the idea that a fanciful geocentric cosmology could not possibly be a valid description of reality. Beginning with the work of Copernicus, Kepler, Galileo, and Newton, it became obvious that both reason and observations clearly pointed to a heliocentric picture of the solar system in which all the planets orbit the sun. It is the sun and not Earth that dominates the solar system.

The sun dominates the solar system simply because it is so massive. Almost 99.9 per cent of the mass of the entire solar system is found in the sun, which means that only one tenth of 1 per cent is left over for all the planets combined. The planets, therefore, can be regarded as microscopic impurities in the vast vacuum of space surrounding our star.

By applying Newtonian mechanics to the observed orbits of the planets and their satellites it is possible to deduce the masses of the objects in the solar system. For example, it turns out that the mass of the sun is 333,000 times the mass of Earth. Everything else in the solar system adds up to a total mass of less than 450 Earths. Since the sun is so massive, it is also the largest object in the solar system. The sun's diameter is 864,000 miles, or about 109 times the diameter of Earth. This means that the sun is so large that $1\frac{1}{3}$ million Earths could fit inside it.

From observing the stars in the sky, we have come to realize that our sun is a very typical star. Temperatures on the sun range from 6,000°K (11,000°F) at the surface to more than 12 million degrees K at the sun's center. The nuclear fires at the sun's core produce 500 sextillion horsepower, which provides all of the heat and light for the rest of the solar system. In virtually every sense, the sun does indeed dominate the solar system.

Chapter 4 The Solar System

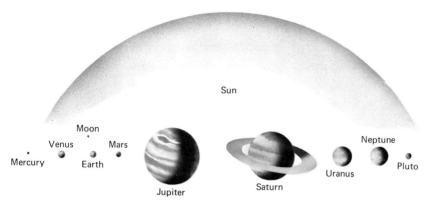

Figure 4-1. *The Sun and Planets.* This scale drawing shows the relative sizes of the nine planets and the sun.

It would be practically impossible for an imaginary astronomer from some advanced race of creatures in some distant part of the universe to know about the existence of the planets. Only if he were to travel toward our star in his spacecraft would his careful scrutiny reveal some tiny pieces of matter in orbit about the sun. If he were to come within a few trillion miles of the solar system, he would first see Jupiter, the largest of all the planets. Then he would notice Saturn. Only with very great difficulty might he discover Earth.

The nine known planets are scattered around the sun in a wide range of orbits. The planet nearest the sun, Mercury, has an average distance from the sun of only 36 million miles (0.39 AU). The most distant planet, Pluto, is $3\frac{1}{2}$ billion miles (39.46 AU) from the sun. As seen from Mercury, the sun appears very large in the sky and, as you might expect, the daytime temperatures on this planet are very high. Lead and tin would melt and flow like water under the noontime Mercurian sun. From Pluto, however, the sun would look like just a bright star in the sky. At this enormous distance from the sun, the temperatures are so low that even the air (if there were any) would be frozen, covering the ground like snow. In general, the planets closest to the sun are the warmest, whereas those farthest from the sun have the lowest surface temperatures.

The masses of the planets also cover a wide range. Jupiter is by far the most massive and Mercury the least. Since it is convenient to express the masses of planets in terms of the mass of Earth (Earth = 1), we find from Newtonian mechanics that the mass of Jupiter is 318 Earths, whereas the mass of Mercury is only $\frac{1}{20}$ that of Earth. Jupiter is more massive than all of the rest of the planets combined. It is also the largest of all the planets, with a diameter of 86,000 miles. Mercury is the smallest. Mercury's diameter is 3,000 miles, approximately equal to the distance between Boston and Los Angeles.

As might be expected, in view of the different masses and sizes of the planets, the strength of gravity on their surfaces varies from planet to planet. You would weigh more on Jupiter than on Mercury. This effect is called the *surface gravity*,

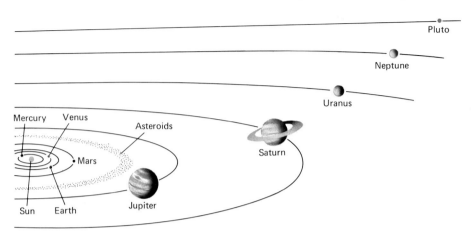

Figure 4–2. *The Planetary Orbits.* This scale drawing shows the relative sizes of the orbits of the planets. Mercury, Venus, Earth, and Mars are crowded close to the sun. By contrast, the orbits of the outer planets are spread out.

a measure of the strength of the force of gravity. For example, Jupiter has the highest surface gravity of any of the planets. You would weigh 2.64 times more on Jupiter than you do on Earth. We, therefore, say that the surface gravity of Jupiter is 2.64. You would weigh the least on Mercury, which has a surface gravity of only 0.39. Thus, if you weighed 150 pounds on Earth, you would weigh only 59 pounds on Mercury, but you would tip the scales at 396 pounds on Jupiter.

The surface gravity of a planet is directly related to the escape velocity needed to leave the planet. Astronauts of the future traveling to the planets will find it much easier to blast off from Mercury than from Jupiter. Due to its low surface gravity, Mercury has the lowest escape velocity of any of the planets—only 9,600 miles per hour. Any object that achieves this speed in an upward direction will leave the planet. Jupiter, on the other hand, has the highest escape velocity because it has the largest mass and the greatest surface gravity. An object must have a speed of at least 129,000 miles per hour in order to overcome Jupiter's enormous gravity. By comparison, the escape velocity from Earth is 25,000 miles per hour. This is the speed that the Apollo astronauts and unmanned interplanetary satellites must achieve from Cape Canaveral in order to start their journeys into deep space.

The escape velocity and the surface temperature of a planet have a profound effect on the planet's atmosphere. In order to see why this is so, first think about what exactly is meant by *temperature.*

All matter, whether it is solid, liquid, or gas, consists of atoms. These atoms are constantly in motion. At a low temperature, the average speed of the atoms in a substance is low. But at a high temperature, the average speed of the atoms is high. If you apply heat to an object, you are putting energy into the object, which

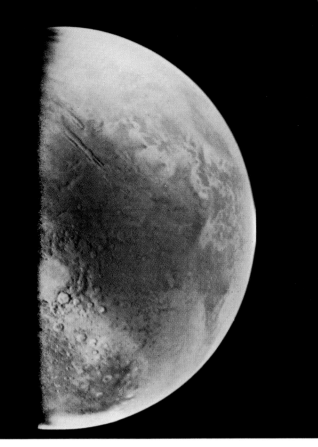

Plate 1. *Mars from Viking.* As the Viking spacecraft approached Mars in June, 1976, this dramatic color photograph was sent back to Earth. Just below the center of the picture is the large impact basin called Argyre. North of Argyre, the "Grand Canyon" of Mars, called Vallis Marineris, can be seen near the terminator. (*JPL; NASA*)

Color plates credited to the Hale Observatories are reproduced with permission. Copyright by the California Institute of Technology and Carnegie Institution of Washington.

Plate 2. *The Martian Surface.* Viking 1 landed on Mars on July 20, 1976, and sent back this scene of the planet's surface. The Martian soil consists of a reddish fine-grained material while some of the larger rocks have a blue-black hue. The horizon is about 1.8 miles away. In transmitting this first color photograph from Viking 1, the blue of the sky may have been overemphasized. Scientists believe the Martian sky has a pinkish tinge, caused by floating dust particles. (*JPL; NASA*)

Plate 3. *Jupiter.* This photograph taken with the 200-inch Palomar telescope is an excellent example of the best earth-based view of this giant planet. The Great Red Spot is shown faintly on the upper left-hand portion of the planet. (*Hale Observatories*)

Plate 4A. *Jupiter from Pioneer 10.* This was the view of Jupiter from Pioneer 10 when it was 1,550,000 miles from the planet. The shadow of Io, one of Jupiter's four largest satellites, is seen on the planet's surface. (*NASA*)

Plate 4B. *Jupiter from Pioneer 11.* Jupiter's Great Red Spot appears prominently in this photograph taken by Pioneer 11 when the spacecraft was 660,000 miles from the planet. (*NASA*)

Plate 4C. *The North Pole of Jupiter.* On December 12, 1974, Pioneer 11 passed over the north pole of Jupiter. This photograph was taken when the spacecraft was 750,000 miles from the huge planet. (*NASA*)

Plate 5. *The Orion Nebula.* This nebula (also called M42 or NGC 1976) is just barely visible to the naked eye in the constellation of Orion. New stars are being formed in this huge cloud of gas. (*Hale Observatories*)

Plate 6. *The Horsehead Nebula.* The constellation of Orion also contains this nebula (also called IC 434) just south of the star Zeta Orionis. A large, cool cloud of gas in the shape of a horse's head obscures some of the nebulosity. (*Hale Observatories*)

Plate 7. *The Trifid Nebula.* This nebula (also called M20 or NGC 6514) is in the constellation of Sagittarius. Dark, obscuring clouds of gas can be seen against the background of the nebula. The red light is emitted by hot hydrogen atoms while the blue light is reflected from dust in the nebula.
(*Hale Observatories*)

Plate 8. *The Rosette Nebula.* This nebula (also called NGC 2237) is in the constellation of Monoceros. Many very young stars are found scattered throughout this nebula.
(*Hale Observatories*)

will result in increasing the average speed of the atoms of which the substance is composed. For example, if a block of ice is heated, the water (H_2O) molecules will begin vibrating faster and faster until the speed is so high that the ice can no longer retain its solid form; it will begin to melt. If you continue to heat the resulting water, eventually the speed of the water molecules will be so great that the molecules will leave the surface of the water; the water will begin to boil. This is what is really meant by temperature. The temperature of a substance is a measure of the average speed of the particles (atoms, molecules, electrons, and so on) that make up that substance. This is why scientists prefer to use the so-called Kelvin temperature scale, as opposed to Centigrade or Fahrenheit scales. The Centigrade and Fahrenheit scales are based on familiar phenomena such as the melting and boiling points of water or the temperature of the human body. But the Kelvin scale is directly associated with the average speeds of the atoms in a substance. For example, at $0°K$ all motion stops. This is the lowest possible temperature; it is called *absolute zero*.

To complete this brief discussion of temperature, the following table relates some typical temperatures from the Kelvin, Centigrade, and Fahrenheit scales.

	Kelvin (K)	Centigrade (C)	Fahrenheit (F)
Absolute zero	0°	−273°	−460°
Water freezes	273°	0°	32°
"Room temperature"	293°	20°	68°
Water boils	373°	100°	212°

Throughout the rest of this book we will frequently make use of the Kelvin scale in discussing temperature.

The planets near the sun have higher surface temperatures than those farther away due to the heating effect of the sun's light. This means that the average speed of atoms in the atmosphere of a planet near the sun is higher than the average speed of similar atoms in the atmosphere of a planet far from the sun. If this average speed is near the escape velocity, the planet will lose its atmosphere. Gases are retained by a planet only if the average speed is much less than the escape velocity. That is why Mercury has no atmosphere. The daytime temperatures are so high ($610°K = 640°F$) and the escape velocity is so low (9,600 miles per hour) that any gases around Mercury easily leave the planet. Earth is more massive and cooler than Mercury and thus Earth is able to hold onto most of its gases except for the lightest: hydrogen and helium. Jupiter is even more massive and cooler than Earth, and as a result, no gases can escape from Jupiter, not even hydrogen and helium.

These effects are, in part, responsible for the vast differences between the

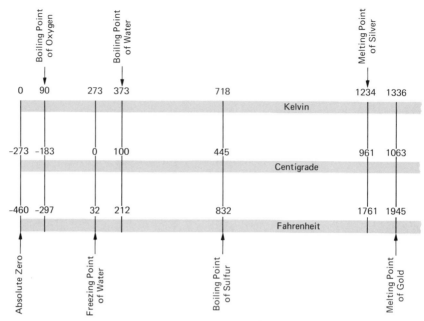

Figure 4-3. *Different Temperature Scales.* This drawing illustrates the relationship between the three most-commonly used temperature scales. Scientists prefer to use the Kelvin scale.

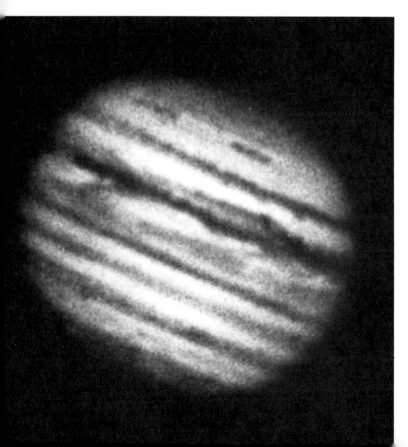

Figure 4-4. *Jupiter.* Jupiter is the largest planet. Its mass is greater than all the other planets combined. (*Lick Observatory*)

inner and outer planets. The inner planets (Mercury, Venus, Earth, and Mars) are all small with hard, rocky surfaces. They are called the *terrestrial planets*. The outer planets (Jupiter, Saturn, Uranus, and Neptune) are all very large and massive. They do not have solid surfaces but, rather, consist of hydrogen, helium, methane, and ammonia in liquid and gaseous form. They are called the *Jovian planets*. If any of these planets has a solid surface, it must be in the form of a small, rocky core not much bigger than Earth buried at its very center. The Jovian planets are so far from the sun that they have been able to retain many of the gases out of which they were originally formed. If a planet like Jupiter were very near the sun, the sun's heat would have caused many of its gases to "evaporate" into space. If a planet like Mercury were very far from the sun, it just might be captured by one of the giant planets and become a "moon." Indeed, the largest satellites of Jupiter, Saturn, and Neptune are about the same size as Mercury.

Only the outermost planet, Pluto, does not fit into this scheme of terrestrial and Jovian planets. Pluto is very small, about the same size as Mars, and has a rocky surface with no detectable atmosphere. In addition, Pluto's orbit is unusual. The orbits of the other eight planets lie in almost exactly the same plane, very near the plane of the ecliptic. But Pluto's orbit is tilted substantially (by 17°) out of the plane of the rest of the solar system. This suggests that Pluto might not

Figure 4–5. Pluto. Even with the most powerful telescopes, Pluto always looks like a dim star. It is never brighter than 15th magnitude. (*Lick Observatory*)

really be a planet at all. Perhaps at one time in the distant past, Pluto was one of the moons of Neptune and some catastrophic event occurred that allowed Pluto to escape and go into orbit about the sun. This idea is supported by the fact that Pluto's orbit is more highly elliptical than any of the other planetary orbits. It is so elliptical that Pluto's perihelion lies *inside* the orbit of Neptune; sometimes Pluto is closer to the sun than Neptune. Finally, one of Neptune's moons, Nereid, has a very elliptical orbit. The orbits of virtually all the other moons in the solar system about their parent planets are very nearly circular. Nereid, with its highly elongated orbit, is a striking exception. In addition, Neptune's other moon, Triton, goes around the planet in *retrograde*, or backward, compared to the direction in which most moons in the solar system go about their planets. Perhaps the same catastrophic events that cast off Pluto also caused the high eccentricity of Nereid's orbit and the retrograde motion of Triton.

The sun is the largest object in the solar system. The planets are the second largest objects in the solar system. Next in line are the satellites, or moons, of the planets. There are 34 known satellites distributed among six of the planets. Only Mercury, Venus, and Pluto do not have moons. Of these 34 satellites, seven could be called giant moons; each is approximately the same size as the planet Mercury. The remaining 27 satellites are very small, typically several hundred miles in diameter or less. Four of the giant moons belong to Jupiter. They are Io, Europa, Ganymede, and Callisto, which Galileo discovered with his telescope in 1610. Saturn, Neptune, and Earth each possess one giant moon.

In addition to the sun, planets, and satellites, there are numerous other objects in the solar system. Thousands of asteroids, which are nothing more than huge rocks, circle the sun between the orbits of Mars and Jupiter. Countless billions of smaller rocks, frequently no bigger than grains of sand, also orbit our sun. They are the meteoroids that blaze forth as "shooting stars" when they collide with Earth's upper atmosphere. In addition, there are large chunks of ice in our solar system. When one of these interplanetary icebergs comes near the sun, the ice is vaporized and we see a comet in the sky. All of this "extra" matter in space does not really amount to very much. All of the asteroids, meteoroids, and comets lumped together would probably amount to only one or two good-sized moons.

During the eighteenth and nineteenth centuries, the study of the solar system dominated much of astronomy. Using telescopes, astronomers spent a large percentage of their time examining the planets and, as a consequence, discovered new satellites, asteroids, and comets. However, with the advent of the space program all this changed. It is almost a waste of time, for example, to look at Mars through a telescope, because what can be seen is vastly inferior to the superb pictures sent back by Mariner 9. As a result, the planets and the solar system no longer belong to the astronomer. As the astronomer turns his telescope to the stars and galaxies, the study of the planets is being taken over by other scientists. The geologist and the geophysicist examine the moon rocks and the surface features of Mercury and Mars; the meteorologist is intrigued by the cloud formations and patterns on Venus and Jupiter.

Until recently, scientists could study only one planet up close. The only rocks

Figure 4–6. *The Earth.* From studying the planets we will better understand the history and evolution of our own earth. (*NASA*)

(aside from meteorites) that could be analyzed in laboratories came from Earth. The only wind patterns and ocean currents that could be observed were those here on Earth. Before the space program, trying to learn the history and evolution of the earth was analogous to trying to formulate Darwin's theory of evolution after examining just one cat or dog. But with the wealth of information that has come from the unmanned interplanetary probes, we can soon expect major breakthroughs in our understanding of the mysteries of the solar system and thereby gain profound insight into the nature of Earth.

4.2 Our Moon

Earth's nearest celestial neighbor, the moon, stands in dramatic contrast to our own planet. Earth has an atmosphere of nitrogen, oxygen, and other gases; the moon has no atmosphere at all. More than half of the surface of Earth is covered with vast oceans; there is absolutely no water or ice on the moon. Earth has a magnetic field; the moon has none. Earth has a molten iron core; the entire moon consists of solidified rock. Earth is teaming with life; the moon is barren, desolate, and sterile.

It might, therefore, seem that the moon is a very uninteresting place. It is a dead world consisting of nothing more than a lot of rocks. The lunar daytime

Figure 4–7. The Moon. Through a small telescope it is possible to see many different types of lunar features such as maria, craters, and mountain ranges. (*Lick Observatory*)

lasts for 2 weeks, during which the sun blazes down on the barren surface of the moon heating it to almost 400°K. Without any atmosphere to retain the sun's heat, during the 2-week-long lunar night, the temperature drops to about 100°K. Meteoroids bombard the moon, producing craters that appear as pockmarks on its surface. It is an incredibly hostile environment.

Studying the moon offers a great wealth of information. Earth is constantly changing. Rain, wind, and other natural forces are continuously reshaping Earth's surface. Even the continents are in motion. Yet, the moon is dead; very little has changed since its creation billions of years ago. In other words, a record of the earliest days of the solar system is preserved on the lunar surface. It is, therefore, clear to geologists that any fundamental understanding of the creation of the solar system can come from the moon, but not from Earth. Natural forces have long since eradicated all traces of the primordial Earth. For example, when you pick up a rock from the ground, chances are that the rock is very young, having been formed only a few hundred million years ago. Only with very great difficulty can geologists find mineral samples dating back as far as $3\frac{1}{2}$ billion years on Earth. The moon is covered with rocks $4\frac{1}{2}$ billion years old; they are almost as old as the solar system itself. Indeed, it is from such samples brought back by the Apollo astronauts that geologists are supported in their belief that the true age of the sun and all the planets is really $4\frac{1}{2}$ billion years.

Looking at the moon through a small telescope, we find that there are basically three different types of features on the lunar surface. The largest features we can see are huge flat areas called lunar seas, or *maria*. This peculiar term comes from astronomers who first looked at the moon through telescopes and believed that there might be oceans covering large portions of the lunar surface. These astronomers of the seventeenth century gave the maria fanciful names such as Mare Serenitatis (the "Sea of Serenity"), Mare Tranquilitatis (the "Sea of Tranquility"), and Mare Imbrium (the "Sea of Showers"). Of course, the maria are dry land. Analysis of samples brought back by the Apollo astronauts indicates that all the maria are very similar. They are covered with very porous rocks and dust. This layer of rock and dust is known as the *regolith*. The top few inches of the regolith are like sandy soil, and the next few feet down consist of larger broken fragments of rock, below which is the *lunar bedrock*. Samples from the maria are of two basic types: there are *igneous rocks*, which were formed from cooling lava, and *breccia*, which are rocks formed when smaller fragments are compressed together under great pressure.

One surprising recent discovery about the moon is that *all* the maria are located on the "front side" of the moon. Between August 1966 and August 1967, the United States sent five lunar orbiters to the moon for the purpose of photo-

Figure 4–8. A Mare. The large flat areas on the "front" side of the moon are called *maria*. This photograph taken by the Apollo 8 astronauts shows the Sea of Tranquility. (*NASA*)

Figure 4-9. The "Hidden" Side of the Moon. This photograph of the "back" side of the moon was taken by Orbiter II when it was 900 miles above the lunar surface. Note how heavily cratered the "back" side is, compared to the "front" side which has large maria. (*NASA*)

graphing the entire lunar surface. Aside from a few low-quality photographs resulting from earlier Soviet flights, man had never seen the "far side" of the moon, which is eternally hidden from our Earth-based view. These highly successful spacecrafts sent back thousands of pictures of both sides of the moon and to everyone's surprise the far side is covered *only* with craters. There are no large flat regions at all!

It was also discovered at this time that the orbits of these spacecrafts around the moon were perturbed slightly in passing over the maria, indicating that there must be large concentrations of matter buried below the surfaces of the lunar seas. These hidden chunks of matter are called *mascons,* a contraction of the term *mass concentrations.* It is reasonable to suppose that these mascons are gigantic iron meteorites that hit the moon sometime in the very distant past and created the maria. Perhaps during this ancient period, much of the moon was still molten. When a gigantic meteorite struck the moon creating a huge hole, lava flowed into the resulting cavity covering the meteorite and giving rise to the smooth appearance of the lunar seas.

Of course, the second obvious type of feature seen on the moon's surface are *lunar craters.* From Earth it is possible to identify thirty thousand such craters, many of which have been named after famous scientists. These craters range in size from 1 mile to slightly over 100 miles in diameter. They were all formed by the violent impact of meteoroids bombarding the moon over billions of years.

Figure 4-10. *The Moon from Apollo 16.* This view of the moon, taken by astronauts in lunar orbit, shows parts of both the "front" and "back" sides. Notice that the front side (on the right) has maria while the back side (on the left) is heavily cratered. (*NASA*)

Figure 4-11. *Two Views of a Crater.* The advantages of doing astronomy from space are dramatically illustrated by two photographs of the crater "Alphonsus." The view on the left was taken with an excellent earth-based telescope. The view on the right was taken by the cameras on-board the Ranger IX spacecraft. (*NASA*)

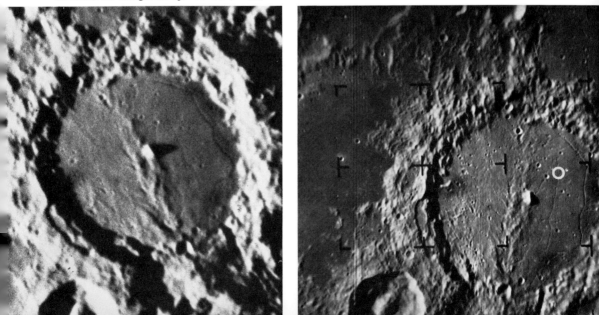

Figure 4–12. A Typical Moon Rock. At first glance, a lunar rock looks like ordinary rocks found here on earth. (*NASA*)

From Earth, even with the most powerful telescopes, it is impossible to see craters whose diameters are less than about 1 mile across. Only by going to the moon could we possibly hope to see finer details of the lunar surface. In 1964 and 1965, the United States sent several Ranger spacecrafts to the moon to return the first close-up photographs. In the few minutes before crashing (as planned) into the moon, the Rangers sent back hundreds of pictures that showed objects as small as 1 foot across. Figure 4–11 shows an excellent Earth-based photograph of the crater Alphonsus along with a corresponding photograph from Ranger 9. Features totally invisible from Earth show up with remarkable clarity. The moon is covered with millions of very small craters that never could be seen from earth.

The next real surprise with regard to craters came from the lunar rocks returned by the Apollo astronauts. If you hold a moon rock in your hand, at first glance it looks like any other rock you have ever seen. Closer inspection, however, reveals that moon rocks seem to sparkle in sunlight. Examination with a powerful microscope discloses that the sunlight is being reflected off of numerous tiny glass craters that cover the surface of the rock. These microscopic craters are called *zap craters;* it is believed that they were formed by tiny grains of interplanetary dust violently striking the exposed surface of the moon rock (the underside of a moon rock is free from zap craters). Since the moon has no atmosphere, dust from outer space directly hits the lunar surface with speeds of thousands of miles per hour. At such high speeds, the force of impact on the surface of a rock is so great that a small area of the rock melts for a fraction of a second and turns into glass, just long enough to form a tiny impact crater. Thus, the moon is covered with literally billions of craters ranging in diameter from more than 100 miles to only a few thousandths of an inch across.

Mountain ranges are the third most prominent feature on the moon. These ranges are named after ranges here on Earth (the Alps, Apennines, Carpathians, and so on); the highest lunar peaks rise to elevations of 25,000 feet or more.

Chapter 4   The Solar System 91

There do not seem to be any lunar mountains quite as high as Mount Everest in the Himalayas.

There are, of course, numerous other types of features on the lunar surface (valleys, hills, crevasses, and clefts) that it would be fascinating to study. But man's firsthand examination of the moon seems to have come to an end. The Apollo program has been terminated and no future exploration of the moon is planned. New discoveries about the moon will now be made by scientists examining photographs and analyzing moon rocks. And already the lunar samples have begun to yield some of their secrets. The rocks brought back by the astronauts are quite similar to certain types of earth rocks, such as basalt, which is solidified lava. But there are some noticeable differences. All earth rocks contain a small amount of water; the moon rocks are completely dry. In addition, the moon rocks do not have as many "volatiles" as Earth rocks, those chemicals that easily evaporate at moderately high temperatures are lacking in samples from the moon. Similarly, those chemicals that do not evaporate easily, especially those containing titanium, are found to be much more abundant in

Figure 4-13. A "Zap" Crater. Examination of moon rocks with electron microscopes reveals that they are covered with tiny craters only a few thousandths of an inch in diameter. (*NASA*)

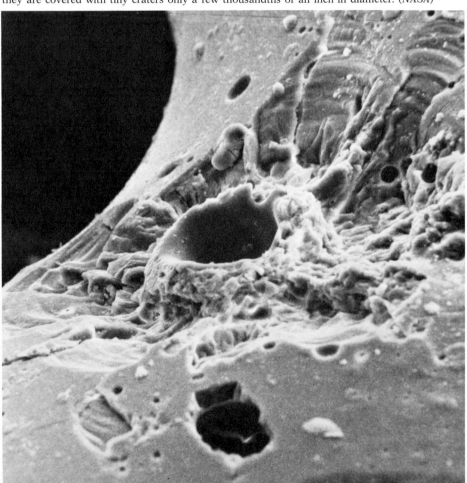

the moon rocks than in Earth rocks. All this evidence points to the fact that the moon must have been created at a higher temperature than Earth. Yet, details of the moon's formation remain a complete mystery. Scientists still do not understand why Earth and the moon, which are so very close in space, seem to have such very different histories. Hopefully, we now have enough information, enough samples, and enough photographs to start fitting the pieces together. Only then will man truly begin to understand the nature of that small part of the universe that he occupies.

4.3 Mariner and Mars

From the pyramids to Palomar man has always turned to the skies for clues to his origin and destiny. The bright planets wandering among the constellations of the zodiac captured imaginations, and these globes were peopled first with ancient gods and later with the exotic creatures of science fiction. In the twentieth century books, magazines, movies, and television and radio shows have used the theme of a coming invasion from "outer space" with almost prophetic certainty. People imagine that they can see flying saucers piloted by little green men. Yet, ironically, man is now invading other planets only to find barren and hostile environments, almost totally incapable of supporting any forms of life at all. We are alone, terrifyingly alone, on a tiny planet orbiting a typical star in one corner of the universe.

Our current understanding of the solar system holds that all the planets had a common origin. About $4\frac{1}{2}$ billion years ago, dust and rocks surrounding the newly born sun were pulled together under their mutual gravitational attraction in a process called *accretion*. The debris left over from the birth of the sun accreted in spherical conglomerations we call the planets. But in spite of their common birth, each planet has its own distinctive character. Earth is teeming with life. Mercury is baked by the sun. Saturn has a beautiful system of rings. And Mars, which is perhaps most like our own planet, is a cold, dry, windblown desert. This incredible diversity alone makes planetary exploration worthwhile; it is only from such exploration that we can ever hope to understand why the planets, which had a common birth, turned out so differently.

Mars is one of the smallest planets in the solar system. Its diameter is only one half that of Earth and its mass is approximately one tenth that of our globe. With such a small mass and size, the surface gravity on Mars is only 0.38, which means that a 150-pound man from Earth would weigh only 57 pounds on the Martian surface. Yet, in spite of these differences, Mars is more like Earth than any other planet. Although it takes almost 2 years to go once around the sun, the Martian day lasts for 24 hours, 37 minutes, and 23 seconds. In addition, Mars' axis of rotation is tilted by 25° from its orbital plane (we recall that Earth's axis is tilted by $23\frac{1}{2}$° to the ecliptic), so that Mars has seasons that superficially seem like those here on Earth.

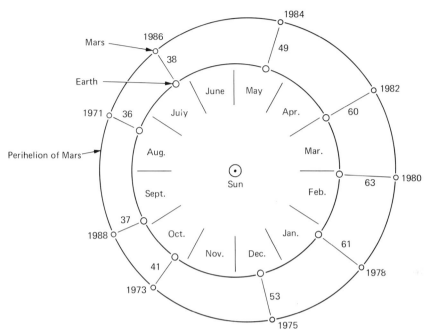

Figure 4–14. *The Orbits of Mars and Earth.* Due to the ellipticity of their orbits, the distance between Mars and the Earth varies from one opposition to the next. Favorable oppositions occur near the perihelion of Mars. Distances are given in millions of miles.

Mars is a very difficult planet to observe. In view of its small size, the best views are obtained only at oppositions when Mars and Earth are close together. This happens once every 2 years. But not all oppositions are equally favorable for observations. Since Mars' orbit is elliptical, oppositions sometimes occur when Mars and Earth are far apart, and sometimes when they are close together, as shown in Figure 4–14. At "favorable" oppositions (for example, 1971, 1988, and so on), the distance between Mars and Earth is about 36 million miles, and Mars appears through a telescope to have an angular diameter of 25 seconds of arc. At "unfavorable" oppositions, the Earth-Mars distance can be as large as 63 million miles, at which times Mars appears so small that it is almost impossible to distinguish any surface features, even with the most powerful telescopes.

When the astronomer trains his telescope on Mars near the time of a favorable opposition, he sees a fuzzy reddish sphere. White areas cap both the north and south poles, while dark formations are seen in the equatorial regions. An excellent Earth-based photograph showing these features is shown in Figure 4–15. When spring comes to a Martian hemisphere, the polar cap shrinks in size and the dark areas, which look almost green against the red surface of the planet, become more pronounced. Little wonder that earlier astronomers imagined that Mars might be an abode of life!

On rare occasions, when Earth's atmosphere is very clear and steady, astron-

Figure 4–15. Mars As Seen from Earth. This is an excellent example of one of the best photographs ever taken of Mars by earth-based telescopes. Note the polar caps and dark markings near the equator. (*Hale Observatories*)

omers can get exceptionally fine views of the Martian surface. It was during one such moment of good visibility, in 1877, that the Italian astronomer Giovanni Schiaparelli saw something he called *canali* on Mars. These features appear to be straight lines crisscrossing the surface of the planet; the Italian word used by Schiaparelli was erroneously translated as "canals," suggesting that they might have been constructed by intelligent beings for the purposes of irrigation. For almost a century following the "discovery" of canals, the existence of intelligent life on Mars was considered to be almost certain. The clever Martians had dug the canals to transport water from the melting snow caps to the lush vegetation near the equator. Today, of course, we realize that the polar caps are composed primarily of dry ice (solid carbon dioxide, CO_2), and that there are no canals on Mars.

Figure 4–16. A Drawing of Mars. Sometimes astronomers feel that they can record more details of what they see by making a drawing instead of a photograph. Both the photograph and the drawing shown here were made on the same night. This drawing shows a few "canals" reported by some astronomers. (*Lick Observatory*)

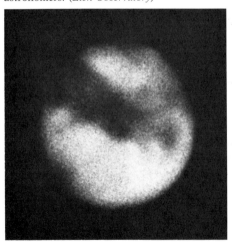
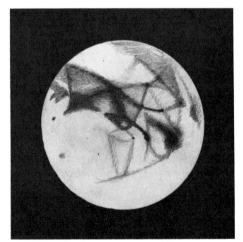

Figure 4–17. A View from Mariner 6. This mosaic of photographs was fashioned from seven pictures taken by Mariner 6. These photographs cover an area on Mars 2500 miles long and 400 miles wide just south of the Martian equator. (*NASA*)

The lesson to be learned from the imaginary canals is that studying Mars from Earth is extremely difficult. Even as late as the 1950s, astronomers were reporting discoveries that have turned out to be mistakes. Mars is so small and far away that it is virtually impossible to make any extensive studies of this planet from Earth. Even with the most powerful telescopes, the best view of Mars we can ever hope to obtain is comparable to a naked-eye view of our moon. The only approach is sending spacecrafts to this mysterious planet.

In the mid-1960s, the United States launched a small spacecraft, Mariner 4, toward Mars. As Mariner 4 flew past Mars in July 1965, at a distance of 6,000 miles, its television cameras took twenty-two pictures of the Martian surface that were transmitted back to Earth. To everyone's surprise, these photographs showed numerous craters; Mars looks more like our moon than Earth. Since the Martian craters showed little erosion, we must conclude that there is very little air or water on Mars, suggesting that this planet is geologically and biologically dead. The case for life on Mars received a further setback when instruments aboard Mariner 4 failed to detect a magnetic field around Mars. Earth has a magnetic field that gives rise to the so-called Van Allen belts. This magnetic field and the resulting Van Allen belts protect us from the lethal, charged particles that are constantly streaming out of the surface of the sun. Without a magnetic field to deflect the charged particles, these deadly particles directly strike the Martian surface. It seems that none of the familiar life forms on Earth, with the possible exception of viruses and bacteria, could endure such a bombardment of radioactivity.

The next breakthrough in Martian exploration came in July and August 1969, when Mariner 6 and Mariner 7 flew past the red planet at distances of about 2,100 miles. In addition to craters, photographs transmitted back to Earth showed chaotic terrain composed of jumbled ridges and valleys unlike anything seen on either Earth or the moon. There were also great flat regions where all traces of craters or hills had been mysteriously eroded.

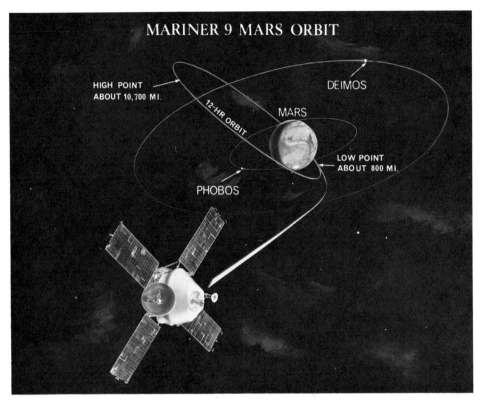

Figure 4–18. The Orbit of Mariner 9. Mariner 9 was launched from Earth in May of 1971 and arrived at Mars six months later. From orbit above Mars, this spacecraft was able to photograph most of the Martian surface. The orbits of Phobos and Deimos are also shown in this artist's rendering. (*NASA*)

Mariners 4, 6, and 7 photographed only 10 per cent of the Martian surface as they cruised past the planet at fairly high speeds. The obvious next step was to place a spacecraft in orbit about Mars for the purpose of extensive photography and mapping. Only a long-lived picture-taking satellite could provide a complete coverage of the planet.

In November 1971, Mariner 9 was placed successfully in an elliptical orbit about Mars, as shown artistically in Figure 4–18. Unfortunately, at the time of Mariner's arrival, Mars was embroiled in one of its planetwide dust storms. No surface details at all could be seen as the storm raged across the planet, and so Mariner's cameras were trained on the two Martian moons, Phobos and Deimos. The resulting photographs revealed that these two moons are small, heavily cratered chunks of rock. Phobos, shown in Figure 4–19, is oblong in shape with a diameter of about 10 miles. Deimos is only about 5 miles in diameter.

As the dust storm cleared early in 1972, Mariner 9 began sending back thousands of superb photographs of remarkable features never before seen by man. Late in January, for example, Mariner 9 sent a series of pictures revealing that there are volcanoes on Mars! The largest Martian volcano, shown in Figure 4–20,

Chapter 4 The Solar System 97

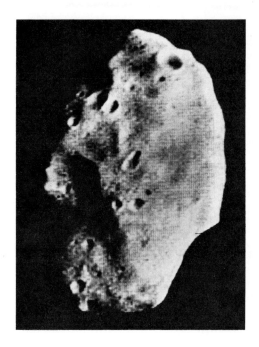

Figure 4-19. Phobos. Phobos is the larger of the two Martian moons. It is only about 10 miles from end to end. The craters on Phobos indicate that it has been struck many times by interplanetary debris. (*NASA*)

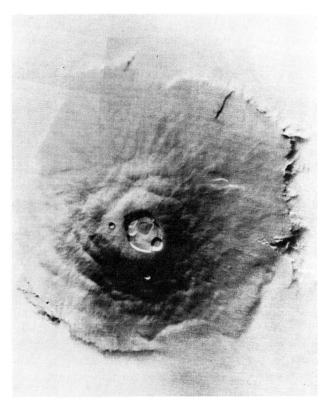

Figure 4-20. Olympus Mons. The largest volcano in the solar system is on Mars. Olympus Mons is 300 miles across at the base and rises 14 miles above the surrounding plains. (*NASA*)

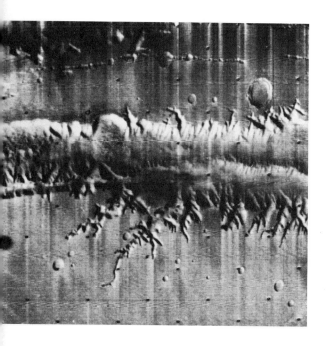

Figure 4–21. *The Martian Canyonlands.* Photographs such as this from Mariner 9 revealed the existence of vast canyons on Mars. The area in this photograph is about 300 miles long and 235 miles wide. (*NASA*)

is called Olympus Mons. It is more than 370 miles wide at the base; if it were on Earth, Olympus Mons would fit just barely between Los Angeles and San Francisco. This immense volcano rises to a height of $14\frac{1}{4}$ miles above the surrounding Martian plains. The volcanic caldera at its summit is 40 miles across, big enough to hold the entire state of Rhode Island. Olympus Mons is more than twice the size of the volcano that formed the Hawaiian Islands, measured from the bottom of the Pacific Ocean.

In addition to having the largest volcano in the solar system, Mars also has the most extensive canyons. The "Grand Canyon" of Mars, a portion of which is shown in Figure 4–21, is about 75 miles wide and 4 miles deep. The entire canyon is thousands of miles long; it would stretch all the way across the United States if it were here on Earth.

One very puzzling aspect of the canyon lands is that there seems to be some evidence of erosion, as if water had flowed on Mars in the distant past. This prospect of water on Mars is further supported by photographs sent back in July 1972, which seemed to show a river bed, as seen in Figure 4–22. This is puzzling because there is now very little water on Mars. Measurements from the Mariner spacecrafts indicate that all the water on Mars put together could not fill a good-sized lake here on Earth. If there were enough water on Mars to erode canyons and fill rivers in the distant past, where could it possibly be now? It is certainly not in the Martian atmosphere in the form of water vapor. The instruments aboard Mariner showed that the Martian air consists entirely of carbon dioxide. One possibility is that there may be a lot of frozen water under the polar caps. The surfaces of the polar caps are made of dry ice. Yet, from studying the way in which the dry ice polar caps evaporate with the approach of spring, some scientists believe that there probably is ordinary water ice under the dry ice.

Chapter 4 The Solar System 99

The subject of water on Mars is important, not only for explaining the features that look like riverbeds, but also in considering the possibility of life. It is now obvious that there are no advanced forms of life on Mars whatsoever. But could there possibly be very simple forms of life such as viruses or bacteria? Life as we know it requires water for survival, and perhaps there are biologically active regions near the Martian polar caps. The answers to questions and speculations about life on Mars can only come from landing a spacecraft directly on the

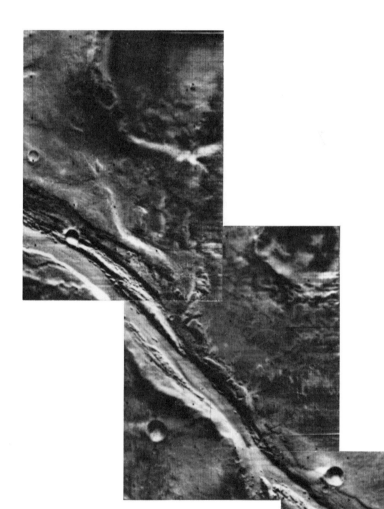

Figure 4–22. *An Ancient Riverbed?* These three photographs show a feature that looks like a dry riverbed. The portion shown here is 45 miles long and is located just north of the equator. (*NASA*)

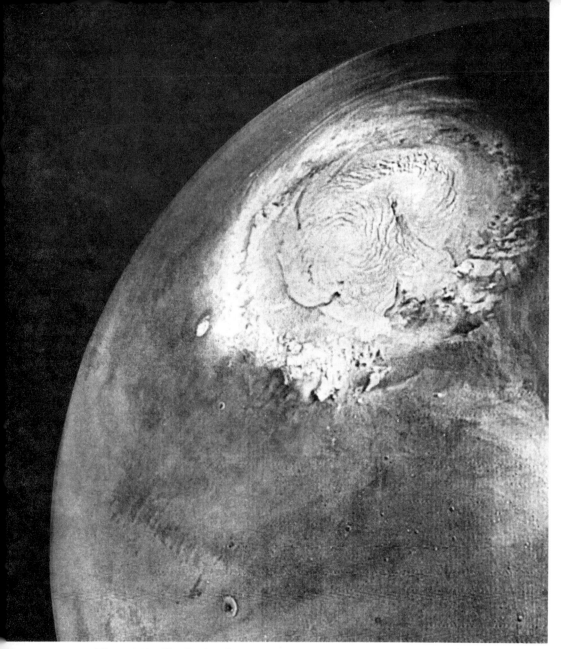

Figure 4–23. *The North Polar Cap.* In August of 1972, Mariner 9 took this fine photograph of Mars' northern polar cap. At this time it was late spring in the northern hemisphere and much of the dry ice had evaporated. (*NASA*)

Martian surface. Such a mission is planned for 1976, when the Viking Lander will parachute through the thin Martian atmosphere and settle down on the surface of the planet. Instruments aboard the Viking Lander will analyze the Martian soil looking for any clues of life.

Although most scientists feel that Mars is a totally dead world like our moon, there is a slim chance that they are wrong. If Viking is successful in detecting any life forms whatsoever, we will have made one of the most profound discoveries in the history of mankind. We now know of only one planet in the universe on which life has appeared. Are we unique? Is it possible that life might exist on other planets in orbit about other stars? With current information, we simply do not have enough information to give intelligent answers to these and similar questions. But *if* life is found on Mars, then we are certainly not unique. Then there must be a high probability for the appearance of life forms whenever conditions allow. Even more importantly, modern biology teaches that evolution favors the development of intelligence. If two creatures evolve on a planet and one is smart while the other is stupid, chances are that the more intelligent creature will be the one to survive. He will be better able to cope with his environment, find food, and shelter his young. It is, therefore, clear that *if* we find any kind of life on Mars, the probability of the existence of intelligent beings—even superior to man—living elsewhere in the universe is extremely high! If, as we expect, Viking reports back that Mars is barren and sterile, then all we can do is continue to explore and to wonder.

Figure 4–24. *The Viking Mission.* This artist's rendering of the Viking mission shows the Lander which will parachute down to the Martian surface. The main spacecraft will stay in orbit above the planet for the purposes of photographing (like Mariner 9) and for relaying transmissions back to Earth. (*NASA*)

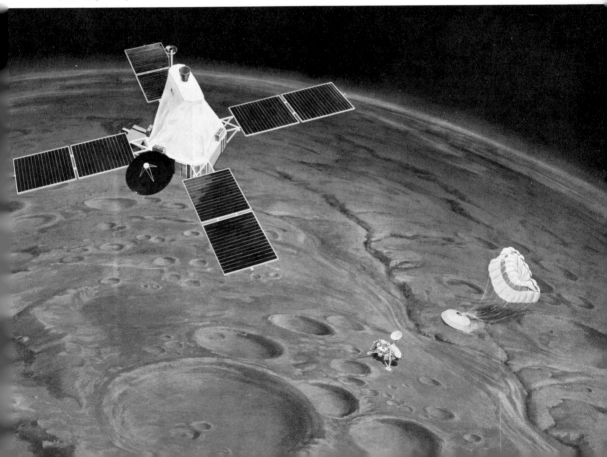

4.4 Venus and Mercury

The morning and evening "stars" that herald the rising sun or the coming night are always the planets Mercury and Venus. These two "inferior" planets—so named only because their orbits lie inside the earth's orbit—are often among the brightest objects in the sky. At maximum brightness, Venus is exceeded in brilliance only by the sun and moon; on a dark night this planet can even cast a shadow. Mercury never gets as bright as Venus. At maximum, Mercury's apparent magnitude is approximately the same as that of Sirius, the brightest appearing star in the sky.

In certain respects, Venus resembles Earth. The mass and size of Venus are very nearly equal to the mass and size of Earth. As a result, the surface gravity on these two planets is approximately the same. But that is where the similarities end. Venus is covered with an incredibly dense, hot, thick atmosphere that completely obscures the surface of the planet. Looking at Venus through a telescope, all that can be seen is featureless, shimmering clouds that eternally hide the rest of the planet from view. As a result of the flights of Mariner 2 (in 1962), Mariner 5 (in 1967), and the two Soviet spacecrafts Verena 4 (in 1967) and Verena 7 (in 1970), it was discovered that the Venusian atmosphere consists almost entirely of carbon dioxide with traces of water, hydrochloric acid, and sulfuric acid. The atmospheric pressure on the surface of the planet is about one hundred times greater than the atmospheric pressure here on Earth. The temperature at the bottom of this crushing atmosphere is an incredible 900°F, hotter even than the surface of Mercury. A plausible explanation of this high Venusian temperature has to do with the so-called *greenhouse effect*. On a sunny day, a greenhouse or a car left with the windows rolled up has inside temperatures much higher than the outside temperature. By analogy, Venus' dense atmosphere absorbs and traps much of the sun's radiation, which heats the planet to very high temperatures.

Since the surface of Venus is eternally hidden from view, up until very recently we had no information concerning Venus' rotation; we had no idea how long a "day" is on Venus. To remedy this, astronomers in the early 1960s bounced radar signals off the planet. These radar waves easily penetrated the thick atmosphere and were reflected back to Earth from the planet's surface. Examining these "echoes," they found that Venus rotates very slowly, once in 243 days. Even more surprising, they found that this rotation is *retrograde,* in comparison to the other planets. Earth, for example, rotates from west to east, whereas Venus rotates from east to west. The sun rises in the west and sets in the east on Venus. Actually, of course, you could never see the sun from the ground on Venus. The "air" is so thick that very little sunlight ever reaches the surface; it would almost be like trying to see the sun from the bottom of the Pacific Ocean.

The planet Mercury is strikingly different from its nearest neighbor. The planet closest to the sun is also the smallest, both in mass and size. In view of its low

Figure 4–25. Venus. This photograph of Venus was taken with the world's largest telescope, the 200-inch on Palomar Mountain. Even with the best equipment, astronomers on earth can see only a featureless, shimmering cloud-cover. (*Hale Observatories*)

surface gravity and proximity to the sun, Mercury does not have an atmosphere. In many respects, this tiny planet seems to resemble our moon.

Mercury's orbit is about half the size of Venus' orbit; the average distance between Mercury and the sun is only 36 million miles. This means that, as seen from Earth, Mercury never appears very far from the sun. At maximum elongation, it is only 28° from the blinding solar disc that makes observations of this tiny planet very difficult. Indeed, many professional astronomers, including Copernicus, have never seen Mercury, even though it is often among the brightest objects in the sky. In addition, since this planet is so very near the sun, it must have a very short orbital period. A "year" on Mercury lasts for only 88 days. And, finally, careful telescopic observations of faint features on the surface of the planet tell us that Mercury's period of rotation is $58\frac{1}{2}$ days. A Mercurian day lasts for exactly two thirds of a Mercurian year.

Telescopic observations of the two inferior planets from Earth are most unrewarding. On Venus all we see is a shimmering, featureless cloud cover, whereas Mercury is so small and far away that it is impossible to detect any surface details. As with Mars, the only possible procedure is picture-taking space probes.

Figure 4–26. The Flight of Mariner 10. This artistic rendering shows Mariner 10 as it journeyed on its historic mission to Venus and Mercury early in 1974. (*NASA*)

Astronomers realized that, during 1974, Mercury and Venus would be positioned in such a fashion that one spacecraft could fly past both planets. A probe could be aimed at Venus in such a way that after passing the cloud-covered planet, it would continue on to Mercury. In doing so, the flight time to Mercury would be very short because Venus's gravitational field gives the spacecraft an extra boost halfway through the journey.

Figure 4–27. Venus. This remarkable photograph of Venus was taken by Mariner 10 when it was 450,000 miles above the planet. Mariner 10 took nearly 3500 photographs of the cloud-covered planet. (*NASA*)

Figure 4–28. Crescent Mercury. This photograph of Mercury was taken by Mariner 10 when it was 124,000 miles above the planet. The resemblance to our moon is striking. (*NASA*)

This ambitious mission was undertaken on November 2, 1973, with the launch of Mariner 10 from Cape Canaveral. After several months of gliding silently through the interplanetary void, Mariner 10 encountered Venus early in February 1974. The television cameras were turned on and Mariner 10 sent back nearly 3,500 incredible pictures from distances down to 3,600 miles above the planet. Features in the clouds as small as 300 feet across could be seen. These photographs, such as the one shown in Figure 4–27, reveal a very turbulent atmosphere. Although Venus rotates once every 243 days, its entire atmosphere revolves completely around the planet once every 4 days! From this observation we can conclude that the wind speeds on Venus are upward of 200 miles per hour. This explains why the radar astronomers who bounced signals off the planet found the surface to be very smooth. The rapidly circulating thick atmosphere has worn down any mountains or valleys that might have been on the planet.

Instruments aboard Mariner 10 found that, unlike Earth, Venus has virtually no magnetic field. Additional instruments measured the temperature of the Venusian clouds (255°K) and discovered that there is very little difference in temperature between the day and night sides of the planet. After about a week of frantic activity, virtually all these instruments were shut down as Mariner 10 continued to coast towards its rendezvous with Mercury.

Figure 4–29. Craters on Mercury. This photograph of Mercury's northern limb was taken at a distance of 49,000 miles from the planet. The linear distance across the bottom of the photograph is about 365 miles. (*NASA*)

In passing Venus, Mariner 10 received an extra midcourse acceleration from the planet's gravitational field, thereby reducing the flight time to Mercury to a few weeks. Encounter with the innermost planet occurred late in March 1974. As the time of closest approach (430 miles above Mercury on March 29) neared, Mariner's cameras were turned on and the space probe preceded to send back 2,400 remarkable photographs showing surface features as small as 500 feet across. To the amazement of scientists around the world, the surface of Mercury looks exactly like our moon! As shown in Figures 4–28 and 4–29, Mercury is a heavily cratered world. Indeed, only careful examination by experts could distinguish the Mariner photographs of Mercury from pictures of unfamiliar parts of the lunar surface. In addition to discovering many lunarlike craters, Mariner 10 also photographed large flat areas that strongly resemble the maria on the moon. Figure 4–30 shows about half of one of these Mercurian maria near the *terminator*, the demarcation line between day and night, This feature is called the Caloris basin, from the Latin word meaning "hot." The comparison with the moon is strengthened by the observation that one side of Mercury is covered only with craters, and all the maria seem to be located on the other side.

Although the surface of Mercury bears a striking resemblance to the moon, it now seems that the interior of the planet is very much like Earth. Instruments aboard Mariner 10 detected magnetic disturbances above the planet, which means that Mercury may have a permanent magnetic field, like Earth. In addition, observations of the mass and size of the planet tell us that the average density of Mercury is about 60 per cent greater than the average density of the moon. Thus, a large part of Mercury must be made out of material denser than the lunar rocks, and perhaps a substantial iron core exists at the planet's center.

Mankind has visited the morning and evening stars. During a few short weeks in 1974, we learned more about these two worlds than all the knowledge that had

Figure 4–30. The Caloris Basin. This photograph of the region near the terminator shows a maria-like feature 800 miles in diameter. It has been named "the Caloris basin" and is very similar in size and appearance to Mare Imbrium on our moon. (*NASA*)

accumulated about them since the dawn of recorded history. Mariner 10 made Earth-based observations virtually obsolete. Our next great advance will come with orbiting satellites above Mercury and Venus and, later, with probes that land on their surfaces. It is clear from the hostility of the environments we have encountered that we should not even begin to hope to detect any life forms.

4.5 The Jovian Planets

With almost prophetic insight, ancient man named Jupiter after the king of the gods. Jupiter, which is often among the brightest objects in the sky, is by far the largest and most massive planet in the solar system. With a mass 318 times greater than Earth's, this vast planet contains more matter than everything else in the solar system combined, excluding the sun. Although it takes 12 years to complete one orbit of the sun, Jupiter has the shortest period of rotation of any of the planets. A "day" on this planet lasts for only about 9 hours and 50 minutes. In view of its rapid rate of rotation, Jupiter has a somewhat flattened appearance. The equatorial diameter is about 12,000 miles greater than the distance from pole to pole.

Perhaps because of its huge gravitational field, Jupiter has far more moons than any other planet. Fourteen known satellites orbit this planet. The four *Galilean satellites*, mentioned earlier, are roughly the same size as our moon, whereas the remaining 10 satellites are very small, 100 miles in diameter or less.

Through a telescope, Jupiter appears to be striped with light and dark cloud bands parallel to its equator. In addition to these bands, one of the most striking surface features is the *Great Red Spot*, which has been observed for several centuries. Up until very recently, the origin and nature of this feature has been a total mystery. An excellent Earth-based photograph is shown in Plate 1.

In December 1973, Pioneer 10 passed within 81,000 miles of Jupiter and sent back a great deal of information about this massive planet. Pioneer 11 flew even closer to the planet (27,000 miles at *periapsis*, or point of closest approach) 1 year later, confirming many of its predecessor's observations. As a result of these unmanned flights, we now know that Jupiter is almost entirely a liquid planet composed primarily of hydrogen and helium with some methane and ammonia. Jupiter's seventeen relatively permanent *belts* and *zones*, which give the planet its characteristic appearance, seem to be huge cyclones stretching all the way around the planet. Due to Jupiter's rapid rotation, the speed of the atmosphere around the planet is about 22,000 miles per hour. In view of this velocity, systems of highs and lows similar to weather conditions on Earth are stretched out in long bands around the planet. The grey-white zones are cloud ridges of rising atmosphere circling Jupiter and looming 12 miles above the cloud belts. Conversely, the orange-brown cloud belts are masses of descending atmosphere about 12 miles deep. Jupiter seems to have a very turbulent interior, much hotter

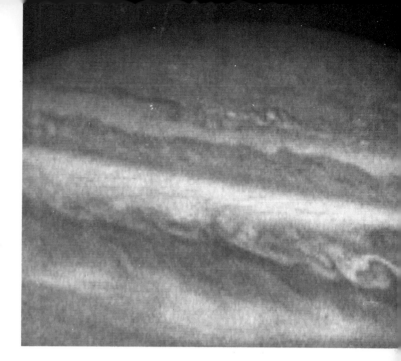

Figure 4–31. *Jupiter's Cloud Tops.* Never-before-seen details of weather patterns on Jupiter are shown in this photograph from Pioneer 10. This view was taken from a distance of 616,000 miles in December 1973. (*NASA*)

than previously thought. The temperature just below the surface of the clouds is about 10°F and rises steadily to an estimated 54,000°F at the planet's center, six times higher than the temperature of the surface of the sun. In many respects, Jupiter may be thought of as an "unborn star," a sphere of gas not quite massive enough to ignite thermonuclear fires at its center, such as those that power our sun.

A surprising result from the Pioneer flights was the discovery that Jupiter has an incredibly intense magnetic field. Huge radiation belts, similar to the Van Allen belts around the earth, circle the planet. As a result, Jupiter is a source of particle radiation, the only one in the solar system besides the sun.

Finally, the Pioneer observations indicate that the Great Red Spot is the vortex of an intense storm that has been raging for hundreds of years. It is 25,000 miles

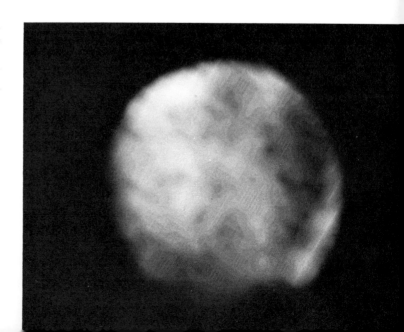

Figure 4–32. *Ganymede.* This photograph of Ganymede, one of Jupiter's four largest moons, was taken by Pioneer 10 at a distance of 467,000 miles. Ganymede has a diameter of 3,270 miles and is believed to consist of water, ice, and rock. (*NASA*)

long and 6,000 miles wide (for comparison, Earth's diameter is about 8,000 miles), and consists of whirling clouds that tower some 5 miles above the surrounding cloud deck.

In addition to examining Jupiter, the Pioneer probes also observed the moons of Jupiter. All four Galilean satellites have an average surface temperature on the sunlit side of about $-230°$F. The two inner Galilean moons (Io and Europa) seem to be made entirely of rock, like our own moon, whereas the outer two (Ganymede and Callisto) probably consist mostly of ice. The innermost large moon, Io, has an atmosphere, as suggested by earlier Earth-based observations. Every time Io orbits Jupiter, the satellite spends 21 hours in Jupiter's shadow. When Io emerges from the shadow, for about 10 minutes it is extremely bright; it is the most reflective object in the solar system. Io then begins to turn from white to its pronounced orange color. It is possible that during the freezing Jovian night, methane snowflakes cover the surface of Io and then evaporate with the return of sunlight.

In performing its mission, Pioneer 10 was catapulted out of the solar system by Jupiter's gravitational field. The spacecraft is now on a path toward the constellation of Taurus. It will arrive in the vicinity of Aldebaran in about 2 million years. The trajectory of Pioneer 11 was chosen much more auspiciously. Just as the gravitational field of Venus was used to direct Mariner 10 toward Mercury, Pioneer is continuing on a $4\frac{1}{2}$ year journey to Saturn. Arrival time is scheduled for September 1979.

Saturn, the second-largest planet in the solar system, was the most distant planet known to ancient man. At 886 million miles from the sun, Saturn takes

Figure 4-33. The Flight of Pioneer 11. Pioneer 11 was launched from earth on April 6, 1973. It flew past Jupiter on December 3, 1974, and is scheduled to arrive at Saturn in September of 1979.

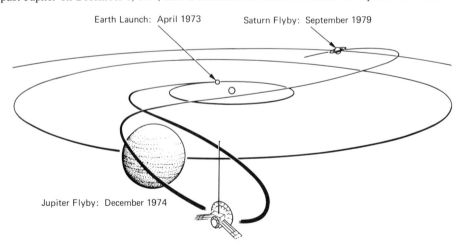

Figure 4-34. Saturn. This earth-based photograph of Saturn shows the magnificent ring system along with cloud belts on the planet's surface. (*Lick Observatory*)

$29\frac{1}{2}$ years to complete one orbit. Through a telescope, its unique *ring system* makes Saturn one of the most impressive sights in the sky. A very fine Earth-based photograph is shown in Figure 4-34.

Saturn's ring system was first discovered by Christian Huygens in 1655. The system actually consists of at least three concentric rings, the outer ring having an outside diameter of 171,000 miles and the inner ring having an inside diameter of 88,000 miles. In between the three major rings there are gaps about 8,000 miles wide.

The rings are not solid sheets. From the earth we can see bright stars shining through them. In addition, the inner portions of the inner ring rotate around the planet much faster than the outer portions of the outer ring. These facts strongly suggest that the rings are made up of billions of tiny particles, probably the size of gravel. Whether these particles are made of rock or ice or a combination of both will be answered only after Pioneer 11 completes its mission.

The planet itself has a mass ninety-five times greater than Earth's and is about 72,000 miles in diameter. Except for its rings, Saturn looks very much like Jupiter. For example, dark and light cloud bands parallel to Saturn's equator are seen on the planet, although they are not nearly as pronounced or distinct as those on Jupiter. In view of Saturn's rapid rotation (a "day" on Saturn lasts for slightly more than 10 hours), these cloud bands are probably cyclonic wind patterns circling the entire planet. Its atmosphere is composed primarily of hydrogen and helium with some methane, just like Jupiter's. However, any ammonia that might be present has been frozen out of Saturn's atmosphere due to its low temperature of only $-230°F$.

Figure 4–35. Uranus. The planet Uranus is shown here with three of its five satellites. (*Lick Observatory*)

At almost a billion miles from the sun, Saturn is the most distant planet that can be easily seen with the naked eye. Yet, beyond Saturn lie two more Jovian planets, Uranus and Neptune, which are remarkably similar to each other. Uranus and Neptune are almost exactly the same size; they each have a diameter of about 30,000 miles. Uranus's mass is about fifteen times that of Earth, whereas Neptune has a mass of seventeen earths. Both are composed primarily of hydrogen and helium with some methane. As with Saturn, any ammonia we might have expected has been frozen out of their atmospheres due to the very low surface temperatures: $-300°F$ for Uranus and $-350°F$ for Neptune. In addition, both planets look very similar through a telescope. They both appear as small, greenish disks, with no surface features or markings. Excellent Earth-based photographs are shown in Figures 4–35 and 4–36.

One unique feature of Uranus is that its axis of rotation is tilted over so far that it lies almost in the plane of its orbit. The seasons on Uranus are, therefore, very strange. When it is summer in Uranus's northern hemisphere, the sun is almost directly over the north pole, and much of the southern hemisphere is in total

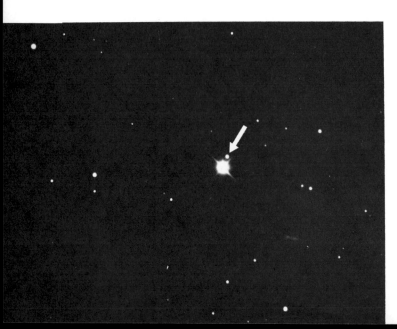

Figure 4–36. Neptune. Neptune and its largest moon, Triton, are shown in this earth-based photograph. (*Lick Observatory*)

Chapter 4 The Solar System

darkness. Half a Uranian year later (Uranus takes 84 years to go once around the sun), the situation is reversed, with the sun almost directly above the planet's south pole. By comparison, Neptune's axis of rotation is inclined by 29° to its orbital plane, which is similar to the inclinations we found for Earth ($23\frac{1}{2}°$) and Mars (25°).

Aside from their rotation periods (11 hours for Uranus and 16 hours for Neptune) and a few moons (five for Uranus and two for Neptune), virtually nothing else is known about these planets. We had an extraordinary opportunity to send a spacecraft on an ambitious mission to *all* four Jovian planets in the 1970s. However, due to budget cuts, this "Grand Tour" mission was abandoned. As a result of this shortsightedness, the outer reaches of our solar system will remain shrouded in mystery for decades to come.

4.6 Interplanetary Matter

As we have seen, our sun contains 99.86 per cent of the mass of the solar system. All nine planets and thirty-four moons combined account for an additional 0.136 per cent. The remaining 0.004 per cent is made up of countless trillions of particles orbiting our sun.

In the mid-1700s, an astronomer named Titius discovered a curious arithmetical progression that seems to give the distances of the planets from the sun in astronomical units (AUs). To obtain this progression, write down the sequence of numbers: 0, 3, 6, 12, 24, and so on. Add 4 to each number. Then divide the resulting sums by 10. It is indeed curious that this little mathematical game produces numbers very nearly equal to the actual distances of the planets from the sun, as shown in the table here.

Titius' Progression	Planet	Planet's Actual Distance (in AU)
(0 + 4)/10 = 0.4	Mercury	0.39
(3 + 4)/10 = 0.7	Venus	0.72
(6 + 4)/10 = 1.0	Earth	1.00
(12 + 4)/10 = 1.6	Mars	1.52
(24 + 4)/10 = 2.8	?	
(48 + 4)/10 = 5.2	Jupiter	5.20
(96 + 4)/10 = 10.0	Saturn	9.54
(192 + 4)/10 = 19.6	Uranus	19.18
(384 + 4)/10 = 38.8	Neptune	30.6
(768 + 4)/10 = 77.2	Pluto	39.4

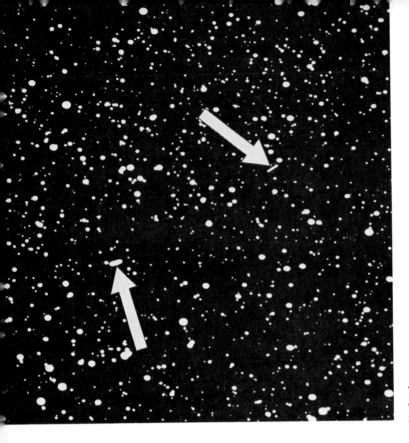

Figure 4-37. *Two Asteroids.* The blurred trails of two asteroids appear on this 1908 photograph. (*Yerkes Observatory*)

Due to peculiar historical circumstances, this arithmetical progression invented by Titius is today known as *Bode's law*. It was not discovered by Bode and most certainly is not a "law." Bode's law does not work well for Neptune and Pluto. However, Uranus was discovered a few decades after the invention of Bode's law. This seventh planet fit so nicely into the arithmetic progression that everyone began to believe that there really might be something very important about this sequence of numbers.

One striking feature of Bode's law is that it "predicts" the existence of a planet at 2.8 AU from the sun. Of course, back in the eighteenth century, it seemed as though there was nothing but empty space between the orbits of Mars and Jupiter. Astronomers, therefore, began searching the skies for the "missing planet" that was supposed to be at 260 million miles from the sun.

In view of the acceptance of Bode's law, no one was terribly surprised when in January 1801, the Sicilian astronomer Giuseppi Piazzi discovered a small object orbiting the sun at 2.77 AU. Clearly this must be the "missing planet." It was named *Ceres*, after the protecting goddess of Sicily. In addition, Ceres must be very small because even through a powerful telescope it looks like a dim star slowly moving among the constellations of the zodiac.

Astronomers began being surprised when, in March 1802, Heinrich Olbers discovered a second starlike object moving in an orbit similar to Ceres's orbit. This second object was named *Pallas*. Then, in 1804, a third object, *Juno*, was found. And, in 1807, *Vesta* was discovered. All had orbits in the gap between Mars and Jupiter.

By the early 1800s, it was, therefore, clear that instead of a "missing planet," there were actually many very small planets, or *planetoids*, orbiting the sun at distances from 2 to 4 AU. Today these objects are more commonly called *asteroids* or *minor planets*, and the region of the solar system they occupy (between the orbits of Mars and Jupiter) is called the *asteroid belt*.

With the invention of astronomical photography in 1891, which allowed astronomers to take time exposures of the stars, asteroids were discovered literally by the hundreds. About two thousand asteroids have been identified and their orbits have been calculated and published. In addition, thousands more have been observed, but no one has bothered to track them and go to the trouble of calculating their orbits. It has been estimated that there might be as many as 100,000 asteroids orbiting the sun.

The first four asteroids discovered in the early 1800s are by far among the largest of the minor planets. Their sizes are listed in the following table.

Asteroid	Diameter (in miles)
Ceres	488
Pallas	304
Vesta	248
Juno	118

There are probably a few hundred asteroids with diameters larger than 25 miles. But most minor planets are not much bigger than 1 mile across.

Many years ago it was popular to believe that the asteroids might really be pieces of a planet that exploded or broke up for some unknown reason. Actually, the total mass of all the asteroids combined is probably less than $1/500$ of the mass of Earth. Lumped together, all the asteroids would make up the equivalent of one small moon. (For comparison, Earth's moon has a mass of $1/81$ that of Earth.)

While objects with diameters bigger than 1 mile and with orbits usually between those of Mars and Jupiter are called asteroids, there are countless trillions of much smaller particles scattered throughout the solar system. These tiny particles are called *meteoroids*. When one of these meteoroids, which typically may be only the size of a grain of sand (about $1/100$-ounce) collides with Earth, it burns up in the atmosphere, producing a "shooting star" or meteor. The high speed of the meteoroid through Earth's upper atmosphere results in so much friction that the particle is vaporized in a fraction of a second by the intense heat. In addition, very many smaller particles called *micrometeorites* are constantly striking Earth. These micrometeorites are typically the size of a piece of dust and are too small to produce meteors. Although it may sound incredible, between 50 and 100 tons of meteoritic material fall on Earth *every day!*

On very rare occasions, very large meteoroids collide with Earth. In spite of

Figure 4–38. *An Iron Meteorite.* When an iron meteorite is cut, polished, and etched, a characteristic crystalline structure is seen. This crystalline structure, called *Widmanstätten figures*, is shown in this photograph of a piece of an iron meteorite. (*Griffith Observatory*)

Figure 4–39. *A Stony-Iron Meteorite.* A slice of a stony-iron meteorite is shown in this photograph. The dark regions are made of the mineral olivine. (*Griffith Observatory*)

Figure 4–40. A Stony Meteorite. A piece has been cut off of this stony meteorite to reveal tiny flakes of iron and nickel contained in the rock. (*Griffith Observatory*)

their high speed through the air, they are so big (a few ounces up to several tons) that they do not completely vaporize. The object that survives the flight and lands on the ground is called a *meteorite*.

With some luck and quite a bit of skill, it is not terribly difficult to find meteorites of which there are basically three types. The most commonly found meteorites are the *irons*, or *siderites*. As their name suggests, they are composed of about 90 per cent iron, with the rest being mostly nickel. They are found most easily with commercially available metal detectors or magnets. The *stony irons*, or *siderolites*, are rarer and, again as their name suggests, consist of a combination of iron and rock. The *stones*, or *aerolites*, are usually among the rarest of all meteorites. They are composed mostly of stony material with small flecks of iron and nickel. They are rare only because they often look like ordinary rocks, especially after having been eroded slightly by the wind and rain. Nevertheless, we have reason to believe that most meteorites are made out of stone.

It is well known that the chances of observing a meteor on a given night vary throughout the year. The best time to observe meteors is during a so-called *meteor shower*. At the time of a meteor shower, Earth is actually passing through a swarm of meteoroids in orbit about the sun, as shown in Figure 4–41. If the meteoroids are strewn uniformly along the orbit, the resulting shower will be roughly the same from year to year. A good example of this type of shower is the *Perseids*, so named because they seem to come from the constellation of Perseus. Every year around August 11, especially if there is a clear moonless night, you

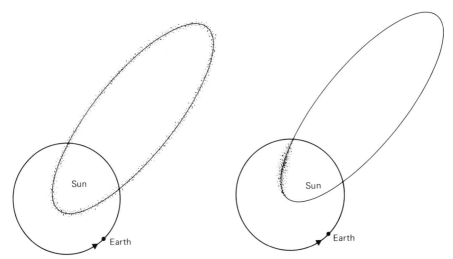

Figure 4–41. *Meteor Showers.* Meteor showers occur when the earth in its orbit around the sun intercepts a swarm of meteoroids also orbiting the sun. Meteor showers vary from year to year depending on how the meteoroids are distributed along their orbit.

can see up to sixty meteors per hour. If, however, the meteoroids in orbit about the sun are bunched up together in one part of the orbit, then the shower can be very different from one year to the next, depending on whether Earth hits or misses the cluster of meteoroids. A spectacular example of this type of shower is provided by the *Leonids*, which appear to radiate from Leo around November 16. At 33-year intervals, Earth passes directly through the meteoroid swarm. In 1833, and again in 1866, as many as 200,000 meteors could be seen in only a few hours. The last good Leonid shower was in November 1966 when up to 140 meteors per second were reported by some observers. On the "off" years, when Earth misses the Leonid swarm, this shower can be most disappointing.

Unfortunately, the term *meteor* often is confused with the word *comet*. Recall that a meteor is produced by a small rock from interplanetary space plunging through Earth's atmosphere at a high speed. The rock is vaporized by the resulting friction with the air in less than a second, which results in the flash of light we call a meteor. A comet, on the other hand, can be seen night after night, slowly moving against the background of stars as it journeys along its orbit around the sun.

Comets have been seen for thousands of years. The bright *head* of the comet followed by a long, flowing *tail* stretching across the sky can provide one of the most spectacular of all celestial phenomena. Actually, very few comets become so bright that they can be seen with the naked eye. Astronomers discover about a dozen comets a year, usually by accident, but most remain so faint that they can be viewed only through a telescope.

When an astronomer discovers a comet, it usually looks like a dim, fuzzy star; indeed, the word *comet* comes from a Latin root meaning "hairy star." As the

Chapter 4 The Solar System

comet approaches the sun, usually along a highly elliptical orbit, the hazy object brightens and develops a long tail. The comet reaches maximum brightness when it is closest to the sun and then fades as it returns to the depths of interplanetary space. This evolution of a comet is shown in a sequence of photographs in Figure 4–42.

Current theories on comets argue that the *nucleus* of a comet contains a large piece of ice consisting of frozen water, methane, and ammonia, as well as

Figure 4–42. *Halley's Comet.* This sequence of photographs shows the changing appearance of Halley's comet from April 26 to June 11, 1910. This comet is scheduled to return in 1986. (*Hale Observatories*)

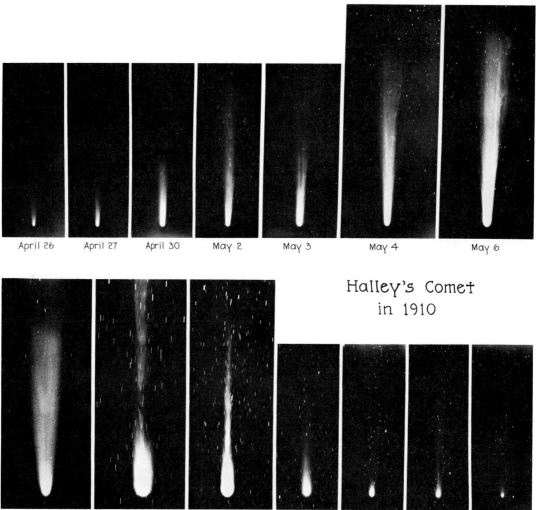

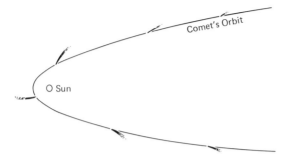

Figure 4–43. *A Comet Passes the Sun.* As a comet swings past the sun, the solar wind causes the comet's tail to point directly away from the sun at all times.

meteoritic material such as small stones and rocks. This so-called dirty iceberg is typically less than 50 miles across; as it approaches the sun, these ices begin to melt and vaporize. The gases from the nucleus produce a small, round, nebulous glow called the *coma*. As the comet gets closer to the sun, the rate of vaporization increases and the sun's radiation pressure and the *solar wind* push the gases out into a long flowing tail, thereby giving the comet its characteristic appearance.

Figure 4–44. *Mrkos' Comet.* Changes in the structure of the tail of a comet are shown in this series of photographs taken on August 22, 24, 26, and 27 of 1957. (*Hale Observatories*)

COMET MRKOS

AUGUST 22   AUGUST 24   AUGUST 26   AUGUST 27

1957

Photographed with the 48-inch schmidt telescope.

This so-called solar wind actually consists of a continuous rain of atomic particles constantly ejected from the sun into space. The pressure from the sun's light and the solar wind is so strong that the comet's tail always points away from the sun, as shown schematically in Figure 4–43. A comet's tail, blown by the solar wind, can be stretched out to lengths of up to 100 million miles, yet the tail itself contains very little matter. Hypothetically, you could take the entire tail of a comet, squeeze it into a suitcase, and walk off with it. In 1910, Earth passed through the tail of Halley's comet with no noticeable effects whatsoever.

In this section we have discussed the odds and ends of the solar system. We have seen that objects of all sizes and shapes roam through the void between the planets. The largest objects are the asteroids. The smallest are the micrometeorites and atomic particles of the solar wind. Yet, in spite of all these various objects, interplanetary space is really quite empty. Interplanetary space is emptier than the best vacuum a scientist can create in his laboratory. The probability of a meteoroid puncturing the spacesuit of an astronaut is totally negligible; he is really quite safe from such dangers. Pioneer 10 and Pioneer 11 both passed directly through the asteroid belt without mishap. In fact, as far as we know, they did not even get close enough to an asteroid to take a good photograph. Most of the matter in the solar system is located in the sun. By comparison, everything else in the solar system can be considered to be microscopic impurities in the vast emptiness of space and time.

Questions and Exercises

1. Briefly discuss the distribution of mass in the solar system.
2. How many planets are there? List them in order from the sun.
3. List the planets in order of decreasing size, with the largest first and the smallest last. (Hint: Refer to the tables at the back of the book.)
4. List the planets in order of decreasing mass.
5. List the planets in order of decreasing surface gravity. On which planets would you weigh most nearly the same as you do here on Earth?
6. Which planet has the shortest period of rotation? Which has the longest?
7. Which planet has the most moons? Which planets have the fewest?
8. Using a compass (and assuming that the orbits of the planets are circles) make a scale drawing of the orbits of the planets around the sun.
9. How many moons are known to exist in the solar system? Of these, which are most nearly like our own moon? What are their names and to which planets do they belong?
10. What is meant by escape velocity?

11. Why and how do the surface temperature and surface gravity of a planet determine whether that planet can have an atmosphere?
12. What is the difference between the terrestrial and the Jovian planets?
13. Discuss one reason why going to the moon may tell us more about the history of Earth.
14. Describe the lunar maria.
15. What is the regolith?
16. In what way does the appearance of the "front" side of the moon differ from the appearance of the moon's "far" side?
17. About how many craters on the moon can be seen through telescopes from Earth?
18. What is a zap crater?
19. In what ways is Mars like Earth?
20. When is the best time to observe Mars?
21. Describe the kinds of surface features discovered on Mars as a result of the Mariner flights.
22. Describe the moons of Mars.
23. Do some research to discover the meanings of the Greek words *phobos* and *deimos*. Why are these two names appropriate for the companions of Mars, the god of war?
24. What is Olympus Mons?
25. Compare the heights of Mount Everest and Olympus Mons.
26. Which planet has the highest surface temperature, Mercury or Venus?
27. Briefly describe Venus' atmosphere.
28. How did astronomers discover Venus' period of rotation?
29. What role does the "greenhouse effect" play in Venus' atmosphere?
30. Briefly describe the surface of Mercury.
31. What are the Galilean satellites? Who discovered them? To which planet do they belong? What are their names?
32. Describe the appearance of Jupiter through a telescope.
33. Which planets in the solar system are known to have a magnetic field?
34. Briefly describe the Great Red Spot.
35. Briefly describe Saturn's rings.
36. Contrast and compare Uranus and Neptune. In what ways are they similar and in what ways are they different?
37. What is Bode's law?
38. What is an asteroid?
39. Where is the asteroid belt?
40. What is the difference between a meteoroid, a meteor, and a meteorite?
41. What causes meteors?
42. About how much meteoritic material falls on the earth every day?
43. Briefly describe the three basic types of meteorites.
44. What is a meteor shower?
45. Describe the appearance of a typical comet.
46. How does the appearance of a comet change as it orbits the sun?

The Physics of Light

5.1 Electromagnetic Radiation

Except for meteorites and a few hundred pounds of moon rocks, *everything* known about the universe beyond the realm of earth comes through light. The nature of the planets, the properties of the stars, and the motions of the galaxies are known to us only because we receive light from these objects. This statement is at the same time both trivial and profound. Obviously, if these objects did not emit light, their existence would not be known. Yet, as we will learn in this chapter, the light that we do receive contains a great deal of information. The atoms that emit the light, as well as the physical conditions surrounding these atoms (temperature, pressure, magnetic fields, and so on), leave an indelible mark on this light, which is transmitted to us through the void of space. Therefore, through understanding the properties of light and how light interacts with matter, scientists can analyze this radiation and, thereby, discover the nature of its source. In this way, the most distant objects in the universe reveal their secrets to the astronomer.

Modern understanding of light began with the pioneering work of Sir Isaac Newton. In the mid-1600s, Newton discovered that, when a beam of white light is passed through a glass prism, the white light is broken up into the colors of the rainbow, as shown in Figure 5–1. This demonstrates that white light actually is composed of all the colors of the rainbow combined. In addition, from the numerous experiments performed during the eighteenth century, it was realized that light exhibits many phenomena normally associated with *waves*. For example, if you shine a beam of light on the edge of a razor blade, careful examination of what happens as the light passes the sharp edge of the razor shows dark and light patterns similar to the ripples set up behind the edge of a pier in the ocean being pounded by water waves. The analogy between water waves and the behavior of light is so striking that physicists have concluded that light must be a *wave phenomenon.*

As scientists began to understand what was going on in experiments with light shining on razor blades, past thin wires, and through pin holes and slits, they realized that they could talk about the *wavelengths* of various colors of light. In ocean waves, the wavelength is simply the distance between successive wave

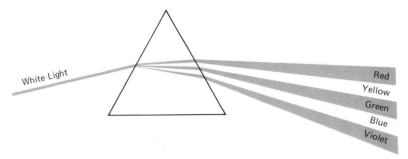

Figure 5–1. A Prism. When a beam of white light passes through a prism, the light is broken up into the colors of the rainbow. This resulting rainbow is called a *spectrum*.

crests. Similarly, the distance between successive crests in light waves is the wavelength of the light, as shown in Figure 5–2. Wavelength is usually symbolized by the Greek letter lambda (λ). And just as the size of wavelengths in the ocean is determined by the properties of water, so the wavelength of a particular beam of light is determined by the color of the light. But, whereas the distance between wave crests in the ocean is very large, the distance between crests in a light wave is extremely small, so small that a new "unit of measure" was invented to express these tiny distances. To see why this was necessary, we must realize that it would be absurd for someone to say that he is 0.00104-mile tall; the mile is not a convenient unit of measure to express one's height. It would be far more reasonable to say that your height is $5\frac{1}{2}$ feet; the foot is a convenient unit of measure. It turns out that the wave lengths of visible light cover the range from 0.000016-inch (violet) to 0.000028-inch (red). To express the wavelength of light in inches is just as inconvenient as expressing your height in miles. To solve this problem, physicists invented a new unit of measure called the *angstrom* (abbreviated Å). One angstrom is equal to one ten millionth of a millimeter, or about four billionths of an inch. In terms of the angstrom, visible light has wavelengths ranging from about 4,000 to 7,000 Å. Light of a particular wavelength reacts with the cells in the retina of the eye to give the sensation of a particular color. In general terms, colors and their corresponding wave-length ranges are given in the table here.

Color	Wave-Length Range (Å)
Violet	3,900–4,400
Blue	4,400–5,000
Green	5,000–5,600
Yellow	5,600–5,900
Orange	5,900–6,400
Red	6,400–7,400

Chapter 5 The Physics of Light

Questions that come to mind are "What is at wave lengths shorter than 3,900 Å?" and "What is at wave lengths longer than 7,400 Å?" Complete answers to these obvious, yet profound, questions were not available until only 100 years ago, when scientists found that visible light constitutes only one small portion of the entire so-called *electromagnetic spectrum*.

During the eighteenth and early nineteenth centuries, physicists performed a variety of experiments dealing with electricity and magnetism. Up until the early 1800s, it was believed that electricity and magnetism were two very separate phenomena. Electricity had to do with Leiden jars, galvanic cells, and Ben Franklin's kite; magnetism was related to strange iron rocks that always pointed to the North Pole when suspended from a string. However, in several extremely important experiments, Michael Faraday and Hans Oersted discovered that there must be a very intimate relationship between these very different phenomena. For example, it was discovered that whenever an electric current flows through a wire, a magnetic field appears around the wire. Similarly they discovered that when a wire is moved through a magnetic field, such as that which exists between the poles of a horseshoe magnet, an electric current flows in the wire.

Another major breakthrough in the understanding of electricity and magnetism occurred in the mid-1800s as a result of the brilliant work of James Clerk Maxwell. Maxwell was to electricity and magnetism what Newton was to gravitation and mechanics. Maxwell showed that all of the experiments with and all our knowledge of electricity and magnetism could be explained by four simple equations, appropriately called *Maxwell's electromagnetic field equations*. These four equations fully express the intimate relationship between electric and magnetic fields first discovered by Faraday and Oersted. Maxwell's equations are among the most important equations in all of physics.

Shortly after Maxwell discovered his four equations, it was realized that they could be combined to give a new set of equations that describe waves of energy traveling at 186,000 miles per second. These wave equations, derived from Maxwell's equations, actually described all the properties of light. In fact, Maxwell's brilliant analysis of the phenomena of electricity and magnetism actually enabled him to predict the existence of light!

The beauty and power of Maxwell's equations cannot be overemphasized. From four simple equations, everything dealing with electricity and magnetism can be completely understood. Furthermore, Maxwell's equations can be used to tell us everything we might want to know about light. We will return to the work

Figure 5–2. Wavelength. Light is a wave phenomenon. The wavelength of a particular color of light is the distance between successive crests in the light wave. This distance is usually symbolized by the Greek letter λ (lambda).

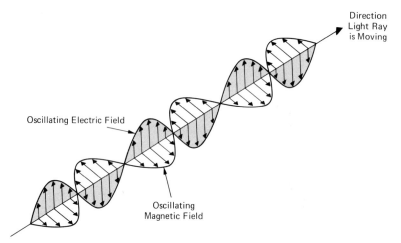

Figure 5–3. *Electromagnetic Waves.* As a result of the work of Maxwell, the true nature of light as a wave phenomenon became clear. A light wave actually consists of oscillating electric and magnetic fields.

of Maxwell in Chapter 11 because, as we will see, Dr. Einstein proved that the entire special theory of relativity logically follows from Maxwell's equations.

Maxwell's description of light reveals that a beam of light actually consists of simultaneous oscillating electric and magnetic fields, as shown schematically in Figure 5–3. As a light wave moves through space, the strengths of the electric and magnetic fields that make up the light vary, just as water waves cause the surface of the ocean to go up and down. Since, according to Maxwell, light is made up of electric and magnetic fields, light is appropriately called *electromagnetic radiation*.

One interesting feature of the Maxwellian description of light is that it involves *no* restrictions on wavelength. It was, therefore, apparent that electromagnetic radiation should be possible at *all* wavelengths, from thousands of miles down to fractions of an angstrom, not just in the range from 4,000 to 7,000 Å. The reason why such radiation cannot be seen is that the human eye does not respond to anything but visible light. Physicists, therefore, set about the business of trying to detect and discover electromagnetic radiation at all wavelengths—radiation that might be called invisible light.

In a series of experiments during the last half of the nineteenth century, many of the exotic types of radiation were discovered. For example, at wavelengths longer than red light, just beyond 7,000 Å, is a form of invisible light called *infrared radiation*. Hot objects such as a kitchen stove emit large amounts of infrared radiation. Although the human eye cannot see this form of light, human skin does respond to infrared radiation giving rise to the sensation of warmth. Infrared radiation covers the range from about 7,000 to 100,000 Å.

At wavelengths shorter than those of violet light, just beyond 4,000 Å, another form of invisible light called *ultraviolet radiation* exists. The human eye does not respond to this type of light, but ultraviolet radiation can be very destructive. If you look at a source of ultraviolet radiation, although nothing can be seen or felt, you will rapidly go blind. This radiation destroys the cells in the retinae of your eyes. Ultraviolet radiation is used to sterilize instruments in hospitals. Lying on the beach during the summer, the ultraviolet radiation from the sun causes chemical reactions to occur in exposed skin, producing a sunburn. Ultraviolet radiation covers the range from about 4,000 down to 100 Å.

In the 1880s, Hertz discovered very long wavelength radiation called *radio waves*. In essence, there is no difference between a common flashlight and a radio transmitter. The former is a source of visible light waves, whereas the latter is a source of invisible radio waves. Just as the human eye detects visible light, radios and television sets detect radio light (of course, additional machinery in radios and TVs convert the invisible radio waves into audible sounds and pictures).

Finally, around the turn of the century, the work of Wilhelm Konrad Roentgen and Mme. Marie Curie led to the discovery of X rays (100 down to 1 Å) and gamma (γ) rays (wavelengths shorter than 1 Å). These forms of invisible light are unique in that they easily pass through matter. When a doctor "takes an X ray," he is literally shining a beam of light through the body. Denser parts of the body, such as bones and teeth, show up as shadows in photographs.

All these forms of light, ranging from the very short wavelength gamma rays to the longest wavelength radio waves, together make up the *electromagnetic spectrum*. The only real difference between all these types of radiation is their wavelength. They all behave according to the same physical laws; they all are described by Maxwell's equations.

A drawing of the electromagnetic spectrum, as shown in Figure 5–4, reveals that visible light constitutes only a very small fraction of the entire spectrum. An overwhelming percentage of all types of electromagnetic radiation is totally invisible to the human eye. This has profound implications for the astronomer. Until very recently, everything known about the universe was based entirely on visible light. When we look through a telescope at the planets or stars, we are seeing *only* the visible light from these objects. Yet, why shouldn't stars also emit radio waves, X rays, and ultraviolet radiation? Clearly, scientists have been missing a very great deal! All understanding of the universe was based entirely on visible light, which is a very small part of the entire electromagnetic spectrum. Scientists, therefore, began to realize that this understanding would be dramatically affected if somehow they could see what the universe looks like in X rays, radio waves, and other wavelengths far removed from those of visible light.

Of course, until the time of Maxwell, scientists simply did not know that these other types of light existed. After some considerable experimentation, physicists found that there was an additional complication that would further hamper the efforts of astronomers. The earth's atmosphere is *opaque* to most of the electromagnetic spectrum. Visible light easily gets through the air we breath. Air is also

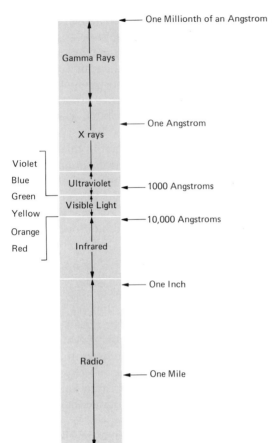

Figure 5–4. The Electromagnetic Spectrum. The electromagnetic spectrum is the complete array of all types of electromagnetic radiation. Note that visible light constitutes only a very small fraction of the entire range of the electromagnetic spectrum.

transparent to certain radio waves. But virtually no other types of radiation can get through the 10-mile-high layer of air above our heads. For example, if the human eye could see only X rays, the sky would look completely black. The X rays from stars and galaxies cannot get through all that air. Of course, this is a very fortunate situation. By being opaque to most of the electromagnetic spectrum, our atmosphere protects us from many of the deadly radiations. For example, the sun emits great quantities of ultraviolet radiation. But a layer of ozone gas high in the earth's atmosphere absorbs almost all of this type of light. If this ozone layer were suddenly to disappear, life on the earth would be destroyed in fewer than 24 hours. The earth's surface would literally be sterilized by the sun's ultraviolet rays.

Thus, the earth's atmosphere is a protective blanket shielding all life from deadly radiations. Simultaneously, this protective blanket of air has kept astronomers in ignorance. From the earth's surface astronomers cannot see what the universe looks like in X rays, gamma rays, and infrared and ultraviolet radiation. But the advent of the space program has given mankind the ability to place scientific equipment in orbit, many miles above the earth's atmosphere. Telescopes have been built to detect and record exotic radiations from the planets, stars, and galaxies. By placing these telescopes in space, we literally have been given a new set of eyes with which to view our universe. As a result, virtually every idea and concept about the universe will be profoundly affected. As the views of the X ray universe, the infrared universe, and the ultraviolet universe are pieced together with our familiar visible universe, our understanding of the cosmos and the field of astronomy will change more rapidly and dramatically than ever before in history.

5.2 Atoms and Spectra

Man's modern understanding of light began when Newton passed a beam of white light through a glass prism and found that the light was broken up into the colors of the rainbow. This rainbow of colors, ranging from violet at 4,000 Å to red at 7,000 Å, is called a *spectrum*. Studying the spectra of stars and galaxies is one of the most important tools the astronomer has for unlocking the secrets of the universe. In view of the importance of spectra, scientists have designed and constructed instruments to aid in the examination of spectra. A device used to observe spectra visually is called a *spectroscope;* a device that produces a photograph of a spectrum is called a *spectrograph*.

In the early 1800s, William Wollaston was observing the spectrum of the sun when he noticed that there were some thin dark lines among the colors. Several years later, this discovery was confirmed by Joseph Fraunhofer, who found a total of six hundred such lines. A high-quality photograph of the solar spectrum taken with a modern spectrograph is shown in Figure 5–5.

At first these so-called *spectral lines* were a complete mystery. What was their meaning? What could cause them? Wollaston's idea that the lines might designate the boundaries between various colors did not appear to be a satisfactory explanation.

A major breakthrough in understanding spectra occurred when physicists realized that spectral lines could be produced artificially in the laboratory. If a beam of white light is passed through a bottle of gas, the spectrum of the light passing through the gas will contain spectral lines. Furthermore, each gas produces its own unique pattern of spectral lines. The spectrum of light passing through hydrogen gas, for example, shows a pattern of spectral lines unlike the pattern produced by any other gas. The hydrogen gas somehow manages to absorb light from the spectrum at specific wavelengths, producing a distinctive pattern of dark lines among the colors. It logically follows that if this same

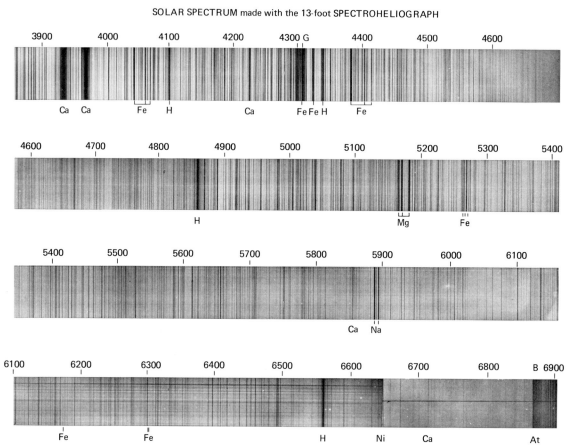

Figure 5-5. The Solar Spectrum. The spectrum of the sun shows many dark lines. These so-called spectral lines are caused by the chemical elements in the sun's atmosphere. (*Hale Observatories*)

pattern of lines exists in the solar spectrum, there must be hydrogen in the gases that make up the atmosphere of the sun. Clearly, mankind now had the ability to discover what the stars are made of; the science of *spectral analysis* was born.

The fundamental idea behind spectral analysis is very simple. Pass white light through jars containing various familiar gases, examine the resulting spectra, and record all the distinctive patterns. Obviously, there are some chemicals that are not normally gases, such as calcium and potassium. In such cases, simply heat a small amount of the solid until it vaporizes and then use the vaporized chemical in experiments. The end results of these experiments are contained in a lot of books that describe the patterns of spectral lines produced by various chemicals. When an astronomer takes a spectrum of the sun or a star, he can use these reference books to *identify* spectral lines with known chemicals. He, therefore, concludes that these chemicals are present in the sun or star. In the solar spectrum shown in Figure 5–5, a person familiar with the patterns of spectral lines produced by certain elements recognizes these lines and realizes that hydrogen (H), calcium (Ca), iron (Fe), nickel (Ni), chromium (Cr), strontium (Sr), astatine (At), silicon (Si), and so on must be present in the solar atmosphere. Some of the more prominent spectral lines in Figure 5-5 are identified with their chemical symbols.

During the second half of the nineteenth century, astronomers began applying spectral analysis to the objects in the sky. Each and every chemical leaves its own unique and indelible fingerprint in the form of spectral lines on the light. Therefore, this new knowledge allowed astronomers to discover what the planets and stars are made of. But this new and powerful tool did not tell them *how* and *why* the spectral lines were formed in the first place.

Around the turn of the twentieth century, physicists began to talk seriously about the concept of *atoms*. It seemed quite reasonable to suppose that everything is made up of very small objects called atoms, which probably have diameters of about 1 Å. An atom is composed of three basic types of particles: *electrons*, *protons*, and *neutrons*. In a series of important experiments performed during the first few decades of this century, physicists were able to discover and measure many of the properties of these three *fundamental particles* that are the building blocks of atoms. It turns out that electrons are the lightest of these fundamental particles. Each electron has an incredibly small mass (it would take 30 thousand trillion trillion electrons to weigh one ounce) and carries a negative electric charge. A proton is about two thousand times more massive than the electron (it would take only 15 trillion trillion protons to weigh one ounce) and carries an equal but positive electric charge. Neutrons have nearly the same

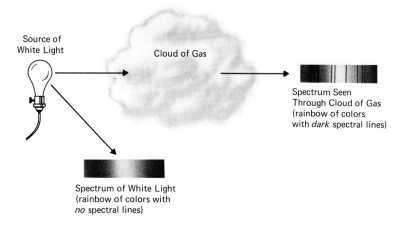

Figure 5–6. *The Formation of Spectral Lines.* The spectrum of white light seen through a cloud of gas contains dark spectral lines. The pattern of spectral lines depends on the chemical composition of the gas.

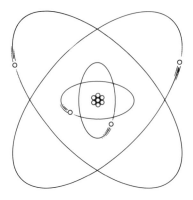

Figure 5-7. An Atom. A nuclear or solar-system model of the atom was first proposed by Rutherford in 1911. Most of the mass of an atom is contained in the nucleus which is composed of protons and neutrons. Electrons orbit the nucleus in circular or elliptical orbits.

mass as protons but have no electric charge at all. The obvious question then becomes "How do these building blocks of matter combine to form atoms?"

In 1911, Sir Ernest Rutherford proposed a "nuclear" or solar-system model of the atom that forms the basis of our current understanding of *atomic structure*. The central idea behind Rutherford's concept of the atom is that, crudely speaking, atoms can be thought of as miniature solar systems. At the center of each atom is the *nucleus*, composed entirely of protons and neutrons. Electrons are in orbit about this tiny massive nucleus, like planets around the sun. Although gravity provides the force that holds the solar system together, it is the electric forces between the negatively charged electrons and the positively charged protons that hold the atom together. Rutherford's solar-system model of a typical atom is shown schematically in Figure 5-7. Under normal conditions, the number of negatively charged electrons equals the number of positively charged protons in the nucleus, so that the atom is electrically neutral. It is the number of electrons in the atom of a particular element that determines the chemical properties of that element. Hydrogen atoms have one electron, helium atoms have two electrons, lithium atoms have three electrons, and so forth.

After Rutherford proposed his model of the atom, it was realized that there were some severe problems. If electrons are in circular or elliptical orbits about the nucleus, they must be constantly undergoing accelerations, just as planets going around the sun are constantly accelerated by the sun's gravitational field. According to Maxwell's equations, charged particles that are accelerated must lose energy by emitting electromagnetic radiation. Therefore, electrons in orbit should rapidly spiral into the nucleus as they emit a flash of light, and Rutherford's atoms, if they exist, should collapse almost at once! This was not good news for Rutherford, who thought he had a pretty good idea of what atoms were like. Fortunately, however, in 1913, Niels Bohr saved the day by proposing a simple idea that offered a way out of this dilemma and led to the modern theory of *quantum mechanics*.

Think for a moment about the solar system. In principle, planets, asteroids, or meteoroids can orbit the sun at *any* distance whatsoever. A spacecraft, such as Mariner 10, can be placed in any orbit just by aiming a rocket in the appropriate direction. Even though there are no fundamental restrictions on orbits about the sun, Niels Bohr came up with the idea that this might *not* be the case with orbits about the nuclei of atoms. Suppose that there are strict conditions on the orbits of electrons in atoms, that only certain orbits are "allowed" and all others are prohibited. In this way, the orbits of electrons about the nucleus are *quantized* because the speed and distance of electrons from the nucleus can have only certain allowed quantities.

Niels Bohr then applied this revolutionary concept of quantization to Rutherford's model of the atom. Bohr assumed that only certain orbits were permitted and that when an electron is in a permitted orbit it does not radiate energy. An electron radiates or absorbs energy only when it "jumps" from one allowed orbit to another allowed orbit. Hydrogen best illustrates Bohr's model of the atom because, with only one electron, it is the simplest of all atoms. Nevertheless, the fundamental ideas can be applied to any atom containing any number of electrons.

According to Bohr, there are certain permitted orbits for the electron in the hydrogen atom. We will call them by such names as orbit 1, orbit 2, orbit 3, and so on in increasing order from the nucleus. Bohr found that the orbit nearest the nucleus (orbit 1) has the lowest energy. As a result, this orbit is called the *ground state* of the atom. If an electron is to go into a higher orbit, it must be supplied with energy to boost it up to that higher orbit. But the electron *cannot* be given any random amount of energy. If the electron is to go from one specific low orbit

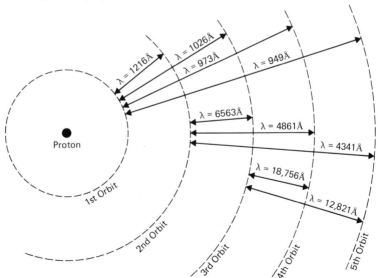

Figure 5–8. The Hydrogen Atom. The hydrogen atom is the simplest of atoms. Its nucleus contains only one proton which is orbited by only one electron. In going from one allowed orbit to another, the electron absorbs or emits light at very specific wavelengths.

to another specific higher orbit, it must be supplied with a very precise amount of energy. This amount of energy corresponds exactly to the difference in the energy of the electron in the lower and higher orbits.

When Bohr calculated the amounts of energy required to boost the electron from the second orbit (orbit 2) to higher orbits (orbit 3, orbit 4, and so on), he found that the needed amounts of energy corresponded to the energy in light of very specific wavelengths. For example, to go from orbit 2 to orbit 3 requires the energy contained in red light at a wavelength of exactly 6,563 Å. To go from orbit 2 to orbit 4, we need the energy of blue light, having a wavelength of exactly 4,861 Å. To go from orbit 2 to orbit 5 requires the energy of light at a wavelength of 4,341 Å, and so forth. These are exactly the same wavelengths at which spectral lines are observed in the spectrum of hydrogen gas! Finally, scientists had an explanation of how and why spectral lines are formed.

Our understanding of the Bohr model of the atom and the formation of spectral lines up to this point can be summarized as follows:

1. An atom contains a nucleus made up of protons and neutrons.
2. Electrons orbit the nucleus of an atom in very specific allowed orbits.
3. When an electron jumps from one allowed lower orbit to another allowed higher orbit, the atom must absorb a very precise amount of energy.
4. If energy is available to the atom from white light, the atom extracts energy from the light at a specific wavelength.
5. If a source of white light illuminates a gas, the atoms in the gas extract energy from the white light at a series of wavelengths, thereby giving rise to the spectral lines we observe.

Figure 5–9. Kirchoff's Laws. A continuous source of white light shows no spectral lines. When white light is passed through a cloud of gas, the atoms absorb light at specific wavelengths resulting in an absorption line spectrum. When one observes this cloud of gas at an angle away from the beam of white light, an emission line spectrum is seen.

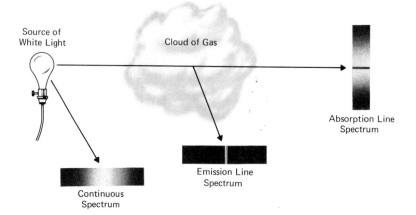

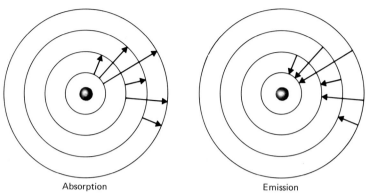

Absorption	Emission

Figure 5-10. Absorption and Emission. When electrons jump from a low orbit to a high orbit, they must absorb energy. When electrons jump from a high orbit back down to a low orbit, they emit energy. This energy is either absorbed or emitted at *very* specific wavelengths, as dictated by quantum mechanics.

Atoms of different chemical elements have different numbers of electrons and, therefore, different allowed orbits. For example, helium with two electrons in orbit about its nucleus has a very different structure from iron atoms, which each have twenty-six electrons. The allowed orbits in the helium atom, therefore, occur at different locations from the allowed orbits in the iron atom. Consequently, it should be no surprise that the spectrum of helium looks nothing like the spectrum of iron. This is why each chemical produces its own unique pattern of spectral lines. The subject of quantum mechanics deals with calculating the structure of atoms and discovering the allowed orbits. Every spectral line can now be understood in terms of electrons jumping from one orbit to another.

The work of Niels Bohr enabled scientists to understand spectral lines. In particular, all the experimental work of Gustav Kirchoff (who formulated some of the most fundamental laws of spectral analysis) can be explained completely. Perhaps the best way to examine and understand these laws is with the aid of the experiment shown in Figure 5-10. In this experiment there is a source of white light, such as a light bulb. Looking at the spectrum of this white light through a spectroscope, we find that the spectrum contains *no* spectral lines at all. We, therefore, say that the hot filament in the light bulb emits a *continuous spectrum.* If this light is shined on a cloud of gas, the spectrum of the light passing through the gas will still show the original, continuous rainbow of colors. But, superimposed on these colors is a series of dark, thin spectral lines. The atoms in the gas *absorb* light at certain specific wavelengths as the electrons in these atoms jump from low orbits to higher orbits. The resulting spectrum is called an *absorption line spectrum.* But as the atoms in the gas absorb light from the continuous spectrum, a lot of atoms find themselves with electrons in very high orbits, far above their ground states. Such atoms are said to be *excited.* It is logical that the atoms prefer to return to their ground states; the ground state is the normal, stable condition for an atom. An atom becomes *deexcited* and returns to its ground state when electrons jump from high orbits back down to lower orbits. In undergoing such transitions, the atoms must lose energy and they *emit* light at certain specific wavelengths. The wavelengths of the light emitted as electrons cascade back down to their ground states and are exactly the same as the wavelengths of the light originally absorbed by the atoms in the first place. In going back to the lower orbit, an electron surrenders exactly the same light that it absorbed. But the atoms have no way of remembering which way the original beam of light from the light bulb was going. Therefore, this light is emitted in all directions. In other words, if the cloud of gas is observed in our experiment at some angle away from the beam of white light, an *emission line spectrum* is seen. Such a spectrum is mostly black, but superimposed on this blackness is a series of bright *emission lines*. The light in these lines comes from the atoms emitting radiation as they become deexcited. The pattern of bright lines in the emission spectrum is exactly the same as the pattern of dark lines in the absorption spectrum.

Whereas atoms give rise to patterns of spectral lines, molecules give rise to bands of lines. Molecules are combinations of atoms such as water, carbon dioxide, and ammonia, which were mentioned in connection with the atmospheres of planets. When dealing with individual atoms, transitions of electrons between various allowed orbits give rise to *line spectra*. However, vibrations and rotations of molecules give rise to *band spectra*. Just as the orbits of electrons are quantized, the manner in which molecules can vibrate and rotate is also quantized. Molecules can vibrate and rotate only in certain specific allowed fashions, and each mode of vibration or rotation gives rise to a huge number of spectral lines having very nearly the same wavelength. These vibrational and rotational spectral lines are bunched together giving the appearance of bands in the spectrum. An example is shown in Figure 5–12.

The central idea behind the quantum theory of molecules is that molecules,

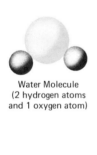

Water Molecule
(2 hydrogen atoms
and 1 oxygen atom)

Carbon Dioxide Molecule
(1 carbon atom and 2
oxygen atoms)

Ammonia Molecule
(1 nitrogen atom and
3 hydrogen atoms)

Figure 5–11. Some Molecules. Molecules consist of atoms. Molecules of water (H_2O), carbon dioxide (CO_2) and ammonia (NH_3) are shown schematically.

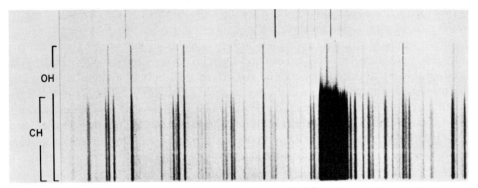

Figure 5–12. Molecular Spectrum. Vibrational and rotational bands due to CH and OH observed in an oxyacetylene blowtorch flame are shown here. (*Lick Observatory*)

which are simply combinations or chains of atoms, cannot vibrate and rotate at just any speed or frequency. Vibrations and rotations can occur only at specific allowed rates. Energy must be added to a molecule in order to speed up the rate at which it is vibrating and rotating. But the molecule will accept energy only in certain specific amounts, amounts precisely equal to the quantity of energy needed for the molecule to go from a slower vibrational or rotational state to a faster vibrational or rotational state. If this energy is available from white light, a spectral line is formed. Detailed calculations of molecules show that there are literally thousands of allowed vibrational and rotational states for each molecule; hence, there are great numbers of lines that together give the appearance of bands.

If a band of lines is seen in the spectrum of a star or planet, then there must be molecules present in the atmosphere of that planet or star. It is the task of the physicist in the laboratory to discover which molecules produce which bands. In this fashion, astronomers discovered that the atmosphere of Venus contains carbon dioxide. A spectrum of the reflected light from Venus showing the carbon dioxide bands appears in Figure 5–13.

Figure 5–13. Carbon Dioxide. This spectrum of Venus shows molecular bands due to carbon dioxide, indicating the presence of this gas in the planet's atmosphere. (*Lick Observatory*)

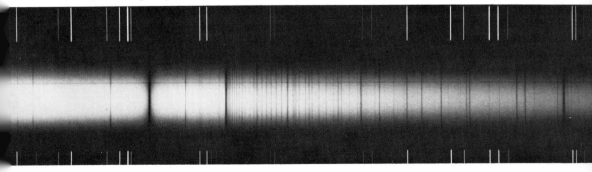

During the last several years astronomers have discovered that there are many molecules in space. In the late 1960s, astronomers began discovering many different kinds of organic molecules in large clouds of gas floating among the stars. These molecules, such as formaldehyde (H_2CO), cyanogen (CN), hydrogen cyanide (HCN), and cyanoacetylene (HC_3N), are usually associated with living matter. Even though no one believes that there might be animals or planets floating around in space, these observations do suggest that the basic building blocks for life may be present in the interstellar gas. Some of the basic molecules out of which we are made may have existed even before the formation of the earth.

5.3 Radiation Laws

The depths of space, stretching for billions of light years in all directions, are populated with stars, galaxies, quasars, and nebulae—the plethora of strange objects that fascinate the modern astronomer. Yet, we are aware of these celestial objects and the universe that they occupy for one and only one simple reason: they emit energy in the form of electromagnetic radiation. If this were not the case, we would have concluded that space and time beyond our earth is an uninteresting void.

So far, we have only begun to scratch the surface of the subject of light. We have seen that the atoms and molecules in a source of light make their presence known through the formation of spectral lines. But why and how do stars emit light in the first place? The complete answer to these questions requires an awareness of the detailed properties of stars based on our understanding of the interaction of matter and energy. We must, therefore, begin by learning how any material object emits light.

Imagine that you have an object, such as a bar of iron, in your laboratory. Suppose this bar of iron is supplied with energy, perhaps with the aid of a blowtorch. As the bar of iron is heated, the temperature of the iron begins to rise; the iron atoms begin to vibrate faster and faster. Soon the bar of iron begins to glow with a dull red color. As more energy is applied in the form of heat, the color and brightness of the light from the iron begin to change. As the temperature rises, the dull red becomes a bright red, then the bright red turns into a blinding yellowish-white, sometimes called white hot. If the bar of iron could be prevented from melting and vaporizing as it approached temperatures of tens of thousands of degrees, the bar would actually glow with an incredibly brilliant bluish color.

This simple experiment embodies some of the most fundamental radiation laws in physics. Such laws reveal how the intensity and color of light emitted by a hot object vary with temperature. As might be expected from this kind of experiment, the chemical composition of the heated object has some effect on the precise nature of the light emitted. A brass bar or an aluminum bar would give slightly different results from an iron bar. Nevertheless, the changes in color and brightness would be nearly the same in all cases.

In order to describe these changes in color and brightness, physicists prefer to imagine that they are dealing with an *ideal* object rather than bars made of iron or brass. With an ideal object, the details of the structure of the atoms out of which the object is composed would not play a role in the results we obtain. The imaginary ideal object invented for this purpose is called a *black body*. A black body is an object that absorbs *all* the electromagnetic radiation that falls on it. This is why it is called black; all light is absorbed and none is reflected. A black body is also a "perfect radiator" in that it emits radiation that depends *only* on its temperature; there is no dependence on the chemical properties of this ideal radiator. The radiation that is given off is called *black body radiation*. Radiation emitted by objects in the real world (bars of iron or the atmospheres of stars) differ from black body radiation only because of their particular chemical and physical properties.

During the second half of the nineteenth century, physicists spent a considerable amount of time trying to study black bodies experimentally. In their experiments they made black bodies out of carbon or covered objects with soot and found the same kind of relationship between temperature and radiation as was described in the observations of a hot bar of iron. For example, the German physicist Wilhelm Wien discovered a way of expressing the manner in which the colors of radiation emitted by hot objects depends on temperature.

Every object that is above absolute zero (0°K) emits some type of radiation. Cool objects emit very long wavelength radiation. For example, the human body at a temperature of about 300°K is a source of infrared radiation. Very hot objects with temperatures in millions of degrees emit very short wavelength radiation, such as X rays. Only if the temperature of an object ranges from about 3,000 to 10,000°K will the object emit primarily visible light.

The radiation emitted by objects above absolute zero covers a range of wavelengths. However, there is a particular wavelength at which *most* of the radiation is emitted. This particular wavelength, called λ_{max}, depends only on the temperature of the black body. As suggested by the experiment with an iron bar, the wavelength at which most of the radiation is emitted depends inversely on the temperature. The higher the temperature, the shorter the wavelength. The lower the temperature, the longer the wavelength. This relationship between the temperature (T) of an object and the wavelength (λ_{max}) at which most radiation is emitted is called *Wien's law*. (See the table on page 140.)

It should be reemphasized that a hot object emits a *range* of wavelengths; Wien's law tells us where *most* of the radiation is emitted. Thus, a bar of iron at 1,000°K gives off a dull red glow; some visible red light is emitted. However, according to Wien's law, most of the radiation from this object is in the invisible infrared at about 10,000 Å.

Physicists, thus, understood why a bar of iron appears to change color (red → orange → yellow → blue) as it is heated to higher and higher tempera-

Temperature (°K) of Black Body	Wavelength (λ_{max}) at Which Most Radiation Is Emitted	Type of Radiation
3°	$1/10$ cm	Radiowaves
300°	$1/1000$ cm	"Far" infrared
3,000°	10,000 Å	"Near" infrared
4,000°	7,500 Å	Red light
6,000°	5,000 Å	Yellow light
8,000°	3,750 Å	Violet light
10,000°	3,000 Å	"Near" ultraviolet
30,000°	1,000 Å	"Far" ultraviolet
300,000°	100 Å	"Soft" X rays
1½ million°	20 Å	"Hard" X rays
3 billion°	$1/100$ Å	Gamma rays

tures. However, a second equally important effect was noticed. As the temperature of the bar of iron increased, the brightness of light given off also increased. Objects at low temperatures emit only small amounts of radiation, whereas objects at high temperatures emit huge amounts of radiation. This relationship between the temperature and brightness of radiation from an ideal black body was formulated first by Josef Stefan in 1879. The resulting *Stefan's law* simply states that the amount of energy emitted by a black body increases as the "fourth power" of the temperature. Thus, according to Stefan's law, if you double the temperature of an object (for example, go from 2,000 to 4,000°K) the amount of energy given off in the form of radiation will go up by a factor of 16, because $2 \times 2 \times 2 \times 2 = 16$. If you triple the temperature of a black body (for example, go from 3,000 to 9,000°K), the black body will give off eighty-one times more energy because $3 \times 3 \times 3 \times 3 = 81$.

From both Wien's law and Stefan's law we now can appreciate why cool objects emit small amounts of reddish light and hot objects emit large amounts of bluish light. Stefan's law tells us about the total amount of energy emitted, and Wien's law tells us at what wavelength most of this energy is concentrated. However, neither of these laws tells us precisely how the radiation from a black body is distributed over *all* wavelengths. For example, according to Wien's law, a black body at 6,000°K emits most of its radiation in the form of yellow light at 5,000 Å. From Stefan's law, the total amount of energy at *all* wavelengths can be calculated; physicists can calculate how many "watts" are given off by the black body. But how much energy is given off by this black body in blue light? What is the intensity of the radiation at 4,000 Å? Or at 8,000 Å? Questions of this type are unanswered by either law. Indeed, by the end of the nineteenth century, it was apparent that from the "classical" understanding of light based on Maxwell's equations, it was *impossible* to answer such questions. A crisis had developed in physics. Classical theory gave absurd answers. The nature of light was once again a mystery.

In 1900, the brilliant German physicist Max Planck resolved this dilemma by proposing that light is *quantized*. According to Planck, light can exist only in discrete packets called *photons*, whose energy depends on the wavelength of the light. The shorter the wavelength of a photon, the higher the energy contained in the photon. An X-ray photon carries a lot more energy than a radiowave photon. Up until the turn of the century, physicists were entirely content to think of light as waves. No one had ever performed any experiments to look for the discrete or quantized nature of light. Yet, the idea that a beam of light actually consists of a stream of very tiny photons, each carrying a specific quantity of energy, proved to be very powerful. Planck was then able to prove both Stefan's and Wien's laws, as well as calculate precisely how much energy is given off at each and every wavelength by a black body at a specific temperature. The final result of Planck's work is the so-called *Planck black body radiation law.*

Perhaps the best way to illustrate the Planck black body radiation law is with the aid of a graph, as displayed in Figure 5–14. This graph simply shows how much energy is emitted at various wavelengths by black bodies at various temperatures. Energy is plotted vertically and wavelength is plotted horizontally, and the effects of both Wien's and Stefan's laws are immediately apparent. At low temperatures, the curves given by Planck's law peak at long wavelengths. At higher temperatures, the peak in the curve (at λ_{max}) moves to shorter wavelengths. In addition, the curves for high temperatures are much higher than the curves for lower temperatures. This is in accordance with Stefan's law, which says that the hotter an object is, the more energy it emits. But, more importantly,

Figure 5–14. *Black Body Curves.* The curves on this graph are drawn according to Planck's radiation law. The precise amount of energy emitted by an ideal black body at a specific temperature and at a specific wavelength is given by these curves.

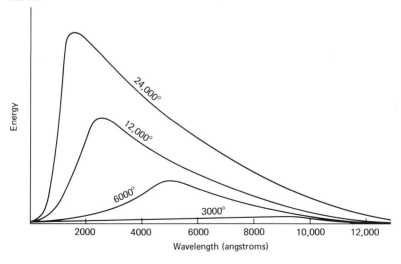

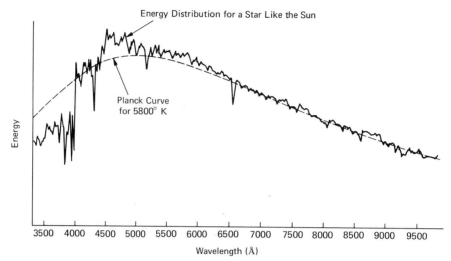

Figure 5-15. *The Energy Distribution of a Real Star.* Stars are not ideal black bodies. The absorption of light by atoms in their atmospheres causes spectral lines. The energy emitted by a real star of a certain temperature does not exactly follow the Planck curve.

Planck's law tells precisely how much energy is emitted at every wavelength by a black body at a particular temperature.

Of course, objects in the real world—such as stars—are not ideal black bodies. Stars are made up of atoms, and these atoms produce spectral lines in the light emitted by the star. Therefore, the actual distribution of energy emitted by a star will not exactly follow a black body curve. Instead, due to the dark spectral lines at various wavelengths, there will be "dips" in the energy distribution from the star, as shown in Figure 5-15. If the star described in Figure 5-15 were a perfect black body, the intensity of radiation at various wavelengths would have exactly followed the dotted Planck curve. However, the atoms in the star's atmosphere have removed energy from this ideal distribution. This is called the *line blanketing effect* and every spectral line causes a little dip or valley to appear in the actual distribution of light from the star.

Using this new understanding of light can be a very powerful tool in astronomy. For example, careful observation of the stars in the sky reveals that they have different colors. Some stars look bluish, whereas others have a reddish tinge. In the constellation of Orion, the star Rigel is noticeably blue, whereas Betelgeuse is very red. Astronomers, therefore, conclude that Rigel must be a hot star and Betelgeuse a significantly cooler one. As we will see in Chapter 6, a technique in astronomy called *photometry* can be used to measure the colors of stars very accurately. From knowing the colors of stars, astronomers can determine their temperatures. This is one more example of how man's understanding of the nature of light can be used to discover the properties of stars trillions of miles away.

5.4 Information from Spectra

The wealth of information contained in the spectra of light received from the stars cannot be overemphasized. Although the objects the astronomer studies are incredibly remote, virtually everything he would want to know has left an indelible mark on the light that reaches his telescope. In a very real sense, the astronomer is limited only by the reliability of his instruments and the depth of his knowledge of the properties of light.

As was pointed out in Section 5.2, atoms and molecules give rise to spectral lines in the light from stars and galaxies. The astronomer can identify spectral lines simply by comparing the pattern of lines in the spectrum of a star or galaxy with patterns of spectral lines obtained from known chemicals in the laboratory. For example, vaporized calcium heated to a high temperature shows two very prominent spectral lines very close together near the violet end of the spectrum. Whenever these lines show up in an astronomer's spectrum, he can be sure that calcium is present in the object he is studying. Thus, from examining spectra, it is possible to determine the *chemical composition* of even the most distant objects in the universe.

Very frequently, it is noticed that a familiar pattern of spectral lines due to some well-known chemical looks slightly peculiar. The spectral lines that the astronomer observes might be shifted from their usual positions, or they might appear very fuzzy and broadened, or a familiar line might be split up into a number of closely spaced lines. In such situations the astronomer is quick to realize that something very unusual is going on with the star or galaxy he is observing. The pathology of the light he receives bears witness to these unusual conditions. At this point, the astronomer turns to the physicist to tell him about the various factors (motion, turbulence, magnetic fields, and so on) that can affect the appearance of spectral lines.

When an astronomer observes the spectrum of some celestial object, he frequently notices that familiar patterns of spectral lines are *not* where they should be. For example, the skilled astronomer is always quick to recognize the prominent two lines of calcium (the so-called H and K lines) in the spectrum of a star or a galaxy. In the laboratory, these two lines *always* show up at wavelengths of 3,968 and 3,934 Å. Yet, in a stellar spectrum, these lines frequently appear in a very different region of the spectrum. Careful scrutiny of other recognizable spectral lines reveals that they *all* have been shifted by corresponding amounts, either toward the red or the blue end of the spectrum. In order to understand what could cause a shift in all the lines in a spectrum, simply appeal to an everyday experience:

> Imagine that you are standing on the sidewalk of a busy street when suddenly you hear the siren of an approaching ambulance or police car. As the vehicle comes toward you, the wail from the siren seems to have a very high pitch. But, after the

ambulance or police car passes you, the pitch of the siren appears to drop significantly. This change in pitch also will be noticed by someone standing near a railroad track listening to the bell or whistle of a passing train. As the source of noise approaches, the pitch seems high; as the source of noise recedes, the pitch seems low. This phenomenon is known as the *Doppler effect*. The sound waves are bunched up in front of an approaching source of noise and, thus, the listener hears a high-frequency pitch. On the other hand, the sound waves are spread out behind a receding source of noise and, thus, the listener hears a low-frequency pitch.

This same sort of phenomenon also occurs with light. If a source of light is coming toward an observer, the light waves will be slightly compressed, as shown schematically in Figure 5–16. If the light from the source contains spectral lines, these lines will appear at wavelengths *shorter* than usual; they will be shifted toward the blue end of the spectrum. Conversely, if a source of light is receding from an observer, the light waves will be slightly expanded, again as shown in Figure 5–16. Spectral lines from this receding source will, therefore, appear at wavelengths *longer* than usual; they will be shifted toward the red end of the spectrum. In other words, if an astronomer finds that all the lines in the spectrum of a star are shifted from their usual "laboratory" positions, he imme-

Figure 5–16. *The Doppler Effect.* The precise wavelength of a particular spectral line is affected by the motion of the source. An approaching source of light has its spectral lines shifted to shorter wavelengths (toward the blue end of the spectrum). A receding source of light has its spectral lines shifted to longer wavelengths (toward the red end of the spectrum).

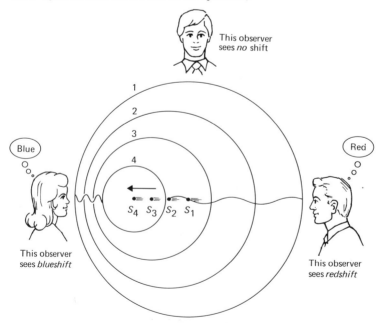

Chapter 5 The Physics of Light

diately concludes that the star is moving. If the lines are shifted to shorter wavelengths, resulting in a so-called *blueshift*, he knows that the star is coming toward him. If the lines are shifted to longer wavelengths, resulting in a so-called *redshift*, he knows that the star is moving away from him. The bigger the shift, the higher the speed.

By way of an example, the spectrum of the quasar known as PHL 1119 shows the familiar lines of hydrogen. The spectral line formed by electrons jumping from the second to fourth allowed orbits in hydrogen atoms is at 4,861 Å in all laboratory experiments. However, as reported in a scientific paper published in 1974, this same line in the spectrum of PHL 1119 appears at 5,437 Å. It has been shifted by 576 Å toward the red end of the spectrum. The astronomer, therefore, concludes that PHL 1119 is moving away from us. Calculations reveal that a shift of this amount corresponds to a recessional velocity of about 12 per cent of the speed of light, or about 670 million miles per hour.

Although shifts in spectral lines are usually due to the motion of the source, many physical processes can occur in the source that will change the appearance of the spectral lines. For example, at around the turn of the century, Pieter Zeeman discovered that the presence of a magnetic field can cause a single spectral line to split up into several spectral lines. If a source of light in the laboratory is placed between the poles of a magnet, the spectral lines from that light will be split into two or more components. The exact number of component lines depends on the details of the structure of the atoms in the source of light. This phenomenon is appropriately called the *Zeeman effect*.

All atoms in magnetic fields exhibit a Zeeman splitting of their spectral lines. For example, in the absence of a magnetic field, the spectrum of sodium is dominated by two very strong lines (the so-called sodium D lines) at 5,890 and 5,896 Å. In the presence of a magnetic field, however, the 5,896 line splits into four lines, whereas the 5,890 line splits into six. As with all atoms, the separation of the components depends on the strength of the magnetic field. The stronger the field, the bigger the separation. By measuring the separation between components of a line split due to the Zeeman effect, a physicist can calculate the strength of the magnetic field acting on the atoms.

The Zeeman effect has important applications to astronomical observations. If an astronomer discovers several closely spaced lines in the spectrum of a star near the same wavelength at which he normally would expect to see only one spectral line, he immediately concludes that a magnetic field must be present. In this way, astronomers have discovered that sunspots possess strong magnetic fields, as do so-called magnetic stars and white dwarf stars. For example, in 1974, a team of astronomers reported their observations of the spectrum of the white dwarf star called GD 90. In place of the familiar hydrogen line at 4,861 Å (second to fourth orbit), they found three lines, (4,793, 4,857, and 4,918 Å) from which they conclude that this star must have an incredibly intense magnetic field, more than ten million times stronger than the natural geomagnetic field of the earth.

The discussion to this point has focused on two effects that can dramatically change the appearance of a spectrum. According to the Doppler effect, the

Figure 5-17. A *Spectroscopic Plate.* The spectrum of a star or galaxy is photographed on a spectroscopic plate. An astronomer is shown holding a typical spectroscopic plate on which the spectrum of a star has been exposed.

motion of a source of light causes a displacement of spectral lines from their usual (laboratory) wavelengths. The higher the speed, the greater the shift. Magnetic fields at a source of light causes Zeeman splitting of spectral lines into two or more component lines. The separation between the components increases with increasing field strength. There are, however, numerous more subtle effects that can alter spectral lines. These effects do not split or shift the lines. Instead, these effects cause the spectral lines to *broaden.* This *spectral line broadening*

Figure 5-18. A *Microdensitometer.* This devise transfers the spectrum from a spectroscopic plate onto a long strip of paper. Astronomers often find it much more convenient to work with the "tracing" rather than with the tiny spectroscopic plate.

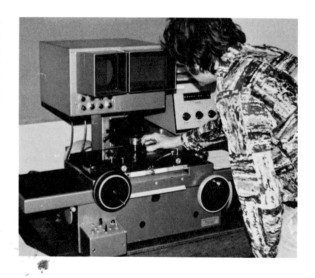

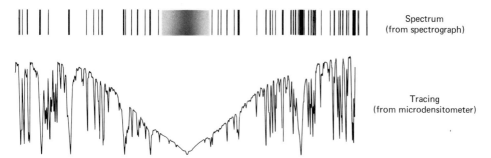

Figure 5–19. *A Spectrum and A Tracing.* A portion of the spectrum of a star is shown here along with its tracing. Notice how the "dips" and "valleys" in the tracing correspond to the features in the spectrum.

results in the lines appearing wide and "washed out." Different physical processes result in different types of line broadening. The astronomer is, therefore, motivated to study the details of the *shapes* of spectral lines.

When an astronomer takes a spectrum of a star with his spectroscope, the final result of his observations is a small photographic plate about the size of a microscope slide. The spectrum on this plate looks like a thin band only a few millimeters wide and perhaps no more than an inch or two long. The spectral lines on this band are frequently so fine and faint that they are invisible to the unaided human eye. Obviously, the astronomer needs some sort of help to examine the details of spectral lines.

To aid in the examination of spectra, astronomers have invented a device called a *microdensitometer*. The central idea behind a microdensitometer is very simple. This machine sends a thin beam of light through the photographic plate containing the astronomer's spectrum. A light-sensitive photoelectric cell measures how much light gets through and converts the light into an electric current. This electric current is used to drive a pen that draws the shape of the spectrum on a long strip of paper that comes out of the machine. In other words, the spectrum on the photographic plate, which is only an inch or two long, is turned into a drawing of the spectrum that may be several feet long. This drawing is called a *tracing* of the spectrum. Spectral lines that were finer than a strand of hair on the photographic plate now show up as dips and valleys as much as an inch wide on the tracing. An example of a spectrum and its tracing is shown in Figure 5–19. From such a tracing, the astronomer now has a permanent record of the precise shapes of the spectral lines in the star or galaxy he is studying.

Even under the most ideal conditions, in the absence of any outside effects, a spectral absorption line is *not* totally black and infinitely fine. Spectral lines have a *natural shape* and a *natural width*. The natural width of a spectral line is about fifty millionths of an angstrom. It is physically impossible for a spectral line to be thinner than this limiting value. However, there are various physical processes that can cause spectral lines to be as much as several angstroms wide.

When an astronomer looks at a star, he is seeing light emitted by very hot gases at the star's surface. For example, the temperature on the sun's surface is about 6,000°K. As we learned in our earlier discussion of temperature (see Section 4.1), the atoms of any object above absolute zero are in motion. The higher the temperature, the higher the speed of the atoms. For example, in the case of hydrogen gas at 6,000°K, the average speed of the hydrogen atoms is about $7\frac{1}{2}$ miles per second. But, these are the same hydrogen atoms that produce the spectral lines we observe. At any given instant, some are moving toward us while others are moving away simply because the gas is at a high temperature. According to the Doppler effect, those atoms moving toward us will try to make spectral lines at slightly shorter wavelengths than usual, while that fraction of the atoms moving away will try to form the spectral lines at slightly longer wavelengths than usual. The final result is that the spectral line is broadened. Since this phenomena is caused by the motions of the atoms in the hot stellar gases, this type of broadening is called *thermal Doppler broadening*. In the case of hydrogen atoms at 6,000°K, for example, this thermal Doppler broadening causes the spectral line at 6,563 Å (second to third orbit) to have a width of about $\frac{1}{4}$ Å.

We have seen that the atoms in the hot gases of a star are constantly in motion. The average speeds of such atoms are typically quite high. For example, hydrogen atoms at 6,000°K have an average speed of $7\frac{1}{2}$ miles per second (equals 27,000 miles per hour). But, if we follow an individual atom in such a stellar atmosphere, we find that it does *not* get very far. The reason for this is that the atoms are constantly bumping into each other. In other words, although the speeds of the atoms are high, they do not cover much ground because of frequent collision that constantly change the directions in which the atoms are moving. During such collisions, the orbits of electrons around their nuclei are perturbed slightly. As a result of such perturbations, the resulting spectral lines from these colliding atoms will be broadened. This type of broadening mechanism is called *collisional broadening* or *pressure broadening*. The rate at which atoms in a gas collide depends on the pressure of the gas. If the pressure is low, then collisions will not occur very often. But, if the pressure is high, the atoms are packed more closely and collisions occur more frequently. As a result, the width of a collisionally broadened line is determined by the gas pressure in the atmosphere of the star. The greater the pressure of the gas, the greater the widths of the spectral lines. From examining the collisional broadening of spectral lines, the astronomer has a means of measuring the atmospheric pressure on the surface of a star.

There are many other physical conditions on stars that can broaden spectral lines. For example, suppose the surface of a star is violently boiling; huge blobs of hot gas are constantly rising and falling in the star's atmosphere. Due to the Doppler effect, rising bubbles will give rise to lines that are slightly blueshifted, while descending bubbles will produce a slight redshift in the lines. The combination of these two effects results in an over-all broadening of the spectral lines. This mechanism is called *turbulence broadening*.

A final example of line broadening mechanisms that will be discussed here deals with the rotation of stars. Astronomers have every reason to suppose that all stars rotate. Some stars rotate slowly, such as the sun, which takes about 25 days to go once about its axis. Other stars rotate much more rapidly, perhaps as fast as once in a few hours. Look at a rotating star; the atoms on one side of the star are coming toward you while the atoms on the other side are moving away from you. The approaching atoms will try to make lines that are slightly blueshifted while the receding atoms will try to make lines that are slightly redshifted. Together they produce lines that are *rotationally broadened*. The faster the star rotates, the broader will be the lines. From measuring the widths of rotationally broadened lines, astronomers can tell how fast a star is rotating.

There are all kinds of effects that can change the shapes of spectral lines. It is important to realize that each effect broadens a spectral line in a very specific and characteristic way. A rotationally broadened line looks very different from a line that is broadened by turbulence. Different effects produce different shapes or *line profiles*. In reality, when an astronomer examines a stellar spectrum he is seeing a combination of many effects. The shape or profile of a particular line may be due partly to rotation, partly to turbulence, partly to pressure broadening, and so forth. He is then faced with the task of deciding how much of each of these effects comes into play. If he is successful in sorting out the contributions of these various effects, he obviously has learned a great deal about the nature of the star.

The starlight received here on earth contains an incredible wealth of information. With patience and care, this light can be analyzed and deciphered to discover some of the most intimate details of even the most remote objects in the universe. Astronomers are limited only by their own skill and resourcefulness.

Questions and Exercises

1. What happens when a beam of white light passes through a prism?
2. What is an angstrom?
3. Approximately what range of wavelengths corresponds to visible light?
4. What is meant by the electromagnetic spectrum?
5. Who was James Clerk Maxwell?
6. List the various types of radiation that make up the electromagnetic spectrum.
7. Briefly discuss the effects of the earth's atmosphere on the astronomer's efforts to observe stars at various wavelengths.
8. What is a spectral line?

9. Briefly discuss the reasons why, using spectral analysis, astronomers are sure that there is some calcium in the sun's atmosphere.
10. What is an atom?
11. Where are electrons, protons, and neutrons found in an atom?
12. Compare and contrast electrons, protons, and neutrons.
13. How does Bohr's model of the atom account for spectral lines?
14. What is a continuous spectrum?
15. Discuss the differences between an absorption line spectrum and an emission line spectrum.
16. How does the appearance of the spectrum of molecules differ from that of individual atoms?
17. What is a black body?
18. What does Wien's law tell us?
19. What does Stefan's law tell us?
20. Who was Max Planck?
21. What is a photon?
22. How does Planck's black body radiation law tell us *more* than Wien's law and Stefan's law combined?
23. What is the Doppler effect?
24. How can astronomers use the Doppler effect to learn something about the motions of stars or galaxies?
25. What is the Zeeman effect?
26. What is a tracing of a spectrum?
27. Discuss how the profiles of spectral lines can be used to tell us about the conditions in the atmosphere of a star.

The Astronomer's Tools

6.1 Refracting Telescopes

Up until the beginning of the seventeenth century, everything known about the universe was based on visual observations made with the naked human eye. In 1610, however, the entire course of astronomy was dramatically and permanently changed when Galileo pointed his telescope toward the stars. Suddenly craters and mountains on the moon, spots on the sun, four satellites going around Jupiter, and the phases of Venus were made visible. These discoveries constituted the first refreshing influx of new astronomical data in almost 2,000 years. Since the time of Galileo, the telescope has been the single most important tool that the modern astronomer has had at his disposal.

A telescope does three things: it magnifies, it resolves, and it gathers light. Magnification is the property most commonly attributed to it. It is well known that when looking through a telescope, distant objects appear larger. This, of course, is the primary advantage in using a telescope. This *magnifying power* tells us how many times larger a telescopic image is compared to the naked-eye view. For example, to the naked eye, the diameter of the moon is $\frac{1}{2}°$. If a telescope gives an image of the moon 20° in diameter, we say that the magnifying power of the telescope is forty, written as 40X, because the moon appears forty times larger through the telescope. For example, if a telescope has a power of 300X, images seen through this telescope will appear three hundred times larger than to the naked eye.

Sometimes commercially available telescopes sold in department stores are advertised according to their "power." Often this constitutes a clear case of misleading advertising. The magnifying power of a telescope can be changed simply by using different *eyepieces*. The more powerful the eyepiece, the higher the magnification of the telescope. It is possible to take a very poorly constructed telescope, equip it with a powerful eyepiece, and arrive at a very high magnifying power. Unfortunately, the customer who buys such a telescope is often severely disappointed. If he looks at Jupiter through his "450X telescope for under $75," he will see a large blob instead of a clear image of the planet. Even though Jupiter will have been magnified, so will all the defects and shortcomings of the

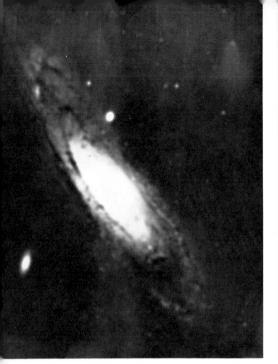 

Figure 6–1. A Demonstration of Resolving Power. These two views of the same galaxy illustrate the effects of resolving power. The view on the left corresponds to a low resolving power. With a high resolving power, as shown on the right, sharp images are obtained. (*Courtesy of Dr. O'Dell*)

telescope. Although a high magnification is frequently desirable, there are numerous other qualities that must be considered in a well-designed telescope.

In addition to magnifying, telescopes also have the ability to *resolve*. The *resolving power* of a telescope tells about the sharpness and clarity of the telescopic image. A telescope with a low resolving power gives fuzzy and indistinct images, whereas one with a high resolving power gives sharp, clear images. For example, through a small inexpensive telescope with a low resolving power, Castor, the second brightest star in Gemini, looks like any ordinary star, much the same as it looks to the naked eye. However, a larger telescope with a higher resolving power reveals Castor to be a *double star*, two stars very close together.

The resolving power of a telescope is expressed as the smallest angle between two stars for which separate, recognizable images are produced. The smaller the angle, the better the resolution. For example, a typical, good amateur telescope has a resolving power of one second of arc (1″). Through this telescope, stars separated by more than 1″ produce clear, distinct images, whereas the images of stars separated by less than 1″ are blurred together.

The resolving power of an individual telescope depends to a large degree on the quality and precision of the telescope's optics. However, due to the wave nature of light, there is a limiting resolving power for all telescopes. No telescope can ever give an infinitely sharp image. Instead, there is a limiting resolving power and no telescope can do better than this limit, no matter how good the optics. This limiting resolving power increases with the diameter of the primary mirror or lens of the telescope. The 200-inch telescope at Mt. Palomar has twice the resolving power of the 100-inch telescope at Mt. Wilson. Theoretically, the Palomar telescope should be able to resolve stars separated by only $\frac{1}{50}$-second of arc. In practice, such things as weather conditions and the stability of the atmosphere prevent the astronomer from ever achieving the theoretical limit.

Finally, telescopes gather light. The primary lens or mirror of a telescope takes all the light covering an area of several square inches or square feet and focuses this light into a small, bright image. It is this focusing (or gathering of light) that results in an increased brightness of the resulting image. Stars that are too faint to be seen with the naked eye easily show up through a telescope. As shown schematically in Figure 6–2, the bigger the primary lens or mirror of a telescope, the more light it gathers. The area of the 200-inch mirror at Mt. Palomar is four times the area of the 100-inch mirror at Mt. Wilson. The Palomar telescope, therefore, gathers four times more light, and its resulting images are four times brighter than the Mt. Wilson telescope. Improved resolution and increased light gathering are the two primary reasons why astronomers build larger and larger telescopes.

There are basically two types of telescopes: those based entirely on the use of lenses, and those based on the use of mirrors. Galileo's telescope, as well as many of the major telescopes built before the twentieth century, were of the first type. Such telescopes are called *refractors* because, by means of lenses, light is

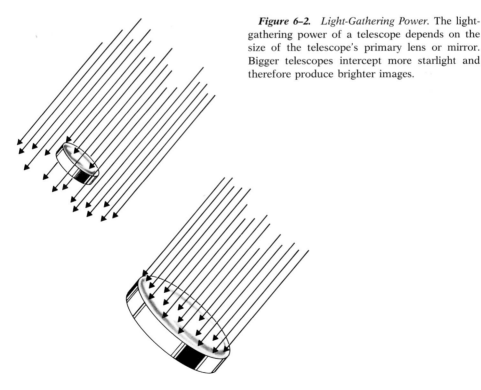

Figure 6–2. *Light-Gathering Power.* The light-gathering power of a telescope depends on the size of the telescope's primary lens or mirror. Bigger telescopes intercept more starlight and therefore produce brighter images.

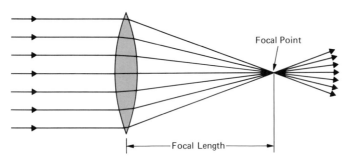

Figure 6–3. *A Convex Lens.* A convex lens focuses incoming starlight to a *focal point.* The distance between the lens and the focal point is called the *focal length.*

bent, or refracted. The refractor consists of a large *objective lens* mounted, for convenience, at one end of a telescope tube. This convex objective acts like a giant magnifying lens and focuses the incoming starlight to a point called the *focal point.* As shown in Figure 6–3, the distance between the lens and the focal point is called the *focal length* of the telescope. At the other end of the telescope, the astronomer usually places an eyepiece with which to examine the image formed around the focal point. The magnifying power of the telescope is determined by the focal lengths of both the objective lens and the eyepiece.

When the astronomer refers to the size of a refracting telescope, he is usually speaking of the diameter of the objective lens. After all, it is the diameter of this lens that plays such an important role in the resolving power and light-gathering ability of a telescope. The largest refractor in the world is the 40-inch refractor at Yerkes Observatory in Wisconsin; the second largest is the 36-inch refractor at the Lick Observatory on Mount Hamilton in California.

No optical system is perfect. Difficulties and imperfections in the design of a telescope appear as distortions in the image. Such distortions are called *aberrations.* The most common aberration found in refractors is *chromatic aberration,*

Figure 6–4. *A Refracting Telescope.* The image formed at the focal point can be examined and magnified with the aid of an "eyepiece." The resulting optical arrangement is called a refracting telescope.

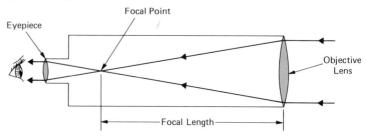

Figure 6–5. The World's Largest Refractor. The 40-inch refractor is located at Yerkes Observatory. It is the largest refractor ever built. (*Yerkes Observatory*)

The objective lens acts a little like a prism in that different colors of light are bent by slightly different amounts in passing through the lens. As a result, different colors of light have slightly different focal lengths. This is a serious defect since the image of a star or planet seen through a telescope with chromatic aberration appears surrounded by a fuzzy rainbow of colors. Chromatic aberration usually is corrected by gluing a second thin lens to the objective lens, as shown in Figure 6–6. The chemical compositions of the two lenses is chosen in such a way that all the colors of light focus at precisely the same focal point.

A second common defect found in refractors is *spherical aberration*. In this type of aberration, light rays passing through different parts of the lens come to focus in slightly different locations. The result is a very fuzzy image that always looks out of focus. Fortunately, the addition of a second lens to correct for chromatic aberration also corrects spherical aberration. Since a second lens corrects for both chromatic and spherical aberration, most refractors have a *two-element objective,* two lenses made out of slightly different kinds of glass.

There are other types of aberrations, such as *astigmatism, coma,* and *curvature of field,* that have the effect of producing slight distortions in the telescopic image. It is the job of the optical engineer to design telescope objectives in which

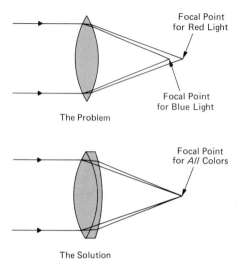

Figure 6–6. Chromatic Aberration. Simple lenses suffer from the fact that different colors of light have slightly different focal lengths. This defect is corrected by adding a second thin lens.

the various kinds of aberrations are eliminated, or at least reduced to acceptable levels. There are, however, some inherent problems with all refractors that no amount of engineering can solve. In order for a refractor to work, light must pass through lenses. In doing so, the glass absorbs some light, causing the image to dim. To make matters worse, glass is opaque to all ultraviolet radiation. Ultraviolet light in the range 3,000 to 4,000 Å does make it through the earth's atmosphere, but it will not pass through glass. An astronomer who tries to observe such radiation through a refractor might just as well have lenses made out of wood. This problem can be circumvented by making lenses out of quartz. Quartz

Figure 6–7. Spherical Aberration. Simple lenses suffer from the fact that light rays entering different parts of the lens have slightly different focal lengths. As with chromatic aberration, this defect is corrected with the addition of a second lens.

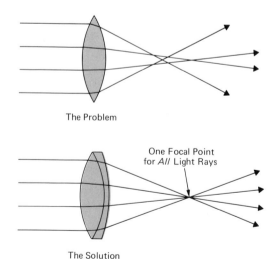

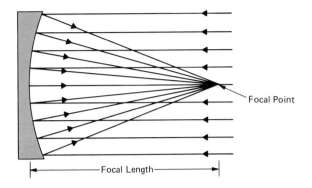

Figure 6–8. *A Concave Mirror.* A concave mirror can be used to focus light rays. As with lenses, the distance between the mirror and the focal point is called the focal length.

is transparent to long wave-length ultraviolet light, but the cost of building such a telescope would be enormous.

A final problem with refractors is that the glass from which objective lenses are made must be of so-called optical quality. It cannot have any of the flaws, such as bubbles, that usually form as the glass cools in the factory. It is extremely difficult to manufacture a slab of optical quality glass much more than three feet in diameter. Yet, astronomers would prefer to have still larger telescopes than those found at Yerkes and Lick. Due to the undesirable absorption of ultraviolet light, as well as to the difficulty of making very large lenses, around the turn of the century astronomers abandoned the idea of building any more major refractors. Instead, they turned to the use of mirrors in the construction of the giant telescopes; these are used at most major observatories today.

6.2 Reflecting Telescopes

As with so many topics and inventions in astronomy, the design of the reflecting telescope originated with the work of Sir Isaac Newton. In his experiments with optics, Newton found that a concave mirror could be used to focus light rays, as shown in Figure 6–8. The incoming starlight is reflected by the concave mirror and converges to a *focal point.* The distance between the mirror and the focal point is called the *focal length.* Such a mirror usually is made out of a slab of glass that has been ground until it is spherically concave. It is then coated with a thin layer of aluminum or silver. Obviously, however, if the astronomer tried to observe the image formed at the focal point by placing his eye or an eyepiece at the focal point, his head would block out the incoming starlight. To circumvent this difficulty, a second mirror called a *diagonal mirror* is placed in front of the focal point to reflect the converging light rays to one side, as shown in Figure 6–9. This allows the eyepiece to be mounted on the side of the telescope for convenient viewing. This type of a telescope is called a *Newtonian reflector.*

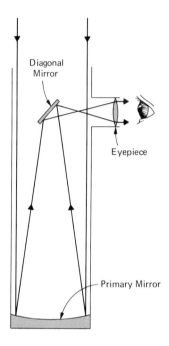

Figure 6-9. *A Newtonian Reflector.* This type of a reflecting telescope makes use of a *diagonal mirror* which reflects the converging light rays to one side so that the astronomer can better view the image with the aid of an eyepiece.

Several advantages of reflectors over refractors are immediately apparent. For example, the glass used in the refractor's objective lens must be of optical quality, but this is not required of glass used in a reflector. Light does not pass through the glass used in a reflector's mirror and, thus, defects inside the glass cannot affect the quality of the image. In addition, a mirror reflects *all* colors of

Figure 6-10. *Spherical Aberration.* Simple concave mirrors suffer from the fact that light rays reflected from different locations on the mirror have slightly different focal lengths. This defect is corrected by making sure the concave surface of the mirror is parabolic.

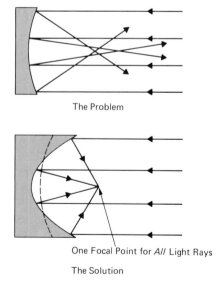

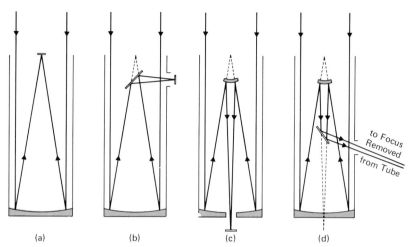

Figure 6-11. Different Types of Reflectors. Four basic different types of reflectors are shown schematically. The four optical arrangements are: (a) prime focus, (b) Newtonian, (c) Cassegrain and (d) Coudé focus.

light in exactly the same way. The reflecting telescope, therefore, does not suffer from any chromatic aberration at all. And, finally, since the incoming starlight does not pass through glass, the glass cannot absorb any ultraviolet light. A reflector is, thus, an ideal telescope for the astronomer who wants to make observations in the wave-length range from 3,000 to 4,000 Å. Any lenses used in his eyepieces or additional apparatus can be made out of quartz.

A defect common to both reflectors and refractors is spherical aberration. As shown in Figure 6–10, light rays reflected from near the center of a spherically concave mirror come to focus at a slightly longer focal length then the light rays reflected from near the edge of the mirror. This defect is corrected by making sure that the concave surface of the mirror is parabolic. With a parabolic mirror, all the light from a distant object comes to focus at the same point, no matter from which part of the mirror that light is reflected.

One difficulty commonly encountered with Newtonian reflectors is that, in practice, the focal point is often high off the ground. This is especially true if the mirror has a very long focal length. Observing with a Newtonian reflector is frequently cumbersome and inconvenient. The astronomer copes with this difficulty by using mirrors to reflect the light rays to a more accessible location, thereby modifying the design of the telescope.

One of the most popular modifications results in the so-called Cassegrain reflector. In the Cassegrain system, the diagonal mirror is replaced with a convex mirror that reflects the converging light rays back toward the primary mirror. A small hole has been drilled through the primary mirror. The light rays pass through this hole and come to focus just behind the primary mirror. This is often a far more convenient location from which to do observations.

160 *Astronomy: The Structure of the Universe*

Another popular modification frequently employed in the case of very large telescopes involves the use of the so-called Coudé focus. Such a system starts off in the manner of a Cassegrain telescope. Light goes from the concave primary to a convex secondary mirror. As the light rays head back toward the primary mirror, the beam is intercepted by a flat, diagonal mirror that sends the light to some distant location in the observatory usually called the *Coudé room*. In the Coudé room, the astronomer can analyze the starlight with complex equipment that is too large or heavy to attach to the side or back of the telescope.

A final alternative arrangement of the reflecting telescope to be discussed here is perhaps the simplest of all. In the case of very large telescopes, such as the Mt.

Figure 6–12. *The 200-inch Telescope.* This view looks "down" the 200-inch telescope toward the primary mirror. An astronomer rides in the observer's cage at the prime focus. (*Hale Observatories*)

Palomar telescope, the mirror is so big that it is possible to place observing apparatus directly at the primary focus without blocking out very much of the incoming light. Such an arrangement is called the *prime focus*. The 200-inch telescope has an *observing cage* at the prime focus in which the astronomer rides while making his observations. Some of the most important spectroscopy and photography in astronomy today is done by astronomers sitting all night long in the observer's cage of the 200-inch telescope. Unfortunately, once an astronomer begins his observations in the cage he cannot interrupt his work. There are humorous stories of observatory assistants subjecting the helpless "caged" astronomer to radio programs of fine acid rock or of all night religious revival meetings.

The world's largest reflector, the 200-inch reflector at Mt. Palomar, has been in operation since the late 1940s. Until that time, the 100-inch reflector at Mt. Wilson had been the largest telescope. Other major reflectors include the 120-inch at Lick Observatory in California and the 107-inch at McDonald Observatory in Texas. In the late 1970s, construction will have been completed on two 150-inch reflectors, one of which is at the Kitt Peak National Observatory in Arizona and the other at the Cerro Tololo Inter-American Observatory in Chile. These twin telescopes, one in the Northern Hemisphere and one in the Southern Hemisphere, will together give astronomers the ability to cover the entire sky with identical instruments.

When an astronomer decides to build an observatory, great care must be exercised in choosing a location, or *site*. The first obvious consideration is weather conditions. An observatory site must be free from rain and clouds for as many days out of the year as possible. Secondly, the sky must be dark. For this reason, observatories often are located far from big cities whose street lights brighten the sky and would prevent astronomers from seeing dim stars and galaxies. Unfortunately, the population is growing and once-remote observatories frequently find themselves in big trouble. For example, when the Mt. Wilson Observatory was constructed in around the turn of the century, Pasadena was a very tiny town and, for all practical purposes, Los Angeles was a million miles away. Today, however, Los Angeles and neighboring cities virtually surround Mt. Wilson and the "light pollution" is so severe that many types of observations astronomers would like to make are physically impossible. Also, retirement communities are springing up in the desert near Mt. Palomar and the lights from San Jose are interferring with observations from Lick, just as the lights from Tucson are degrading the work at Kitt Peak. It is indeed conceivable that virtually all the major observatories in the world will be rendered useless by population growth before the end of the century. This is, however, the least of the problems that will confront our species as we continue to overpopulate our planet.

Finally, no discussion of telescopes would be complete without mentioning an ingenious system that utilizes both a lens and a mirror called the *Schmidt telescope*. The Schmidt telescope is a reflector specifically designed for wide-angle stellar photography. It consists of a spherical mirror with a short focal

Figure 6–13. A Drawing of the 200-inch Telescope. This cutaway drawing of the world's largest telescope shows the prime focus, the Cassegrain focus and the Coudé focus. (*Hale Observatories*)

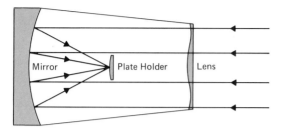

Figure 6–14. *A Schmidt Telescope.* The optical system of a Schmidt telescope employs both a lens and a mirror. Excellent, wide-angle photographs of the sky are made with Schmidt telescopes.

length. To correct for spherical aberration, a thin *correcting lens* is placed at the front of the telescope, as shown in Figure 6–14. In the most famous Schmidt telescope, the 48-inch Schmidt at Mt. Palomar, photographic plates measuring 14 inches square are placed at the prime focus, resulting in spectacular wide-angle pictures of the sky. Shortly after the completion of the 48-inch Schmidt in 1949, a 7-year observing program was undertaken to photograph the entire sky down to a declination of $-33°$ (stars farther south than $-33°$ cannot be seen from Palomar). Every region of the visible sky was photographed twice, once with a red filter and then again with a blue-sensitive plate. The former records reddish (cool) stars, whereas the latter picks up mostly bluish (hot) stars. The initial survey consisted of 1,870 plates; it is the finest and most useful survey of the sky that the astronomer has at his disposal.

Figure 6–15. *The 48-inch Schmidt.* The 48-inch Schmidt telescope on Palomar Mountain has been used to produce excellent photographic surveys of the sky. (*Hale Observatories*)

The telescope is the astronomers' most important tool, and a wide variety of auxiliary equipment has been developed for the purpose of analyzing starlight. Most of the important advances in astronomy today are made by using this auxiliary apparatus in conjunction with the world's major telescopes.

6.3 Spectroscopy and Photometry

In view of the tremendous amount of information contained in spectra, it should come as no surprise that the *spectroscope*, or *spectrograph*, could be called the astronomer's second most important tool. Indeed, spectroscopy takes up about half of the telescope time at the major observatories around the world. An astronomer's spectrograph is simply a device that takes starlight from a telescope, passes it through a prism, and records the resulting colors of the rainbow on a photographic plate. The basic design of a spectrograph is shown in Figure 6–16. The light from the astronomer's telescope comes to focus at the slit of the spectrograph. A *collimating lens* directs the beam of light through a prism and a *camera lens* focuses the spectrum onto the photographic plate. If the object the astronomer is studying is very bright, he will have to expose the photographic plate for only a short time. If the object is dim, he may have to

Figure 6–16. A Spectrograph. The basic optical design of a spectrograph involves a "collimating lens" which directs the incoming starlight onto a prism or diffraction grating. A "camera lens" then focuses the spectrum onto a photographic plate.

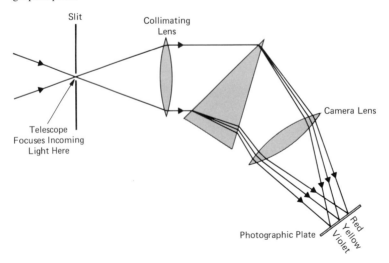

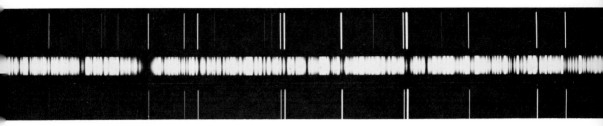

Figure 6–17. A Typical Spectrum. The spectrum of a star is bordered above and below by a comparison spectrum. The lines in the comparison spectrum serve as reference markers from which the astronomer can deduce the wavelengths of the spectral lines in the star's light. (*Hale Observatories*)

expose the photographic film for many hours as the telescope tracks the star or galaxy across the sky.

The spectrum obtained by the astronomer usually looks like a fuzzy band on which spectral lines are superimposed. Obviously, the astronomer wants to identify these lines with some chemicals. However, due to the Doppler effect, these lines can be far from their usual wavelengths. For example, the calcium H and K lines appear in the blue part of the spectrum in laboratory experiments, whereas in the spectra of distant galaxies they frequently are found among the red colors of the spectrum. Since spectral lines often are shifted from their usual wavelengths, it would be very advantageous for the astronomer to have some sort of reference marks so that he could easily tell how big the shift was. To achieve this, a *comparison spectrum* is artificially superimposed on the photographic plate while it is in the spectrograph. This comparison spectrum is usually of iron and appears above and below the astronomical spectrum, as shown in Figure 6–17. Since the wavelengths of the iron lines are well known from laboratory experiments, the astronomer then has convenient reference marks against which he can measure the wavelengths of spectral features in the object he is studying.

Often, the prism in a spectrograph is replaced with a *diffraction grating.* A diffraction grating consists of a shiny mirror on which many parallel grooves have been etched. Such a grating has the same effect as a prism; light that is reflected off of the grating is broken up into the colors of the rainbow. One of the advantages of a grating over a prism is the same as the major advantage reflectors have over refractors. With a grating, light does not have to pass through glass, which absorbs some of the wavelengths astronomers try to study.

As we saw in the last chapter, the spectrum an astronomer obtains after working all night is very small. It looks like nothing more than a faint, fuzzy smear on a small piece of glass. The astronomer must use a microscope to examine the features on his spectrum. As discussed in Section 5.4, the astronomer obtains a *tracing* of his spectrum with the aid of a device called a *microdensitometer.* A microdensitometer essentially consists of a light source (light bulb), a photoelectric cell ("electric eye"), and a chart recorder (mechanical pen

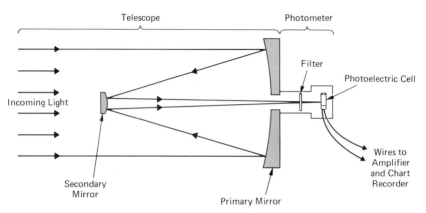

Figure 6-18. *The Design of A Photometer.* At the heart of a photometer is a light-sensitive photoelectric cell or photomultiplier which converts the incoming starlight into an electric current.

and paper). A microdensitometer shines a fine beam of light through the astronomer's spectrum, and a photoelectric cell measures how much light gets through. If part of the spectrum is heavily exposed (for example, the star emits lots of radiation at these wavelengths), then very little light gets through the photographic plate to the photoelectric cell. If a part of the spectrum is very lightly exposed (for example, in the middle of an absorption line), then light easily passes through the photographic plate to the photoelectric cell. The photoelectric cell converts the beam of light into an electric current, which drives the mechanical pen. As the beam of light in the microdensitometer scans across the astronomer's spectrum, the mechanical pen accurately traces out the spectrum on a long strip of paper. It is frequently far more convenient for the astronomer to examine this tracing than the original spectrographic plate. Every dip and valley in the tracing represents a spectral feature, and the shapes of these valleys accurately portray the shapes or profiles of the spectral lines.

Spectroscopy allows the astronomer to investigate details of the light emitted by stars, galaxies, planets, and other interesting objects, but the techniques of *photometry* permit a much broader, more general examination of this light. From our understanding of the nature of light, we realize that cool objects emit primarily long wavelength radiation, whereas hot objects emit short wavelength radiation. Therefore, if you look up into the sky and see a reddish star such as Aldebaran or Betelgeuse, you can be quite sure that it is a cool star having a surface temperature of about 3,000°K. On the other hand, if you see a bluish-white star such as Rigel or Vega, you can be equally sure that it is a hot star with a surface temperature of 10,000°K or higher. Obviously, the colors of stars betray their temperatures. If, somehow, the astronomer could accurately measure the colors of stars, he could immediately infer their temperatures. This is made possible by a device called a *photoelectric photometer*.

At the heart of all photoelectric photometers is an electronic tube that converts light into electricity. This electronic tube, a *photomultiplier,* is similar to common devices called electric eyes, which frequently are used to open doors automatically in department stores. The photomultiplier converts starlight from the telescope into an electric current, which drives a pen on a piece of paper and accurately records the intensity of the starlight. A diagram showing the design of a photometer in connection with a Cassegrain telescope is shown in Figure 6–18.

In the 1950s, astronomers devised a system of standardized filters to be used in photometers by which the colors of stars could be measured. This system, which is in wide use today, is called the *U, B, V system.* The U, B, V system utilizes three colored filters, placed in front of the photometer, that allow only certain specific wavelengths of light to pass through. The U (for ultraviolet) filter is transparent only to ultraviolet light from about 3,000 to 4,000 Å; the B (for blue) filter is transparent only to blue light from about 3,800 to 5,100 Å; and the V (for visible) lets through primarily yellow light from 5,100 to about 6,400 Å. A graph showing how much light is transmitted through each of these standard filters at various wavelengths is given in Figure 6–19. Using these three filters, the astronomer can sample the light from a star or galaxy at specific, standardized bands of wavelengths. For example, suppose the astronomer is observing a reddish star such as Antares in Scorpius with his U, B, V photometer. Since the star appears red even to the naked eye, more light will get through the V filter than the B or U filters. However, if the astronomer observes a hot bluish star such as Alcyone in the Pleiades, more light will come through the U and B filters than the V filter.

Using U, B, V filters, the astronomer can measure how bright a star or galaxy is at specific wavelength intervals. He, therefore, can speak of the *U magnitude, B magnitude,* and *V magnitude* of a celestial object. The V, or *visual magnitude* of a star is usually very nearly the same as the apparent magnitude, which was

Figure 6–19. *The U,B,V System.* The colors of stars are measured by using standard filters along with a photometer. This graph shows the ranges of wavelengths which can pass through three standard filters frequently used by astronomers.

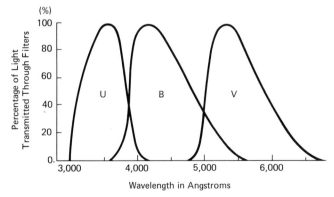

discussed in Chapter 1. They differ by only a very small amount due to the fact that the naked human eye and a photometer with a V filter do not measure exactly the same thing. For example, the apparent magnitude of the bright star Sirius is -1.50, while its V, or visual magnitude is -1.42.

Recalling that the colors of stars tell us about stellar temperatures, astronomers realized that they had the means to measure these colors accurately by simply measuring differences in U, B, and V magnitudes. When an astronomer subtracts one magnitude from another in this way, he obtains the so-called *color index* of the star. The most commonly used color indices are U-B and B-V. There is, then, a direct correlation between the color index of a star and its surface temperature. For example, the reddish star Antares has a visual magnitude $V = +0.92$ and a blue magnitude of $B = +2.72$; more light passes through the V filter than the B filter (recall that the higher the number, the dimmer the star). Thus, Antares has a color index $B-V = +1.8$, which corresponds to a temperature of about $3{,}000°K$. The bluish star Alcyone has $B-V = -0.1$, which corresponds to about $12{,}000°K$. Our sun, with a surface temperature of $6{,}000°K$, has a B-V of 0.6.

Using the techniques of spectroscopy and photometry, astronomers have the ability to analyze the light from distant celestial objects. It is from such analyses that we learn about the nature of the objects in our universe.

6.4 Radio Telescopes

Up until only a few decades ago, everything mankind knew about the cosmos was based entirely on observations with visible light. The visible light to which the human eye responds was the only source of information. Yet, during the nineteenth century it became clear that visible light was only a very small fraction of the entire electromagnetic spectrum. Somehow there had to be many forms of "invisible" light with wavelengths both shorter and longer than those of the visible light. Since the astronomer's telescopes were designed for use with visible light, it became painfully obvious that he was observing only a small fraction of all possible types of radiation from celestial objects. It was quite reasonable to suppose, therefore, that astronomers' ideas and theories about the universe could be very much in error, since these theories and concepts were based only on ordinary optical observations. Astronomers during the early twentieth century were very aware of the limitations they faced. Unfortunately, no one knew what to do about it.

In the early 1930s, Karl Jansky, a young engineer at Bell Telephone Laboratories, was experimenting with very long radio antennas when he noticed that he was receiving static from some unknown source. This static was picked up on his radio receiver and would come in 4 minutes earlier each day. Recalling that stars rise 4 minutes earlier from one day to the next (that is, the solar day and the

Figure 6–20. A Typical Radio Telescope. The 140-foot radio telescope at the National Radio Astronomy Observatory in Greenbank, West Virginia, is an example of one of the finest radio telescopes in operation today. (*N.R.A.O.*)

sidereal day differ by 4 minutes), Jansky correctly concluded that he was detecting radio waves from outer space.

The importance of this discovery cannot be overemphasized. Up until 1931, astronomers only knew about visible light coming from celestial objects. Yet, now radio light also had been detected from cosmic sources. Astronomers then realized that if they could figure out how to build special antennas, which might be called *radio telescopes,* they could pinpoint these cosmic sources of static. Building such telescopes would be like giving new eyes to a blind person. After all, mankind literally had been blind to radio waves from outer space. With Jansky's discovery, astronomers had the hope of being able to "see" the radio universe just as clearly as they could see the visible universe.

The first antenna designed to pick up cosmic radio noise was built by Grote Reber in 1936. However, due to World War II, real progress was delayed until the late 1940s. At the end of the war, astronomers realized that many of the recent advances in electronics and electrical engineering could be directly applied to the problem of building radio telescopes. Radio telescopes began to spring up in England, the Netherlands, Australia, and later in the United States.

The basic idea behind the design of a radio telescope is very simple. A radio telescope is a reflector. It consists of a large parabolic "dish" that causes radio waves to concentrate at a focus. This dish, which reflects the radio waves to the focal point, can be made out of metal or a fine wire mesh. At the focal point there is a radio receiver that converts the radio waves into an electric current. This current is carried through wires to amplifiers and electronic recording devices. If the radio telescope is pointed toward a source of radio noise, a strong "signal" is detected informing the *radio astronomer* that he is "looking" at a bright spot in the *radio sky*. If the radio telescope is not aimed at a source of radio waves, no static comes through the receiver.

Radio waves comprise a very large part of the electromagnetic spectrum. They have wavelengths extending from about $\frac{1}{10}$-inch to thousands of miles. The earth's atmosphere is, however, transparent only to wavelengths between $\frac{1}{10}$-inch to about 100 feet. It is in this range that the radio astronomer makes his observations. Just as you can "tune" your car radio to specific frequencies or wavelengths, the radio astronomer can select the precise wavelength at which he wants to make observations. The resulting view of the radio sky depends critically on the wavelength of the observations. The radio sky at a wavelength of 1 inch looks very different from the radio sky at wavelengths of 1 foot. A radio source that is a strong emitter at $\lambda = 1$ inch might be very weak at $\lambda = 1$ foot, and vice versa. The radio astronomer is, therefore, careful to note the precise frequency, or wavelength, at which he is observing.

Radio telescopes are a lot bigger than their optical counterparts. The largest optical reflector has a mirror 200 inches in diameter, but the world's largest radio telescope has a dish 1,000 feet in diameter. There are several reasons for this. First of all, it is much easier to build a large radio telescope than a large optical telescope. Due to the long wavelength of radio radiation, minor defects in the precise shape of the parabolic dish are unimportant. Secondly, radio astronomers *must* build large radio telescopes. The energy carried by a photon of radio light is much less than the energy carried by a photon of visible light. The dish must, therefore, be large so that enough radio energy can be collected to produce a detectable signal. In addition, in order to obtain good resolution, the radio astronomer has no choice but to make his observations with a large dish. If a radio astronomer tried to make observations with a small telescope, he would find that it would be impossible to pinpoint the locations of radio sources in the sky. Because the wavelength of radio waves is so long, the diameters of radio telescopes must be large to give a clear view of the radio sky. All the major radio telescopes in the world have diameters of more than 100 feet. Some of the most well-known radio telescopes are at the Arecibo Observatory in Puerto Rico, the National Radio Astronomy Observatory in West Virginia, the Jodrell Bank Station and Mullard Radio Observatory in England, and the Radiophysics Laboratory in Australia.

Radio astronomers do not work under some of the restrictions that hamper the observations of ordinary optical astronomers. For example, the sun emits a lot of light but very little radio radiation. Therefore, the radio astronomer can make observations at any time, day or night. In addition, he is not hampered by weather conditions. After all, a radio or television set works just fine, even when it is raining or snowing.

It is important to realize that the radio astronomer does not "see" radio sources with his eyes. Neither does he "hear" radio sources. Rather, his sensitive electronic equipment tells him if his telescope is pointed toward an object from which radio waves are coming. When a radio astronomer discovers a source in the sky, he carefully notes the location (right ascension and declination) and then examines photographs of that region of the sky to see if anything interesting is there. Sometimes radio sources are found to coincide with well-known objects such as nebula and galaxies. Sometimes astronomers working with the most powerful optical telescopes cannot see anything but the dark sky at the precise location of an intense radio source.

Back in the 1950s, when radio astronomers had finally developed reliable telescopes, they began making *surveys* of the sky. After all, no one had any idea what the radio sky looked like. Radio astronomers, therefore, began scanning the sky in a systematic fashion to discover radio sources. Each time a source was discovered, the position would be noted along with such information as the intensity of the radio signal. The final result of such surveys are catalogues in which the known radio objects in the sky are listed, just as the stars in the sky are listed in star catalogues. From catalogues of radio sources, it is possible to get a good idea of what the radio sky looks like. For example, we could draw a map of the radio sky as shown in Figure 6–21. The first drawing in Figure 6–21 shows what the visible sky looks like. It is based on photographs of the entire sky and is drawn so that the Milky Way stretches across the center of the drawing. The

Figure 6–21. The Visible Sky and the Radio Sky. The upper drawing shows the entire visible sky. The drawing is centered about the Milky Way. The lower drawing shows the entire radio sky. Both drawings have the same scale and orientation. (*Lund and Griffith Observatories*)

second drawing shows the entire radio sky to the same scale as the first drawing. Again, the Milky Way prominently stands out. Every bright spot in the radio view represents a radio source, just as every bright spot in the optical view represents a visible star. The optical view is what you see with your eyes, which respond only to visible light. The radio view is what you would see *if* your eyes responded only to radio light.

When a radio astronomer discovers a radio source in the sky, it is possible for him to examine many details of the structure of this source. This is done by moving the radio telescope around so that he can measure the intensity of the radio radiation in the vicinity of the source. In this way, the radio astronomer can discover the shape of a radio source. The results of his observations are best expressed by a drawing with contour lines, as shown in Figure 6–22. Each contour line represents a specific intensity level of radio radiation. In cases in which a radio source is discovered to be associated with a familiar optical object, it is often instructive to draw the contours directly on a photograph of the optical object, as shown in Figure 6–23. In this way the astronomer can readily see if certain optical features correspond to certain radio features.

As a direct result of radio astronomy, many new things have been learned about the universe in which we live. Well-known objects such as nebulae and galaxies have been discovered to emit radio radiation. In order to account for this radiation, the scientist is led to a deeper understanding of these familiar objects. In addition, many new and bizarre objects have been discovered by radio astronomers. In the early 1960s, *quasars* were found, which are just as

Figure 6–22. A Radio Map. The results of the observations of radio astronomers are conveniently displayed in the form of contour maps. Each contour corresponds to a specific level of intensity of radio radiation. This particular map is of a strange galaxy called 3C 411. (*Adapted from Drs. Spinrad, Smith, Hunstead, and Ryle*)

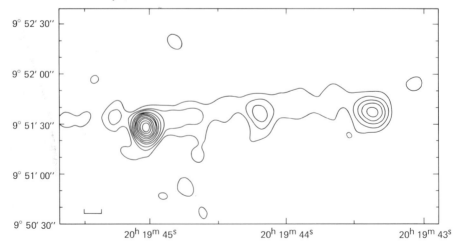

Figure 6–23. Radio and Optical Pictures. It is often enlightening for a radio astronomer to draw his contour maps directly onto the optical photograph of the object he is studying. (*Ohio State University Observatory*)

mysterious and baffling to the astrophysicist today as ever. And, in the late 1960s, radio astronomers discovered *pulsars* giving off regular, rapid bursts of radio noise, blinking on and off and telling us about the recent collapse of a massive star. In discussing these and other topics throughout the rest of this book, we often will turn to the observations of the radio astronomer. Through radio astronomy we are aware of the appearance of the sky in radio light. By synthesizing the radio and visual views of the heavens, our understanding of the universe will be enhanced profoundly.

6.5 IR, UV, and X-Ray Astronomy

In observing the universe, the ground-based astronomer has two options. He can use optical telescopes and make observations of visible light, or he can use radio telescopes and make observations of radio light. These options are possible because the earth's atmosphere is transparent to visible and radio light. The astronomer, therefore, speaks of the "optical window" and the "radio window" in referring to the transparency of the air to these types of radiation. Unfortunately for the astronomer, these are the only two "windows" that exist. The air we breath is opaque to all other wavelengths in the electromagnetic spectrum. For example, water vapor in the air efficiently absorbs almost all incoming infrared radiation; a layer of ozone high in the atmosphere equally efficiently absorbs ultraviolet radiation from outer space. Trying to observe the universe at infrared, ultraviolet, and X-ray wavelengths from the earth's surface is like trying to look through a brick wall. None of these radiations can pass through our atmosphere. The only salvation is to go *above* the earth's atmosphere with airplanes, rockets, balloons, or satellites. Only at high altitudes can the astronomer make observations unhampered by the opaque air.

In between visible light and radio light in the electromagnetic spectrum is a type of light called *infrared radiation*. Infrared light has wavelengths from about 8,000 Å to 1 millimeter. Some of the so-called near infrared radiation (around 10,000 Å) can pass through the earth's atmosphere. Special photographic plates sensitive to infrared light around 10,000 Å can be used to produce pictures at these wavelengths, which lie just beyond the range to which the human eye responds (the human eye does not see light with wavelengths longer than about 7,500 Å). Best results with near infrared photography are obtained from observatories at high altitudes, such as the new 84-inch reflector on Mauna Kea in Hawaii.

At wavelengths longer than 10,000 Å, the earth's atmosphere is opaque to virtually all infrared light, and the astronomer must turn to new techniques. Fortunately, there are certain types of substances that respond to infrared light. For example, when crystals made out of germanium are exposed to infrared light, their electrical properties change. Using this discovery, astronomers realized that they could construct infrared telescopes. A telescope used for "far" infrared observations of the sky is typically an ordinary reflector with a few important alterations. Instead of placing a photographic plate at the focal point, an infrared sensitive crystal is used. Such a crystal is called a *bolometer* and is made of either germanium containing a little gallium or of an alloy of indium and antimony. In order for the bolometer to be sensitive to infrared radiation, the crystal must be cooled down to a very low temperature with the help of liquid helium. When the telescope is pointed toward a star or galaxy emitting infrared radiation, this light is focused on the bolometer. The infrared light is absorbed by the crystal and converted into heat, which raises the temperature of the crystal. With a change in temperature, the electrical resistance of the crystal changes. In short, the infrared astronomer does not actually see anything with his telescope. Instead, he measures the changes in the electrical resistance of a cold germanium crystal at the focus of his telescope. From such measurements, the astronomer can tell whether or not his telescope is pointed toward a source of infrared light. Figure 6–24 shows a simplified design for a typical infrared telescope. The mirrors of the telescope are coated with gold (rather than silver), which more efficiently reflects the infrared. The bottle of liquid helium keeps the bolometer at about 2°K.

Finally, to make observations, the infrared astronomer places all his equipment aboard an airplane or attaches it to a balloon, which enables him to get above the water vapor in the earth's atmosphere. Very successful observations, especially by Dr. Frank Low at the University of Arizona, are carried out in airplanes that fly at altitudes up to 50,000 feet. Such observations have discovered that many stars emit unusually large amounts of infrared light. This seems to suggest that these stars are surrounded by large quantities of dust that might be in the process of condensing into planets. In addition, several galaxies have been discovered to be emitting exceptionally huge amounts of infrared light. The source of this light is not understood; perhaps some violent processes or explosions are occurring at the centers of these galaxies. But, as of the mid-1970s, the source of infrared light from galaxies is one of the baffling mysteries in modern astronomy.

Infrared radiation lies just to the long wavelength side of visible light, and ultraviolet radiation lies to the short wavelength side of visible light. Ultraviolet light covers the wavelength range from 4,000 Å down to about 100 Å. Just as the near infrared light manages to squeeze through the long wavelength side of the optical window, the near ultraviolet (3,000 to 4,000 Å) also gets through the

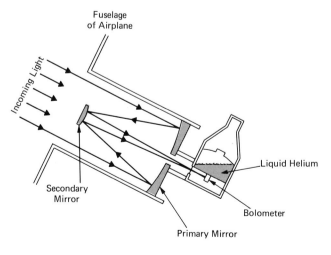

Figure 6–24. The Design of An Infrared Telescope. The basic design of an infrared telescope is shown in this diagram. An ordinary telescope is used to focus the incoming radiation onto a crystal which is sensitive to IR. The astronomer measures the changes in the electrical properties of the crystal.

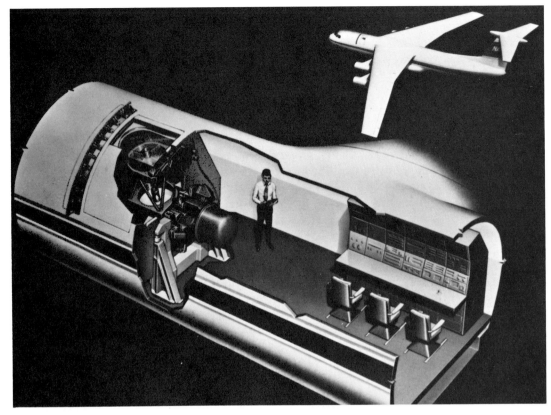

Figure 6–25. The 36-inch Airborne Infrared Telescope. In order to observe the infrared sky, astronomers fly their equipment at altitudes of 50,000 feet. Only at such altitudes is the astronomer above most of the water vapor in the earth's atmosphere. (*NASA*)

Figure 6–26. Infrared Observations of a Nebula. Infrared observations are often expressed in the form of a contour map, just like radio observations. An infrared map of the "Omega Nebula" in Sagittarius is shown here along with a corresponding optical photograph (adapted from Drs. Lemke and Low). (*Hale Observatories*)

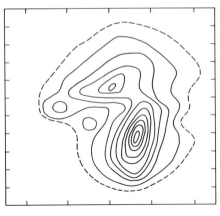

earth's atmosphere. Astronomers are accustomed to making observations in the near ultraviolet from ordinary optical observatories. However, to observe at wavelengths shorter than 3,000 Å, it is absolutely necessary to get far above the earth's atmosphere.

Some of the most successful far ultraviolet observations of celestial objects have been performed by Dr. George Carruthers and his associates at the Naval Research Laboratory, Washington, D.C. These observations have been made from rocket flights, from Apollo missions to the moon, and from Skylab. A schematic diagram of the kind of telescopes used by the ultraviolet astronomer in these space flights is shown in Figure 6–27.

At first glance, the ultraviolet telescope looks like an ordinary Cassegrain. The primary mirror is spherical and has a hole drilled through its center. All the incoming light comes to focus at the focal point. However, a *photocathode* made of potassium bromide is placed at the focal point. This chemical has the property of causing electrons to be emitted when ultraviolet light strikes the photocathode. Photons of ordinary visible light do not carry enough energy to have any effect on the photocathode. But the short wavelength ultraviolet photons have enough energy to shake electrons loose from the atoms in the photocathode. These electrons travel from the photocathode through the hole in the primary mirror and strike a piece of photographic film. The entire telescope is surrounded by a *focusing magnetic,* which focuses the electrons onto the film and insures that the electron image is a faithful reproduction of the ultraviolet part of the optical image.

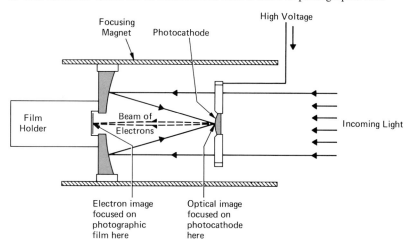

Figure 6–27. *The Design of An Ultraviolet Telescope.* The basic design of an ultraviolet telescope is shown in this diagram. An ordinary telescope focuses the incoming radiation onto a "photocathode." Ultraviolet light causes the photocathode to emit electrons which are in turn focused onto a sheet of photographic film.

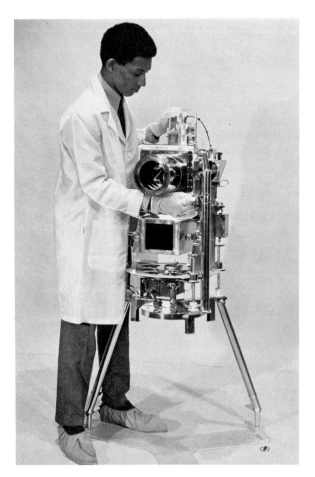

Figure 6-28. An Ultraviolet Telescope. Dr. Carruthers is shown here with the ultraviolet telescope which he designed. This telescope was taken to the moon by the Apollo 16 astronauts. (*NRL*)

The kinds of photographs obtained by Dr. Carruthers's ultraviolet telescopes are incredible. For example, Figure 6–29 shows a photograph of the nearest galaxy, the so-called Large Magellanic Cloud. This photograph was taken by Apollo 16 astronauts from the moon in April 1972. For comparison, an excellent earth-based photograph of exactly the same region of the sky is also shown. In the earth-based photograph we see this galaxy in visible light (4,000 to 8,000 Å), the way it looks to our eyes. In the lunar-based photograph we see this same galaxy in invisible ultraviolet light (1,250 to 1,600 Å). Incredibly hot stars, which emit most of their radiation in the ultraviolet, blaze forth in this remarkable photograph brought back from the moon. Conversely, the cooler stars, which prominently appear in the visible image, are not seen at all in the ultraviolet view.

Another dramatic example of ultraviolet astronomy is shown in Figure 6–30. Two views of Comet Kohoutek are shown here. Both photographs were taken from rocket flights early in January 1974, as Comet Kohoutek passed near the

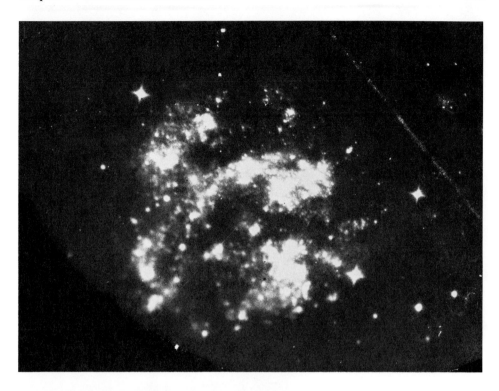

Figure 6-29. The Large Magellanic Cloud. The photograph above shows the appearance of the Large Magellanic Cloud in ultraviolet light. An ordinary optical photograph, shown on the left, demonstrates that there are many stars in the sky that are invisible to the human eye. (*NRL and Lick Observatory*)

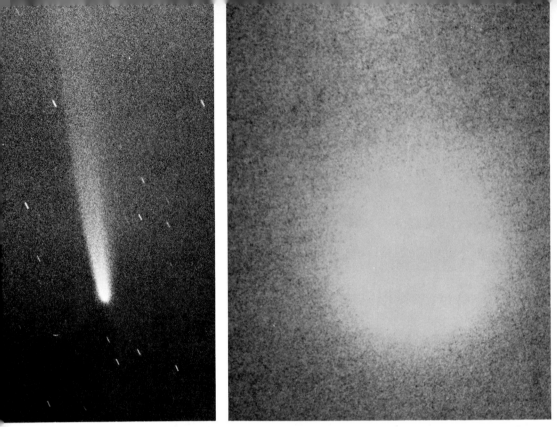

Figure 6-30. Comet Kohoutek. The photograph on the left shows the visual appearance of Comet Kohoutek. The photograph on the right is the ultraviolet view. (*Johns Hopkins University; NRL*)

sun. Both photographs are shown to exactly the same scale. The view on the left is in visible light; it looks like any ordinary comet. However, the ultraviolet view reveals that the head of Comet Kohoutek is surrounded by a huge halo. The halo shown in this photograph is 4 million miles in diameter. In ultraviolet light, Comet Kohoutek is four times larger than our sun, which has a diameter of about 1 million miles.

When we look at these photographs, we are seeing what no man has ever seen before. We are seeing that which the human eye cannot see. We are seeing the invisible universe.

At wavelengths shorter than those of ultraviolet light lies the domain of X rays. No X rays at all get through the earth's atmosphere; all X-ray observations must be done from outer space. Fortunately for the astronomer, there are devices that easily detect X rays. Such devices are called *proportional counters* and are very similar to ordinary Geiger counters. In essence, a proportional counter consists of a bottle of gas. When an X ray passes through the gas, electrons are knocked off the atoms in the gas and electronic equipment is used to detect these electrons. Special care is taken to insure that the proportional counter responds only to X rays and not to other processes that might jar electrons loose from the atoms in the gas.

Since the early 1960s, astronomers have been flying X-ray detectors in small rockets. The rockets go up, look around, and come back down. During the few precious moments above the earth's atmosphere, hopefully the proportional counters in the nose cone of the rocket will detect something. A few bright X-ray sources have been discovered in this way. Of course, the final result of the astronomer's efforts is sometimes a large pile of junk scattered over some remote section of the desert in the southwestern United States. Clearly, there must be a better way.

Unquestionably, one of the most important advances in the history of astronomy centers about the flight of *Explorer 42*. Explorer 42 is a satellite that was launched by NASA into earth orbit on December 12, 1970. The launching site was off the coast of Kenya so that the satellite would go into an orbit directly above the earth's equator. The launch date corresponded to the seventh anniversary of Kenyan independence "in recognition of the kind hospitality of the Kenyan people," and the satellite was promptly christened *Uhuru*, which means "freedom" in Swahili.

Figure 6–31. *Uhuru.* This X-ray satellite was launched in December 1970 from Kenya. Almost 200 X-ray sources in the sky have been discovered during the first four years of operation. (*NASA*)

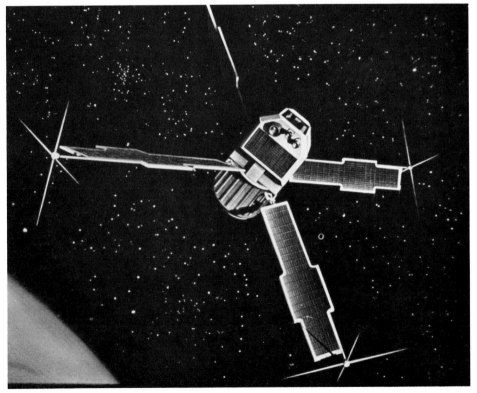

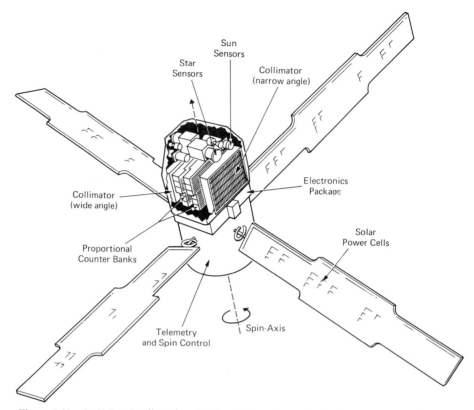

Figure 6–32. *An X-Ray Satellite.* This sketch of *Uhuru* shows the basic arrangement of apparatus used to observe the X-ray sky. As the satellite rotates, the X-ray telescopes on board scan the sky for X-ray stars and galaxies. (*Adapted from R. Giacconi*)

Figure 6–33. *An X-Ray Telescope.* An X-ray telescope consists of a series of proportional counters. This particular X-ray telescope will be used in the High Energy Astronomical Observatory (called "HEAO") scheduled to be launched in the late 1970s. (*NRL*)

Chapter 6 The Astronomer's Tools 183

Uhuru consists of two X-ray telescopes, back-to-back, as shown in Figure 6–32. One telescope is for wide-angle viewing, the other for narrow-angle viewing. *Collimators* select the region of the sky exposed to the proportional counters. If the proportional counters are pointed toward an X-ray source, the proportional counters become activated and data is transmitted back to earth. The satellite spins about its axis once every 12 minutes so that the X-ray telescopes scan the sky looking for X-ray sources. *Uhuru* has operated flawlessly for many years.

From *Uhuru*, mankind obtained its first organized and complete view of the

Figure 6–34. *The X-Ray Sky.* This map of the sky shows 161 X-ray sources observed by *Uhuru*. If our eyes could see X rays, this is what the sky would look like. For comparison, a drawing of the sky in visible light is also shown. (*Adapted from R. Giacconi.*) (*Lund Observatory*)

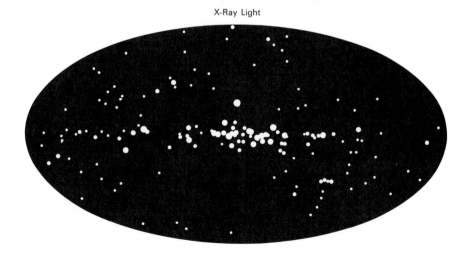

X-ray sky. By 1974, no less than 161 X-ray sources had been discovered. Some sources correspond to nearby stars, whereas others are associated with distant galaxies and quasars. Figure 6–34 shows a map of the X-ray sky drawn in such a way that the Milky Way stretches across the center of the drawing. For comparison, a corresponding drawing of the visible sky is also shown. Those X-ray sources that appear to lie in the Milky Way are believed to be caused by nearby starlike objects, and those sources that lie above and below the Milky Way are generally associated with very distant objects. The kinds of objects that emit these X rays range all the way from clusters of galaxies millions of light years in size to black holes that measure only a few miles across.

In the 1970s, astronomers for the first time became able to observe the heavens at virtually all wavelengths. No longer is the astronomer's work based only on what his eyes can see. The last quarter of the twentieth century will be devoted to bringing together all observations at various wavelengths, synthesizing and unifying the data to give mankind its first truly complete picture of the universe. Virtually every idea, every theory, and every concept of the cosmos will be profoundly affected by this new knowledge.

Questions and Exercises

1. What are the three major functions of a telescope?
2. Contrast and compare magnifying power, resolving power, and light-gathering power.
3. Describe the optical design of a refractor.
4. How is chromatic aberration corrected?
5. Where is the world's largest refractor?
6. Why don't astronomers build any more large refractors at modern observatories?
7. Contrast and compare the various types of reflecting telescopes.
8. How is spherical aberration corrected in reflecting telescopes?
9. What is a Coudé room?
10. What is a Schmidt telescope?
11. Briefly describe the design of a simple spectrograph.
12. What is a photometer?
13. What is the U, B, V system?
14. Discuss the reasons why the color index of a star (such as B-V) is related to the temperature of the star.
15. What is a radio telescope?
16. Compare and contrast a radio telescope and an ordinary reflecting telescope.

17. Discuss one way in which the radio astronomer can express the results of an observation.
18. What is a bolometer and how is it used in astronomical observations?
19. Briefly describe the design of a telescope capable of making observations in the far ultraviolet.
20. What is *Uhuru?*
21. What kind of objects are detected by infrared telescopes?
22. What kind of objects are detected by ultraviolet telescopes?
23. What kind of objects are detected by X-ray telescopes?

The Nature of Stars

7.1 Distances and Magnitudes

In spite of the fact that astronomy is the most ancient of all the sciences, it is remarkable how very recent man's understanding of the universe is. Although man has observed and pondered the cosmos for thousands of years, many of the concepts and ideas in modern astronomy books are only a few decades old. For example, up until 100 years ago, astronomers really did not know where the stars are. They could give the right ascension and declination of a star, and, since stars appear so much dimmer than the sun, they realized that the stars must be trillions of miles away. But the exact locations of stars remained unknown until the mid-nineteenth century.

In 1838, astronomers finally had the required optical and mathematical tools necessary to measure the distances to the nearest stars. The method used in these observations is called *parallax*, and the basic idea is very simple. Imagine

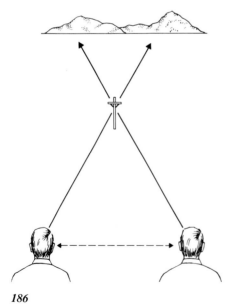

Figure 7–1. *Parallax.* Parallax is simply the effect whereby near objects appear superimposed against different parts of a distant background, depending on the location of the observer.

Chapter 7 The Nature of Stars

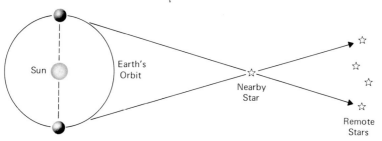

Figure 7–2. Stellar Parallax. As the earth orbits the sun, the location of nearby stars against the background of distant stars will appear to change slightly. From careful observations of these slight changes, astronomers can calculate the distances to nearby stars.

looking at a nearby object, such as a telephone pole, against a distant background of mountains, as shown in Figure 7–1. As you move from one location to another, the nearby object will appear to shift in relation to the distant background. This technique of parallax can be applied directly to the measurement of stellar distances. The earth orbits the sun and, therefore, as seen from the earth, nearby stars should appear to move slightly with respect to distant background stars, as shown in Figure 7–2. In other words, if the astronomer observes a star field containing a nearby star on two separate occasions a few months apart, he will notice a small shift in the location of that nearby star. The bigger the shift, the nearer the star. By measuring the size of this shift over a known period of time, the astronomer can easily calculate the distance to the star.

The problem with parallactic measurements is that all stars, even the nearest ones, are very far away. Therefore, the *parallactic angle* that the astronomer tries to measure is always very small. Even the nearest star, α Centauri, has a maximum displacement of only $1\frac{1}{2}$ seconds of arc. It was not until 1838 that Friedrich Bessel, in Germany, was able to measure the parallax of 61 Cygni. Shortly thereafter, Thomas Henderson at the Cape of Good Hope detected the parallax of α Centauri, and Friedrich Struve, in Russia, measured the parallax of Vega. Today, astronomers have measured the parallaxes of thousands of stars. However, only about seven hundred stars are near enough so that their distances can be determined with reliable accuracy.

Since the distances to the stars are so huge, it is obvious that new "yardsticks" had to be invented to express those distances. To express the distances to the stars in miles or kilometers would be just as inconvenient and absurd as expressing the distance from New York to Chicago in angstroms or millimeters. Two types of yardsticks were invented to meet this need. The first is called a *parsec* (pc). A parsec is defined as the distance you would have to be from the sun so that the radius of the earth's orbit (1 AU) would appear to subtend an angle of 1 second of arc. It turns out that 1 parsec is equal to about 19 trillion miles. Since the definition of the parsec contains angles and the size of the earth's orbit, the parallax of a star is directly related to the distance of the star in parsecs in a very simple fashion. The nearest star, α Centauri, is 1.31 parsecs from the sun.

A second yardstick that also has turned out to be very convenient is the so-called *light year*. A light year is simply how far light can travel in one year. Since the speed of light is 186,000 miles per second, a light year turns out to be equal to about 6 trillion miles. One parsec equals about $3\frac{1}{4}$ light years.

It should be emphasized that the parallactic measurement of stellar distances is the *only* method the astronomer has that gives a *direct* determination of the distances to the stars. If an astronomer carefully and patiently measures the parallax of a nearby star, no one can argue with his answers. The method of parallax is, in principle, both simple and straightforward. However, all other methods of distance determination are *indirect*. As we will see later, to measure the distances to remote stars or galaxies, the astronomer must make a few assumptions in order to crank out answers. Since these assumptions can be called into question, astronomers are constantly reexamining and reevaluating indirect methods.

The measurements of stellar distances during the last century had a profound effect on the course of astronomy. To see why this is so, imagine looking up at the star-filled nighttime sky. If the sky is very clear, faint stars down to an apparent magnitude of 6 can be seen. By way of illustration, suppose attention is focused on one particular star, such as Polaris at the end of the handle of the Little Dipper. Polaris has an apparent magnitude of $2\frac{1}{2}$; it is not a particularly bright star in the sky. Yet, with a little thought, it is realized that just by looking up into the sky gives no indication how bright Polaris *really* is. Polaris might be a very dim star that just happens to be nearby, or it might be an exceptionally bright star at some incredibly remote distance from the earth. Either way, Polaris can still have an apparent magnitude of $2\frac{1}{2}$ as seen by the naked eye. Therefore, just knowing or measuring the apparent magnitude of a star does not tell anything fundamental about the star itself. On the other hand, if astronomers could discover the "real brightness," or so-called *absolute magnitude* of a star, they would then have some basic knowledge about the nature of this star. They would know exactly how much light, how many "kilowatts" of energy this star emits.

When astronomers were finally able to measure stellar distances, they realized that they also could calculate the absolute magnitudes of stars. If an astronomer knows both the apparent magnitude and distance to a star, he can easily calculate the real brightness of that star. For example, Polaris is actually quite far away. The distance to Polaris is 240 parsecs. Since Polaris is easily seen in the nighttime sky with the naked eye, in reality Polaris must be an extraordinarily brilliant star, far more brilliant than our own sun. In fact, from knowing Polaris' apparent magnitude ($2\frac{1}{2}$) and distance (240 parsecs), calculations reveal that this star is almost ten thousand times brighter than our sun. Polaris emits almost ten thousand times more light than the sun.

The physically important quantity associated with stellar brightnesses is the *absolute magnitude* rather than the apparent magnitude. By mutual agreement among astronomers, the absolute magnitude of a star is defined as the magnitude that star would have *if* it were at a distance of 10 parsecs from the earth. For example, the apparent magnitude of the sun is $-26\frac{1}{2}$. However, if the sun were at a distance of 10 parsecs, it would look like a fifth magnitude star. Therefore, the absolute magnitude of the sun is 5. Similarly, if Polaris were at a distance of 10 parsecs, it would appear to have a magnitude of $-4\frac{1}{2}$, a little brighter than Venus at maximum. Thus, the absolute magnitude of Polaris is $-4\frac{1}{2}$.

By way of summary, there is a specific relationship between the distance to a star, its apparent magnitude, and its absolute magnitude. If any two of these quantities are known, the third can be calculated. Apparent magnitudes can be measured by simply looking up into the sky and seeing how bright stars appear. Distances can be obtained directly from parallax measurements. Therefore, beginning in the mid-1800s, astronomers were able to discover the absolute magnitudes of stars.

This is, perhaps, the place to interject a seemingly trivial, yet nevertheless important, consideration. The earth's atmosphere absorbs certain types of radiation and, as a result, when the astronomer measures the magnitude of a star, he is really measuring how bright the star is *after* its light has passed through the earth's atmosphere. In addition, very hot stars and very cool stars emit a substantial fraction of their radiation *outside* the visible range of wavelengths. The astronomer would prefer to know how bright the star is in *all* wavelengths and before any of its light is absorbed by the air. The astronomer, therefore, "corrects" his measurements to allow for these effects. The resulting magnitude he obtains is the so-called *bolometric magnitude*. When the astronomer corrects the absolute magnitude of a star to arrive at a number that expresses the total amount or radiation emitted by the star, he obtains the so-called *absolute bolometric magnitude*. For stars like the sun, this *bolometric correction* is very small because most of the radiation from such stars is in visible light that easily passes through the earth's atmosphere. However, for extremely cool or extremely hot stars this correction can be substantial.

Today, astronomers realize that the absolute magnitudes of stars cover a wide range. The brightest known stars have absolute magnitudes of -10, whereas the dimmest stars have absolute magnitudes as low as $+15$. This range of 25 magnitudes corresponds to a factor of 10 billion in energy output. The brightest known stars emit 10 billion times more light than the faintest known stars. By comparison, our sun—with an absolute magnitude of 5—is a fairly dim star. It emits ten thousand times more light than the dimmest stars, but it would take 1 million suns to shine as brightly as the most brilliant known stars.

In conclusion, the ability to measure stellar distances has enabled astronomers to discover the real brightnesses of stars. This is the first fundamental piece of information needed to understand the true nature of the stars we see in the nighttime sky.

7.2 Stellar Spectra

In studying the properties of stars, spectroscopy is perhaps the single most important technique of the astronomer. After aiming a telescope at a star, the astronomer focuses the starlight on the slit of his spectrograph. He then exposes the stellar spectrum on a spectrographic plate, thereby producing a permanent record of his observations. The dimmer the star, the longer the necessary exposure time. In the case of very dim objects, the astronomer may expose the spectrographic plate continuously for many hours to pick up a few faint spectral lines that will give some information about the nature of the star.

To the naked eye of the casual observer, all stars look very similar. Some are bright, others are dim, but they all look like tiny pinpoints of light against the dark background of the nighttime sky. Stellar spectra, however, reveal that individual stars differ widely. Spectra of certain stars contain lines due to gases such as hydrogen and helium; others show many lines produced by metals, and some stars have spectra dominated by broad bands of molecules such as titanium oxide.

In spite of the myriad of stellar spectra, astronomers believe that all stars have roughly the same chemical composition. Hydrogen, the lightest element, is by far the most abundant element. Between 50 and 80 per cent of the mass of a typical star consists of hydrogen. Helium, the second lightest element, is the second most abundant element. Hydrogen and helium together make up from 96 to 99 per cent of the mass of a star. This leaves less than 4 per cent of a star's mass for all the other elements combined. Of all the remaining "heavy elements," the most abundant are neon, oxygen, carbon, nitrogen, calcium, magnesium, chlorine, argon, silicon, sulfur, and iron.

These facts concerning the chemical composition of stars may seem to present a puzzling paradox. How is it possible that stars such as the sun, composed almost entirely of hydrogen and helium, have spectra dominated by the spectral lines of metals? Indeed, in the sun's spectrum, no helium lines show up at all and hydrogen lines are very weak. On the other hand, lines due to calcium are strong in the spectrum of our sun.

In order for the spectral lines of a particular chemical to appear in the spectrum of a star, some of that chemical must be present in that star's atmosphere. However, the strength of a spectral line depends on factors other than simply how much of that chemical is contained in the star's atmosphere. Spectral lines that are prominent and dark are termed *strong*, whereas faint spectral lines are termed *weak*. Only if conditions are favorable will the atoms of a chemical respond to produce strong spectral lines. If conditions are unfavorable, the resulting spectral lines will be weak or absent altogether.

In the discussion of atomic structure (see Section 5.2) it was noted that spectral lines are formed by electrons jumping from one allowed orbit to another. To form an absorption line, an electron jumps from a low orbit, near the nucleus of the atom, to a higher orbit farther from the nucleus. Such a jump requires some energy and the electron extracts this energy at a specific wavelength from the light emitted by the star. The result is a dark space in the spectral rainbow of colors called an *absorption line*.

Hydrogen is the most abundant element in stars. Yet, only a fraction of the stars in the sky show hydrogen lines in their spectra. To understand why this is so, imagine a very cool star with a surface temperature of 3,000°K. At such a low temperature there is not enough energy to raise electrons in hydrogen atoms from low orbits to high orbits. Therefore, even though there is plenty of hydrogen around, the electrons in the hydrogen atoms remain in their ground states and no spectral lines of hydrogen are formed. Conversely, imagine a very hot star with a surface temperature of 25,000°K. Such a star is so hot and its atmosphere contains so much energy that electrons that normally orbit the nuclei of hydrogen atoms have been torn off. This state of affairs is called *ionization*. An atom that is missing one or more electrons is said to be *ionized*. In hot stars, all the hydrogen is ionized; instead of having hydrogen atoms, there are hydrogen *ions* and loose electrons. Since the electrons are not in orbit about the hydrogen nuclei, they cannot produce spectral lines.

It turns out that only when the temperature of the star is between 7,500 and 11,000°K are conditions favorable for the formation of hydrogen absorption lines. Such a star is not too cool; there is enough energy to *excite* electrons from lower orbits to higher orbits in the hydrogen atoms. Such a star is not too hot; there is not enough energy to ionize the hydrogen atoms altogether.

From this illustration, it is seen that temperature is the primary consideration in determining which absorption lines appear in the spectrum of a star.

The second most abundant element in stars is helium. The helium atom contains two electrons in orbit about its nucleus. However, physicists have discovered that these two electrons are held quite tightly to their lowest orbits; it is difficult to excite helium atoms. Therefore, helium absorption lines are seen in the spectra of relatively hot stars whose temperatures lie between about 11,000 and 25,000°K. The temperature in a star's atmosphere must be at least 11,000°K in order to have enough energy to excite helium atoms. In stars hotter than 25,000°K, the temperature is so high that one of the two electrons in helium atoms is torn away. This is referred to as *singly ionized* helium and it produces a

Figure 7–3. Typical Stellar Spectra. The appearance of stellar spectra can vary widely from one star to the next. The upper spectrum is of a very hot star (τ Scorpii) whose surface temperature is about 25,000°K. The middle spectrum is of Procyon (α Canis Minoris) whose surface temperature is about 7,000°K. The lower spectrum is of a cool star (β Pegasi) whose surface temperature is about 3,000°K. (*Hale Observatories*)

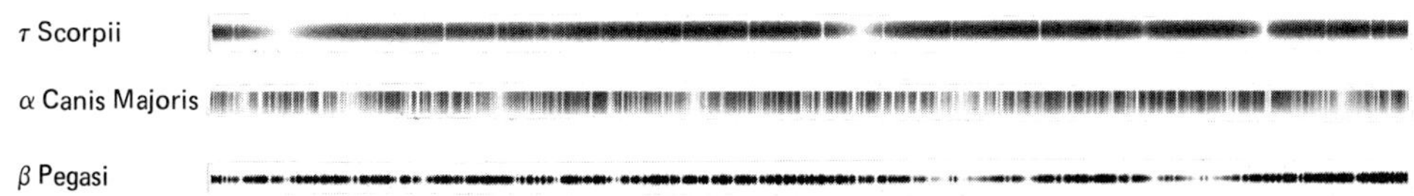

τ Scorpii

α Canis Majoris

β Pegasi

pattern of spectral lines very different from the pattern produced by *neutral* or *unionized* helium. Also, stars with temperatures of more than 25,000°K are so hot that many other types of atoms are ionized. The spectra of such stars show patterns of lines due to *doubly ionized* nitrogen (nitrogen atoms missing two electrons) and *triply ionized* silicon (silicon atoms missing three electrons).

In certain cases it is possible for conditions to be favorable for a very rare chemical to produce spectral lines. For example, the metal titanium is not very abundant in stars. However, if the star is cool, the titanium combines with the oxygen in the star's atmosphere to produce titanium oxide (TiO). Very cool stars with temperatures less than 3,500°K show very strong, broad bands of titanium oxide dominating their spectra. In other words, although such a star is composed mostly of hydrogen and helium, and although titanium is a very rare element, in very cool stars the temperatures are just right for producing molecular spectral bands of titanium oxide. No hydrogen or helium lines are seen in the spectra of such stars; they are far too cool to excite electrons out of their ground status in either of these two abundant elements.

When astronomers first began examining and recording stellar spectra about a century ago, they did not understand the importance of temperature in determining the strength of spectral lines. The nineteenth-century astronomer classified the spectra of stars entirely on their appearance. Letters of the alphabet were assigned to types of spectra, depending on which spectral lines were the most prominent. "A stars" showed the strongest hydrogen lines, "B stars" showed moderately strong hydrogen lines, and so on. However, after the work of Max Planck, Niels Bohr, and other physicists during the first few years of this century, a much deeper understanding was possible. It was realized that temperature is the critical factor in determining the strengths of spectral lines. The final result is that astronomers now recognize seven principal *spectral classes*. These classes still retain their old alphabetical names and are usually listed in order of decreasing temperature as follows: O, B, A, F, G, K, M. The so-called O stars are hot blue stars with surface temperatures in excess of 25,000°K, whereas the M stars are cool red stars with surface temperatures less than 3,500°K. The ordering of letters in this important sequence can be remembered from the first letters in the saying: "Oh, Be A Fine Girl, Kiss Me!"

Almost every star in the sky can be classified into one of these seven categories based on the appearance of its spectrum. This *spectral sequence* from O to M constitutes a continuous progression of stellar temperatures. In going from one class to the next, spectral lines change in strength, in response to temperature. This scheme of *spectral classification* is very important and widely used in modern astronomy. The table on page 193 provides a summary of the spectral features in each of the seven categories.

To supplement this table, Figure 7–4 shows seven stellar spectra, one from each of the seven spectral classes. From examining these spectra, it can be seen how the strengths of spectral lines vary from one class to the next, reflecting the temperatures of stars.

The seven spectral classes can be further subdivided into tenths. For example,

Summary of Spectral Class Features

Spectral Class	Color	Temperature (°K)	Spectral Features	Example Stars
O	Blue	Greater than 25,000°	Singly ionized helium, doubly ionized nitrogen, triply ionized silicon	λ Orionis 10 Lacertae
B	Blue	11,000–25,000°	Neutral helium, singly and doubly ionized silicon, singly ionized oxygen and magnesium; weak hydrogen lines	Rigel Spica
A	Blue	7,500–11,000°	Strong hydrogen lines; some weak lines of neutral and ionized metals	Sirius Vega
F	Bluish white	6,000–7,500°	Moderately strong hydrogen lines; lines of neutral and ionized metals	Canopus Procyon
G	Yellowish white	5,000–6,000°	Very conspicuous ionized calcium; many lines of neutral and ionized metals	Sun Capella
K	Orange	3,500–5,000°	Dominating neutral metals	Arcturus Aldebaran
M	Red	Less than 3,500°	Neutral metals and molecular bands of titanium oxide	Betelgeuse Antares

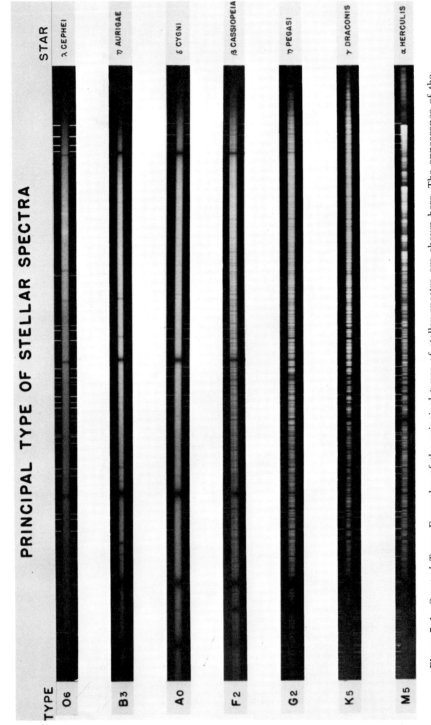

Figure 7-4. Spectral Types. Examples of the principal types of stellar spectra are shown here. The appearance of the spectrum of a star depends on the star's temperature. (*Hale Observatories*)

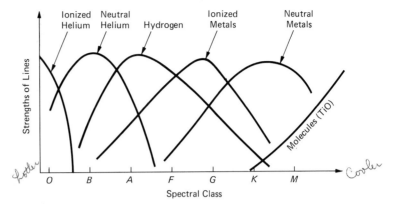

Figure 7-5. *Strengths of Spectral Lines.* This graph shows how the relative strengths of spectral lines vary according to spectral type. The spectra of very hot stars are dominated by helium lines. The spectra of cool stars are dominated by metallic lines and molecular bands.

a B5 star has a spectrum and temperature in the middle of the B range; it is halfway between a B0 and an A0 star. Sirius is an A1 star, which means that its temperature is near the higher end of the range of A-type spectra. Our sun is a G2 star.

From examining the spectrum of a star, the astronomer can make an immediate determination of the surface temperature of that star. All he has to do is see which spectral lines are strong and which are weak. He then compares his spectrum with pictures of spectra such as those shown in Figure 7–4; this gives the spectral class of the star. From understanding how temperature affects the strengths of spectral lines (as summarized in the previous table) the astronomer knows the star's temperature. This is a profoundly important piece of information about any star. Stellar temperatures provide important clues indicating how stars are born, how they grow old, and how they die.

7.3 The Hertzsprung-Russell Diagram

During the first decade of the twentieth century, astronomers realized that their observations were successfully providing some critically important and fundamental data about stars. From distance measurements, astronomers were obtaining the absolute magnitudes of stars; they were measuring the true brightness of stars. From spectral classification, astronomers were obtaining the surface temperatures of stars; they were measuring how hot stars are. As we saw earlier in this chapter, the stars in the sky have a wide range of absolute magnitudes. The dimmest stars have an absolute magnitude of +15, which means that they emit only a ten thousandth as much light as the sun. On the other hand, the brightest known stars have absolute magnitudes of −10, which means that they emit a million times more light than the sun. The range in stellar temperatures is not as striking. The coolest stars we see in the sky have surface temperatures around 3,000°K, whereas the hottest have surface temperatures of about 25,000°K.

In 1911, the Danish astronomer Ejnar Hertzsprung came up with the idea of drawing a graph to compare the behavior of magnitudes and spectral classes of stars. This work was repeated independently by the American astronomer Henry Norris Russell two years later. In many respects, the graph first devised by these two scientists is perhaps the most important diagram in all of astronomy and astrophysics! It is appropriately called the Hertzsprung-Russell diagram.

The Hertzsprung-Russell diagram (often called the H-R diagram) is really nothing more than a plot of the magnitudes of stars against their spectral type. The H-R diagram is always drawn with the magnitude scale on the vertical axis and the spectral class measured on the horizontal axis. Figure 7–6 shows a modern version of the H-R diagram based on many decades of observation. The vertical axis gives the absolute magnitudes of stars (from +15 to −10), and the horizontal axis gives the spectral class of stars (O through M). Each dot represents a star. The location of a dot on the diagram is uniquely determined by the

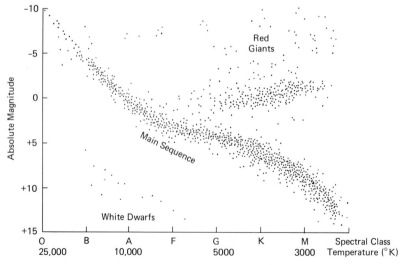

Figure 7–6. *The Hertzsprung-Russell Diagram.* The H-R diagram is simply a graph on which the brightnesses of stars (absolute magnitudes) are plotted against their spectral type (temperatures). Each dot represents a star. Notice that there are three main groupings of stars on the H-R diagram: red giant stars, main sequence stars, and white dwarf stars.

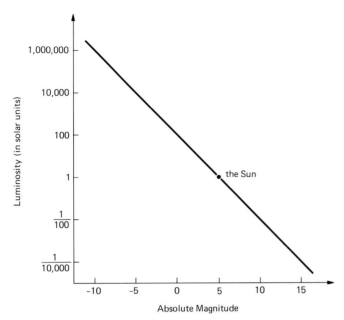

Figure 7-7. *Solar Luminosity and Absolute Magnitude.* Astronomers can speak of the real brightnesses of stars in terms of absolute magnitude *or* solar luminosity. This graph shows the relationship between these two ways of expressing brightness.

absolute magnitude and spectral class of the star that dot is supposed to represent. For example, our sun (spectral type G2 and absolute magnitude +5) is represented by one dot approximately in the middle of the graph.

In drawing a Hertzsprung-Russell diagram, the astronomer has several options. On the vertical axis he can plot absolute magnitude. But the absolute magnitude of a star can be expressed in terms of *solar luminosity:* how many times brighter or dimmer the star is compared to the sun. Figure 7-7 shows the relationship between absolute magnitude and solar luminosity. If the astronomer chooses to measure the brightness of stars on the vertical axis in terms of solar luminosity instead of absolute magnitude, the scale would range from 1/10,000-sun to 1 million suns.

Similarly, on the horizontal axis, the astronomer has several options. Just as in the case of Hertzsprung and Russell, he can use spectral classes (O to M). However, the spectral class of a star is directly related to the star's surface temperature. Therefore, the astronomer can scale the horizontal axis by temperature. In such a case, the temperature ranges from 25,000 down to 3,000°K. There is, however, a third option. Recall that the colors of stars are directly related to stellar temperatures: reddish stars are cool, whereas bluish stars are hot. Indeed the *color index*, B-V, as measured with photoelectric photometers, is related to the (blackbody) temperature of a star, as shown in Figure 7–8. So, the horizontal axis could be scaled according to the color index B-V. Nevertheless, no matter how the astronomer chooses to draw it, the H-R diagram is a graph showing the brightness and temperatures of stars. The H-R diagram is always drawn so that bright stars are near the top, dim stars are near the bottom, hot stars are to the left, and cool stars are to the right.

The strikingly important fact immediately seen from the H-R diagram shown in Figure 7–6 is that the graph is *not* randomly covered with dots. Rather, the dots seem to be bunched together into three groupings. Since each dot represents a star, there are three basic types of stars in the sky. They are called *red giant stars, main sequence stars,* and *white dwarf stars.*

Running diagonally across the H-R diagram is the *main sequence.* Almost every star observable by the naked eye is a main sequence star, including the sun. The main sequence extends from the bright, hot, blue stars in the upper lefthand corner of the diagram down to the cool, dim, red stars in the lower righthand corner.

The second main grouping of stars is the red giants. Every reddish star seen by the naked eye is a red giant. This name very appropriately describes these stars; they are cool and, therefore, reddish. They are also gigantic. Familiar stars such as Arcturus, Betelgeuse, Aldebaran, and Antares are all red giants. Dots representing each of these stars appear in the upper righthand corner of the H-R diagram. If one of these red giants were placed in the center of our solar system, its surface would extend beyond the orbit of Mars!

Finally, there is a third grouping of stars represented by dots in the lower

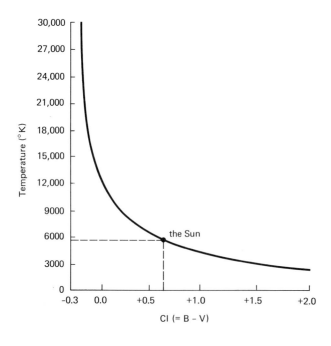

Figure 7–8. Color Index and Temperature. Since the color of a star depends on its temperature, astronomers can speak of a star's temperature or color index (B-V). This graph shows the relationship between these two quantities.

lefthand corner of the H-R diagram. Since they have temperatures of around 10,000°K, they appear whitish. In addition, their dimness indicates their small size. They are, therefore, called white dwarfs. White dwarfs are typically about the size of the earth and are so dim that none can be seen in the nighttime sky without the aid of a telescope.

The data that was used in drawing the H-R diagram shown in Figure 7-6 are based on observations of both the nearest and brightest stars in the sky. This H-R diagram is a powerful tool in astronomy. There is, however, a possible point of confusion that should be cleared up. By way of example, consider the two stars Antares (in Scorpius) and HD 42581 (in Lepus). Both of these stars have exactly the same spectral class; they are both M1 stars and show essentially the same absorption lines in their spectra. However, they are very different. Antares is a red giant with an absolute magnitude of -4.7, and HD 42581 is a main sequence star whose absolute magnitude is 9.33. Antares emits almost a million times more light than HD 42581. One of the primary reasons for this is that Antares is much bigger than HD 42581. Indeed, Antares is almost a thousand times larger than the small main sequence star HD 42581. This, in turn, means that the conditions in the atmospheres of these two stars, where the absorption lines are formed, are very different. Due to its small size, HD 42581 is very compact, and the atmospheric pressure at its surface is high. By contrast, Antares has its matter spread over a much larger volume of space, and therefore, its atmospheric pressure is very low. This effect shows up in the appearance of spectral lines. Although the same patterns of lines occur in the spectra of both stars (they have the same spectral class and, therefore, the same temperature), the shapes and relative strengths of these spectral lines are different in the two stars (they have different luminosities and different atmospheric pressures).

Astronomers have studied how the absolute magnitudes of stars affect the appearance of spectral lines. In the early 1940s, Drs. Morgan, Keenan, and Kellman summarized these effects in a famous catalogue, *An Atlas of Stellar Spectra*. A page from this atlas is shown in Figure 7-9. All the spectra shown in Figure 7-9 are of stars with the *same* spectral class; they are all A0 stars and all have a surface temperature of 10,000°K. But, even though the same spectral lines show up in the three spectra, the appearance of the lines differs from one star to the next. The differences in the appearance of these spectral lines are a direct result of the different physical conditions at the surface of the stars. Small, dense stars with high atmospheric pressure tend to have broad spectral lines. Due to the high pressure, the atoms in the star's atmosphere suffer many collisions and the lines are widened by collisional broadening. On the other hand, giant stars with low atmospheric pressures tend to have sharper lines because collisions between atoms occur less frequently in the rarified gases of these stars.

When an astronomer takes a spectrum of a star, he is in a position to extract a lot of information very quickly. He first of all determines which spectral lines are present. This gives him the *spectral class* of the star. For example, if the astronomer sees strong helium lines, he must have a B star; if the spectrum contains

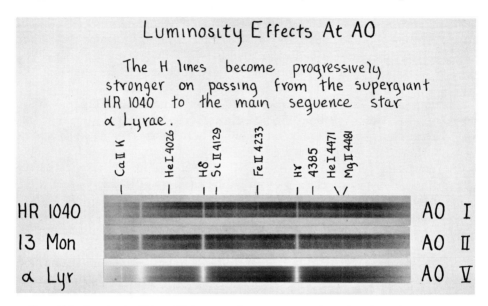

Figure 7–9. Luminosity Effects. All the stars whose spectra are shown here have the *same* spectral type or surface temperature. The spectra appear different, however, because the stars have very different absolute magnitudes due to the fact that they have very different sizes. (*Yerkes Observatory*)

titanium oxide bands, he must be observing an M star. He then examines the relative strengths and widths of these spectral lines. By comparing his spectrum with standard spectra in the Morgan-Keenan-Kellman atlas, he determines the *luminosity class*. This tells him if he is observing a giant star or a dwarf star, or something in between. Since the astronomer now knows both the spectral and luminosity classes of his star, he then has a fairly accurate idea of where this star belongs on the H-R diagram. The punch line is that from the location of the star on the H-R diagram, the astronomer can now easily calculate the distance to the star. This method of distance determination is called *spectroscopic parallax*.

The method of spectroscopic parallax can be summarized as follows:

1. The astronomer takes a spectrum of a star.
2. From seeing which lines are present, he knows the spectral class of the star, which tells him the star's temperature.
3. From examining the strengths and widths of the spectral lines, the astronomer determines the luminosity class of the star. This tells him where the star belongs on the H-R diagram, from which he can read off the absolute magnitude of the star.
4. The astronomer then measures the apparent magnitude of the star just by seeing how bright it appears in the sky.
5. Since he now knows both the absolute magnitude of the star and the apparent magnitude of the star, he calculates the distance to the star.

This is one of the indirect methods of distance determination mentioned earlier. Spectroscopic parallax works well as long as the star under consideration is "normal." If the star is abnormal (for example, if it has an unusual abundance of elements or it ejects shells of matter into space), this method will not work because the stellar spectrum will not fit neatly into the Morgan-Keenan-Kellman scheme.

There is a second powerful method of distance determination made possible by the H-R diagram that is worth mentioning. This so-called *method of main sequence fitting* allows the astronomer to determine the distances to entire clusters of stars without ever going to the trouble of taking a single spectrum.

With the aid of telescopes, astronomers find many clusters of stars in the sky. These star clusters generally fall into one of two types. The *open clusters*, sometimes called *galactic clusters*, are fairly loose groupings of stars. The *Pleiades*, shown in Figure 7–10, is a good example of a typical open cluster. On the other hand, there are *globular clusters* such as M3, shown in Figure 7–11. Globular clusters derive their name from their obvious spherical appearance.

One of the most efficient ways an astronomer can study a cluster is to use

Figure 7–10. An Open Cluster. The Pleiades in the constellation of Taurus is an example of an open or galactic cluster. Young stars are found in open clusters. (*Lick Observatory*)

Figure 7-11. A Globular Cluster. M3 (also called NGC 5272) in the constellation of Canes Venatici is an example of a globular cluster. Very old stars are found in globular clusters. (*Lick Observatory*)

photoelectric photometry. The astronomer attaches his photometer to his telescope and, with the aid of his standard set of filters, begins measuring U, B, and V magnitudes of individual stars in the cluster. From such observations, he can immediately obtain the color index B-V for each star. He then can express the results of his observations in terms of a *color-magnitude diagram*, such as those shown in Figures 7-12 and 7-13. A color-magnitude diagram of a cluster is simply a graph on which the astronomer has plotted the V magnitude of the stars against their color indices. Notice that the resulting color-magnitude diagrams look suspiciously like H-R diagrams. Indeed, recall that the color index of a star is directly related to the temperature and spectral class of that star. Therefore, the horizontal axis of a color-magnitude diagram is the same as the horizontal axis of a H-R diagram. However, the verticle axes in these two types of diagrams are *not* the same. In the color-magnitude diagram, the vertical axis measures V (apparent) magnitude, while the vertical axis of the H-R diagram measures absolute magnitude. This is where the main sequence fitting technique comes in.

The astronomer can draw his color-magnitude diagram of the cluster he is studying on a transparent piece of paper. He can then place the resulting graph

Chapter 7 The Nature of Stars

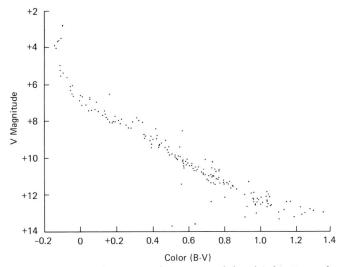

Figure 7–12. *Color-Magnitude Diagram of the Pleiades.* From observing the V magnitude and the color index of each star in the Pleiades with his photometer, the astronomer can draw a color-magnitude diagram for the cluster. (*Adapted from Drs. Johnson and Mitchell*)

Figure 7–13. *Color-Magnitude Diagram of M3.* From observing the V magnitude and the color index of each star in M3 with his photometer, the astronomer can draw a color-magnitude diagram for the cluster. (*Adapted from Drs. Sandage and Johnson*)

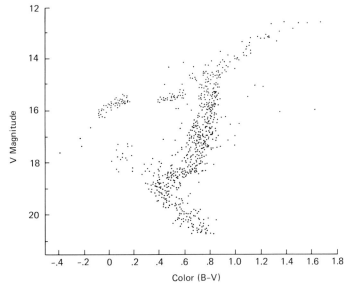

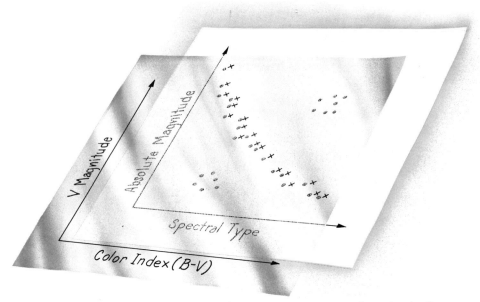

Figure 7-14. *The Main-Sequence Fitting Technique.* An astronomer can determine the distance to a cluster by placing a color-magnitude diagram over a standard H-R diagram such as that in Figure 7-6. He moves the color-magnitude diagram around until the main-sequences on the two diagrams coincide. He then knows which V magnitudes (from the color-magnitude diagram) correspond to the absolute magnitudes (from the standard H-R diagram).

on top of a graph of the H-R diagram. By sliding the transparent piece of paper up and down (in magnitude), the main sequence on the color-magnitude diagram can be made to coincide with, or *fit* directly over, the main sequence on the H-R diagram. The astronomer then knows which apparent magnitudes (from stars on the color-magnitude diagram) correspond to which absolute magnitudes (from standard stars on the H-R diagram). Since he now has both the apparent and absolute magnitudes of the stars in his cluster, the distance to the cluster can be calculated easily.

In examining the color-magnitude diagrams for the Pleiades and for M3, notice that they look very different. The diagram for the Pleiades clearly shows the main sequence, but there are *no* red giants. By contrast, the diagram for M3 shows only a small piece (the lower end) of the main sequence and there are lots of red giants. What is the meaning of these differences? Why are some stars main sequence stars, while others are red giants or white dwarfs? These are among the most important and fundamental questions astronomers have ever asked.

7.4 Our Sun—A Typical Star

The sun is the only star in the sky that astronomers can observe at close quarters. It is the only star on which surface features can be seen easily; and in view of its proximity, more detailed information is available about the sun than any other star in the sky. The sun's spectral class (G2) and absolute magnitude (+5) place it directly on the main sequence in the H-R diagram. The sun is a typical, garden-variety star. Astronomers, therefore, believe that, from studying the sun, a great deal can be learned about main sequence stars in general.

Only the outer layers of the sun can be seen with the naked eye. These outer parts of the sun are called the *solar atmosphere*. Astronomers find it convenient to divide the atmosphere into three distinct regions, which have very different properties. These three layers in the solar atmosphere are called the *photosphere*, *chromosphere*, and *corona*.

Virtually all the light received from the sun comes from the photosphere. Indeed, the word *photosphere* means "sphere of light." Looking at the sun under normal conditions we are seeing the solar photosphere. Figure 7–15 is a photograph of the photosphere.

The first detailed observations of the photosphere date back to the time of Galileo. With the aid of his telescope, Galileo noticed that occasionally there were dark spots on the sun. By observing these so-called *sunspots* from day to day, astronomers were able to figure out how fast the sun rotates. Surprisingly, the sun does *not* rotate like a rigid body such as the earth. Sunspots near the solar equator take 25 days to go once around the sun. By contrast, sunspots at latitudes 30° north or south of the equator take $27\frac{1}{2}$ days to go once around the sun, whereas at latitudes of 75° sunspots take 33 days to return to their starting

Figure 7–15. *The Sun.* Most of the light from the sun comes from the solar photosphere. An ordinary photograph of the sun is a photograph of the photosphere. Sunspots are often seen in such photographs. (*Hale Observatories*)

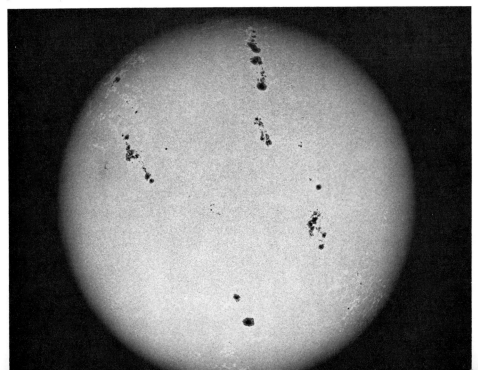

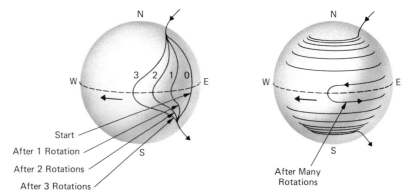

Figure 7-16. *Differential Rotation of the Sun.* By observing sunspots at different locations on the sun, astronomers realized that the sun does not rotate like a rigid body. Rather, the sun rotates faster near the equator than near the poles, as shown schematically in this diagram.

place. At the north and south poles, the rotation period may be as long as 35 days. This phenomenon is called *differential rotation;* the sun rotates faster at the equator than at the poles.

Sunspots are not permanent features on the sun. Rather, they start off as a small blemish on the photosphere called a *pore,* grow to maximum size that can be tens of thousands of miles across, and then fade away. This entire process takes several months. In addition, the number of sunspots seen on the photosphere varies. Sometimes the sun is covered with literally hundreds of spots, whereas at other times no sunspots can be seen.

Since the time of Galileo, astronomers have been counting the number of

Figure 7-17. *The Sunspot Cycle.* The number of sunspots seen on the sun varies with a period of 11 years. Sunspot maxima occurred in 1948, 1959, and 1970. The sun is almost devoid of sunspots in 1954, 1965, and 1976.

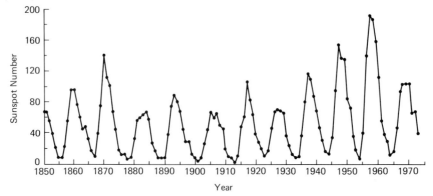

sunspots observed on the solar surface. By the mid-1800s it had become clear that the number of sunspots varies periodically. This phenomenon is called the *sunspot cycle*. Over the course of one sunspot cycle, which lasts about 11 years, the number of sunspots goes from a maximum (hundreds of spots) to a minimum (virtually no spots) back to a maximum again. The time of the greatest number of sunspots is called *sunspot maximum* (in 1948, 1959, 1970, and so on), and the time of the least number of sunspots is called *sunspot minimum* (in 1954, 1965, 1976, and so forth).

Figure 7–18 shows an excellent photograph of a sunspot. This photograph was taken from a balloon flown at a high altitude to overcome the limitations imposed by the earth's atmosphere. It seems that a sunspot consists of two parts: a dark, inner region called the *umbra* surrounded by a lighter region called the *penumbra*.

In spite of their appearance on photographs, sunspots are *not* black. They look dark only in contrast to the neighboring photosphere. From studying the light emitted by sunspots, astronomers have discovered that sunspots are quite cool compared to the average temperature of the solar surface. The normal temper-

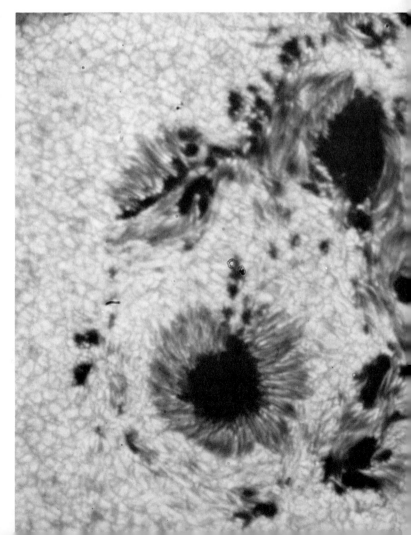

Figure 7–18. *A Sunspot Group.* This photograph of a group of sunspots was taken from a balloon carrying a telescope. The darkest areas of a sunspot are called the umbra, while the surrounding greyish regions are the penumbra. (*Princeton University*)

ature of the photosphere ranges from a minimum of 4,500 to 6,800°K at a depth of about 160 miles. Spectral analysis of the light emitted by sunspots shows them to be about 1,500°K cooler than the surrounding photosphere.

Studying the spectra of light from sunspots also has revealed another important fact. In 1908, the American astronomer George E. Hale noticed Zeeman splitting of the spectral lines in sunspot spectra. This means that there must be very intense magnetic fields associated with sunspots. The average strength of the background magnetic field of the sun is roughly the same as the strength of the earth's geomagnetic field. However, inside a sunspot, the magnetic field can be hundreds or even thousands of times more intense than the average background strength. This discovery provided the first important clue in explaining the cause of sunspots.

According to a theory proposed in the early 1960s by H. A. Babcock, the differential rotation of the sun causes the sun's magnetic field to become very twisted. As a result of the fact that the sun rotates faster at the equator than at the poles, the solar magnetic field becomes so entangled that it erupts through the photosphere. At a location where the magnetic field has ruptured the photosphere, the boiling and bubbling of the sun's atmosphere is impeded. The hot gases in the sun cannot boil as violently as usual in such a location. This area is, therefore, cooler than the surrounding photosphere; a sunspot has formed.

An important instrument for observing the sun is the *monochromer*. A monochromer is simply a device, often consisting of just a filter, that allows the astronomer to look at the sun at *one* specific wavelength. The most fruitful observations are made at wavelengths of 6,563 Å (in the middle of an absorption line of hydrogen) and 3,934 Å (in the middle of an absorption line of singly ionized calcium). When an astronomer uses a monochromer (from the Latin meaning "one color"), *all* the light from the sun is blocked out *except* light at the desired wavelength. It turns out that the light emitted by the sun at 6,563 and

Figure 7–19. Three Views of the Sun. All three photographs shown here were taken on the same day. At the left is a white-light view. The middle photograph shows the sun as seen in the light of the hydrogen atom while the appearance of the sun in the light of the calcium atom is seen at the right. (*Hale Observatories*)

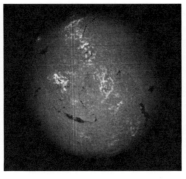
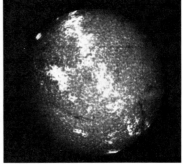

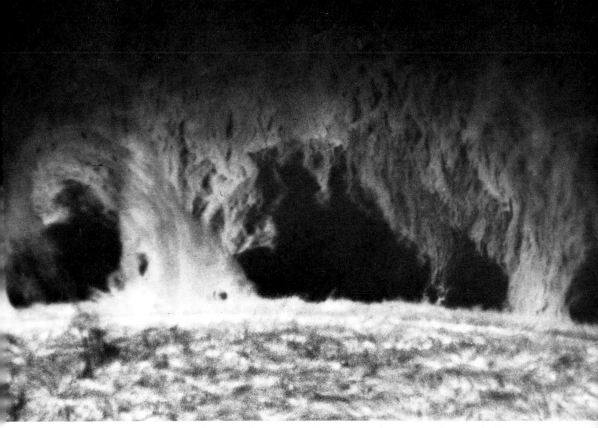

Figure 7–20. *A Prominence.* This magnificent view of a solar prominence was photographed on March 31, 1971, in the red light of the hydrogen atom. The prominence rises to an altitude of 40,000 miles above the solar surface. (*Big Bear Observatory*)

3,934 Å does *not* come from the photosphere. Rather, the light at these wavelengths comes from higher altitudes in the solar atmosphere, from the so-called *chromosphere.* In other words, when the astronomer takes a photograph of the sun through a monochrometer, he gets a picture of the chromosphere, *not* the photosphere. Photographs taken in the red light of the hydrogen atom at 6,563 Å are called *H-alpha photographs* or *H-alpha filtergrams,* whereas those taken in the violet light of the calcium atom are said to be photographed in the *calcium K-line.* Figure 7–19 shows three views of the sun taken on the same day. The first view is just a white-light photograph showing the photosphere. The second and third photographs were taken in the light of hydrogen and calcium, respectively, and show the structure of the sun's upper atmosphere. So-called *plages* (bright areas) and *filaments* (dark areas) are seen.

Photographs of the edge of the sun, or *solar limb,* in the light of calcium or hydrogen atoms are particularly interesting. Such filtergrams show dramatic *prominences* that consist of huge jets of gas gushing up hundreds of thousands of miles out of the solar surface. Prominences can take many different and beautiful shapes that can change very rapidly as the gas rises and falls. Figures 7–20, 7–21, and 7–22 show some different views of typical prominences.

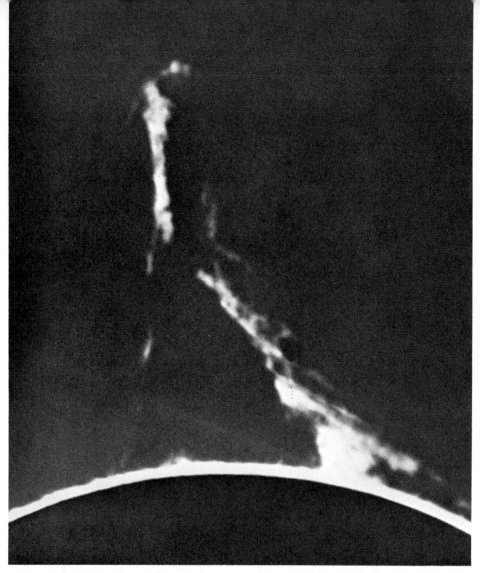

Figure 7–21. A Prominence. This prominence was photographed in 1958 in the violet light of the calcium atom. The prominence extends over half a million miles above the sun's surface. (*Hale Observatories*)

Careful study of the sun at various wavelengths, even in the ultraviolet and X-ray region of the electromagnetic spectrum (see Figure 7–22), reveals that many of the phenomena astronomers observe are closely related. For example, prominences appear bright against the dark sky. But, if an astronomer "looks down" on a prominence and sees it against the sun's disk, by contrast the prominence appears dark and is called a *filament*. Prominences and filaments are the same thing. In addition, plages and prominences (or filaments) often occur over sunspots. The twisted magnetic field that erupts through the photosphere and produces a sunspot also pushes gas thousands of miles above the sun's surface giving rise to prominences. This indicates that sunspots are only *one* phenomenon associated with the life cycle of a so-called *activity center* on

Chapter 7 The Nature of Stars

the sun. As a piece of the twisted magnetic field of the sun bursts through the solar surface, a whole range of phenomena are produced: plages, sunspots, filaments, prominences, and even X-ray bursts. These phenomena, associated with an activity center, typically last as long as 270 days, or ten solar rotations.

Finally, the sun's *corona* constitutes the outermost regions of the solar atmosphere. The corona is seen best at the time of a total solar eclipse when the blinding disc of the sun is blocked by the moon. Spectroscopic studies of the light of the corona reveal that the temperature of this outermost part of the sun's atmosphere is between 1 and 2 million degrees! Solar astronomers believe that noise (so-called *acoustic energy*) from the violent bubbling and boiling of the gases in the sun's surface heats up the thin gases in the corona to incredibly high temperatures. In other words, the sounds produced by the boiling sun give rise to the beautiful delicate corona we see during a solar eclipse.

Although the study of the sun's surface has given astronomers insight into the appearance of a typical star, one of the most important discoveries to come from

Figure 7–22. *The Invisible Sun.* This remarkable photograph was taken by astronauts on board Skylab in 1973. This view shows the appearance of the sun in the light of singly ionized helium in the far ultraviolet ($\lambda = 304$ Å). A beautiful prominence arching hundreds of thousands of miles into space is seen. (*NRL/NASA*)

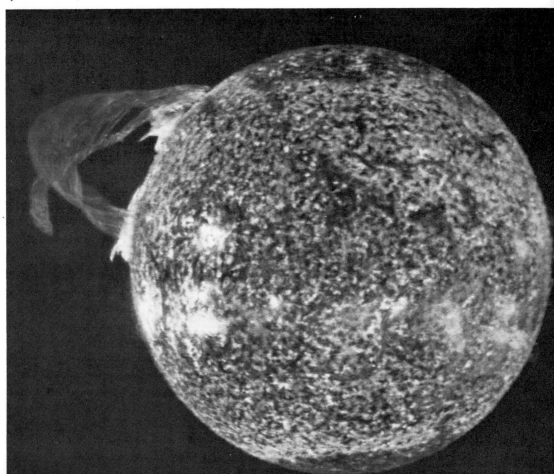

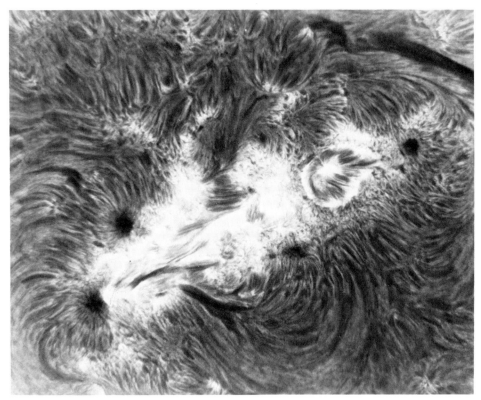

Figure 7–23. *An Activity Center.* Sunspots are only one of many phenomena associated with activity centers. Flares, plages, prominences, and filaments occur around activity centers like the one shown in this photograph. (*Big Bear Observatory*)

the sun has to do with processes that cannot be seen at all. In the 1900s geological evidence began to accumulate that showed that the earth was very old. In fact, it appeared that the earth must be several billion years old. Astronomers and physicists then began to realize that they had a real dilemma on their hands. By any physical process they knew, they could not explain how the sun could be older than a few hundred million years. In other words, it seemed as though the sun must be much younger than the earth. If an astronomer made the reasonable assumption that the sun and the earth were formed roughly at the same time, he was forced to the absurd conclusion that the sun should have burned out long ago. The only logical explanation is that science really could not explain why the sun shines!

Several decades passed before physicists were able to piece together a number of remarkable discoveries in nuclear physics and arrive at a satisfactory explanation for the sun's energy. Recall that early in the twentieth century, Rutherford proposed a model of the atom like one for a solar-system. Electrons in an atom were thought to be in orbit about a dense nucleus like planets around our sun. This model of the atom has been so successful that even today all the properties of matter can be explained in terms of ideas coming from this one simple picture of the atom.

It should be emphasized that all the properties of matter encountered in our everyday experience have to do *only* with the manner in which electrons orbit the nuclei of atoms. For example, the way in which atoms combine chemically to form molecules has to do only with the electrons in these atoms. Such topics have virtually nothing to do with the nuclei of atoms. Nevertheless, with the discovery of radioactivity, physicists began studying the properties of the atomic nucleus. Nuclei are composed of protons and neutrons bound together at the centers of atoms by incredibly strong forces called *nuclear forces*. The nature of these forces, which hold protons and neutrons together, was unknown to science until nuclear physicists began trying to break nuclei apart with machines commonly called atom smashers (that is, cyclotrons, synchrotrons, betatrons, and the like). Scientists had discovered that there must be an incredible amount of energy contained in the nuclei of atoms.

In the 1920s, based on the new awareness of the properties of nuclei, a radically new scheme was proposed to account for the source of energy in the sun. This scheme involved a process called *thermonuclear fusion*, in which the nuclei of light elements combine together to produce heavier elements. In this fashion heavy elements are created by fusing together the nuclei of lighter elements.

The simplest of all thermonuclear reactions involves the transformation of hydrogen (the lightest element) into helium (the second lightest element). Under exceptional conditions of incredibly high temperatures and densities, four hydrogen nuclei can be fused together to produce one helium nucleus. What is so remarkable here is that in the process some matter seems to "disappear." The ingredients weigh more than the end-product. Four hydrogen nuclei weigh a little more than one helium nucleus. In this thermonuclear reaction, a small amount of the mass of the ingredients has been converted into energy, according to Einstein's famous equation $E = mc^2$. In other words, take the amount of matter that seems to have disappeared, multiply it by the square of the speed of light (c^2), and it will equal the amount of energy released by the reaction. Since the speed of light is such a big number, physicists saw immediately that such a reaction should produce an incredible amount of energy.

Further calculations by physicists revealed that the temperature and density at the sun's center should be high enough for this thermonuclear reaction to occur easily. And since the sun is made mostly of hydrogen, there is enough nuclear "fuel" to cause the sun to shine for many billions of years. Astronomers were finally able to explain why the sun shines. Hydrogen is converted into helium at the sun's center with the accompanying release of huge amounts of energy. It is believed that 600 million tons of hydrogen are converted into helium *each second* to produce the sunlight that makes life possible here on earth.

Several decades after astrophysicists conceived of thermonuclear reactions as the source of the sun's energy, scientists discovered that they could make these reactions occur here on the earth. The device that does this is called a *hydrogen bomb.* In the detonation of a hydrogen bomb, hydrogen is converted into helium, with the resulting production of a huge amount of energy. Exactly the same reaction that causes the sun to shine also occurs during the explosion of a hydrogen bomb. Exactly the same physical process that has made the earth a place where life could develop and evolve can be used to reduce our planet to one of the most barren and desolate pieces of rock in the solar system.

Up until the early 1970s, astronomers felt very secure in their understanding of thermonuclear reactions as the source of the sun's energy. Everything fit together very well; everything made a lot of sense. But now there seems to be a big problem that may have some profound implications.

As nuclear physicists continued to explore the nuclear properties of matter, they began discovering a host of *elementary particles.* They found that there were many more particles besides electrons, protons, and neutrons. In breaking up nuclei with their atom smashers, these physicists found that many different particles could be produced (for example, pions, muons, K° particles, and so on). In particular, as nuclear physicists studied the reactions that are supposed to occur in the sun (4H → He) they realized that particles called *neutrinos* should be released in great quantities. Neutrinos are unusual particles and are perhaps most easily compared with photons of light. A neutrino has no mass and no electric charge, just as a photon is massless and chargeless. Also, neutrinos travel at the speed of light, just like photons. However, neutrinos have the unusual property of easily passing through matter. Neutrinos easily pass right through the earth, or even through the sun, with no difficulty at all. Matter is virtually transparent to these tiny particles. Neutrinos have been detected in the laboratory and there is no question that the reactions in the sun should be producing neutrinos. If physicists understand anything about nuclear reactions, every time four hydrogen nuclei are fused into one helium nucleus, two neutrinos should be produced.

Since neutrinos have the unusual property of easily traveling through great quantities of matter, astronomers thought that they might have a method of actually "seeing" the center of the sun. If astronomers could build "neutrino telescopes" they might be able to look directly at the sun's core, since neutrinos travel directly from the center of the sun to the earth with no difficulty at all. Ordinary photons of light take about a million years to get from the sun's core to the solar surface; photons suffer many collisions with the atoms in the sun in making the arduous journey from the core to the surface. By contrast, neutrinos pass directly through the sun, almost as though it were not there.

Beginning in the late 1960s, a team of scientists under the direction of Dr. R. Davis of Brookhaven National Laboratory, built the first neutrino telescope. It consisted of a huge 100,000-gallon tank of cleaning fluid (tetrachloroethylene) buried deep in the earth in South Dakota. The central idea is that, on very rare occasions, a neutrino from the sun should hit the nucleus of a chlorine atom in the cleaning fluid. Nuclear physicists tell us that the chlorine atom should be turned into an argon atom in such a reaction. Although this nuclear

Figure 7–24. A Neutrino Telescope. In theory, this large tank of cleaning fluid should be very efficient in capturing neutrinos from the nuclear reactions in the sun. In fact, very few neutrinos have been detected. The "case of the missing neutrinos" is one of the greatest mysteries in modern astronomy. (*Courtesy of Dr. Davis*)

reaction is very rare (remember, neutrinos pass through almost everything with no interaction at all), there are so many chlorine atoms in 100,000 gallons of cleaning fluid that Davis honestly expected to find some argon gas in his tank. After many years of careful experimentation, Dr. Davis has been able to detect only a small fraction of the argon gas he initially expected. Something is *very* wrong.

If hydrogen is being converted into helium in the sun's center, lots of neutrinos should be bombarding the earth. Some of these neutrinos should produce argon in Davis' experiment. Scientists can calculate exactly how much argon should be produced over a period of time in 100,000 gallons of cleaning fluid. Yet, only one fifth of the anticipated amount of argon has ever been found. What could be wrong?

Although this may sound like a bizarre experiment, the careful observations by Davis seem to be destined to have far-reaching implications. Astronomers are now grappling with novel ideas in an attempt to account for the fact that we do not "see" neutrinos from the sun. For example, Dr. W. Fowler, at the California Institute of Technology, has proposed that perhaps the sun has turned off! Perhaps, for unusual reasons, the thermonuclear reactions at the sun's core stop for short periods of time. This, of course, would explain why Davis does not detect neutrinos; they simply are not being produced. But, in turn, Fowler's ideas tell us that the sun will gradually cool and become dimmer than it is today. This means that we could now be at the beginning of a massive ice age! As the sun cools, the average temperature here on earth will decrease and the polar caps will grow and cover much of our planet. Could this be the explanation of the ice ages that periodically overtake our planet?

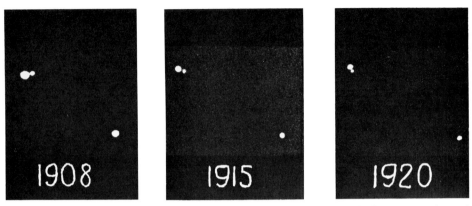

Figure 7–25. Kruger 60. Careful observations over many years often reveals that two stars located near each other in the sky are actually in orbit about their common center. Kruger 60 is a good example of such a binary star system. (*Yerkes Observatory*)

This so-called *neutrino problem* is one of the hotly debated issues of modern astronomy. If nuclear physics is correct in its predictions concerning neutrinos, something very peculiar is going on in the sun. But, perhaps nuclear physics is wrong. Perhaps there is something wrong with Davis' experiment that has gone unnoticed. By the end of the 1970s, understanding the cause of the neutrino problem might produce a new level of insight into the nature of physical reality, a level that could well have profound implications for the future of life on earth.

7.5 Binary Stars

It might come as a surprise that many of the stars seen in the sky are not just individual stars. Rather, about half of the stars in the sky are actually *double stars* or *binaries*, two stars revolving about each other, held together by the force of gravity. Binary stars have played a critical role in astronomy because, from studying the motions of the two stars in a binary, astronomers (under ideal conditions) can determine the masses of these stars. From binary stars astronomers have been able to discover how much matter goes into making up a star.

The existence of binary stars was not certain until 1804. In that year, the astronomer William Herschel succeeded in observing the orbital motions of the two stars that make up Castor (β Geminorum) in the constellation of the Twins. In comparing his observations with earlier ones, Herschel noted that the two stars in this famous binary had changed their orientation, just as if they were revolving about each other. By the end of the nineteenth century, William Herschel's son, John Herschel, had discovered and catalogued thousands of similar double stars.

Chapter 7   The Nature of Stars   217

   The discovery of the first known binaries illustrates an obvious method of finding double stars: the astronomer simply looks for two stars that are very close together. This is done most easily with a good telescope that has a high resolving power. Many of the stars seen in the sky (such as Castor) are resolved into two stars when viewed through a telescope. The astronomer then records the *separation* and *position angle* of the two stars. The separation tells how far apart the stars appear, while the position angle measures the orientation of the two stars in the sky. After a period of time (months or years) the astronomer again observes these stars to see if any changes have occurred. If he is seeing a true binary and not just two stars that happen to be lined up in the sky, one star will appear to be revolving about the other. Figure 7–25 shows a series of photographs of the binary called Kruger 60 taken during a period of several years and illustrating the kind of motions the astronomer sees.
   If the astronomer can actually see both stars in a binary system, he is observing a so-called *visual binary*. Many years of painstaking observations permit astronomers to draw the path of one star about the other. Figure 7–26 is a good example of such a drawing. The important point of such a series of observations is that it enables a determination of the mass of the binary. Just as it it possible to

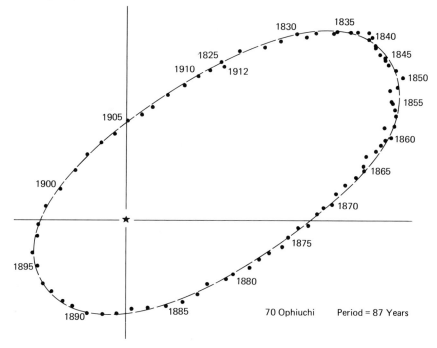

Figure 7–26. *A Binary Orbit.* By measuring the separation and position angle of the two stars in a binary system, the astronomer can draw their orbit. The orbit of 70 Ophiuchi, a visual binary whose period is 87 years, is shown here.

calculate the orbit of the moon around the earth from Newtonian mechanics, the astronomer can apply these same mathematical tools to the orbit of one star about the other in a binary. The final result is that the mass of the double star system can be determined.

Often the astronomer can see only one star in a binary system; the second star might simply be too faint to be seen, or perhaps the two stars are so close that they cannot be resolved. In such cases, the astronomer can be sure that he is observing a double star by one of several techniques.

Astronomers often take spectrograms of stars. On occasion, the astronomer will notice that the spectrum of a star shows *two* sets of spectral lines, indicating that two stars are being observed. In addition, spectrograms taken at different times will reveal that these two sets of lines shift back and forth periodically. This is a *spectroscopic binary*. Although the astronomer may see only one star through his telescope, he is actually detecting two stars in orbit about each other. Due to the Doppler effect, when one of the stars is coming toward him, its spectral lines are shifted toward the blue, while the other star, which is moving away, has its lines shifted toward the red. Half a revolution later, the roles of the two stars are reversed, and their spectral lines have exchanged their locations in the spectrum. Figure 7–27 shows a good example of a spectroscopic binary. In these two spectra, spectral lines can be seen shifting back and forth as the two stars alternately approach and recede from the earth. The whole point of these spectroscopic observations is that the Doppler effect tells the astronomer some-

Figure 7–27. A Spectroscopic Binary. The star Mizar in the Big Dipper (Ursa Major) is a spectroscopic binary. Two spectra are shown. In the lower view all the spectral lines are doubled. One set of lines is from the approaching star (slightly blueshifted) and the other set is from the receding star (slightly redshifted). (*Hale Observatories*)

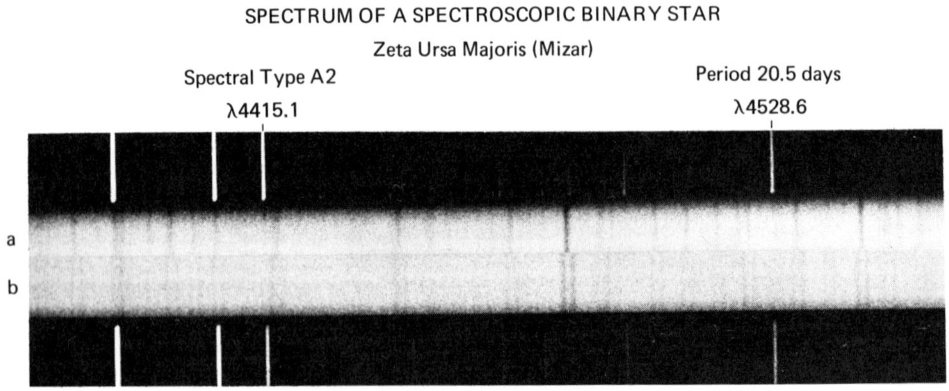

SPECTRUM OF A SPECTROSCOPIC BINARY STAR

Zeta Ursa Majoris (Mizar)

Spectral Type A2   Period 20.5 days
$\lambda 4415.1$ $\lambda 4528.6$

(a) June 11, 1927. Lines of the two components superimposed. (One star is moving to the left while the other star is going to the right)

(b) June 13, 1927. Lines of the two components separated. (One star is coming toward the earth while the other star is receeding)

thing about how fast the stars are moving in their orbits. By applying Newtonian mechanics to the speeds of the stars, the astronomer can then calculate information concerning the actual masses of the stars.

Sometimes an astronomer will see only one set of lines in the spectrum of a star, but that set of lines will be observed to shift back and forth. He, therefore, concludes that he is again seeing a binary, except that the second star is far too dim to be detected. Such a system is called a *single-line spectroscopic binary*. One of the most important stars being studied by astronomers in the 1970s is a single-line binary called HDE 226868. As we will see later, its companion star is a black hole.

And still on other occasions, an astronomer might notice that the spectrum of a star shows two sets of lines that do not appear to shift back and forth. For example, he might see the spectrum of an A star on top of the spectrum of a K star. Since a single star cannot have two different spectra at the same time, he must be observing two different stars. Yet, in such a case, the stars may be so far apart that they move very slowly about each other. Since their motions are so slow, taking perhaps hundreds of years to complete one revolution, the Doppler effect does not produce any noticeable shifting of the lines. Such a system is called a *spectrum binary*.

A second method for detecting unresolved binaries involves a little luck. Imagine observing two stars revolve about each other in a binary, which just happens to be oriented in such a way that the earth lies very nearly in the orbital plane of the binary system. Under such circumstances one star will alternately pass in front of the other star producing *eclipses*. Such a system is called an *eclipsing binary*.

An astronomer knows that he is observing an eclipsing binary (even though he cannot see two separate stars) from the characteristic way in which the brightness of the system varies periodically with time. As shown schematically in Figure 7–28, the total amount of light from an eclipsing binary will show dips when one star passes in front of the other. To record this variation in brightness, the astronomer constructs a *light curve*, which is just a graph showing the magnitude of the system as a function of time. If the two stars in question have different magnitudes, successive dips in the light curve will have different depths. When the dim star is in front of the bright star, the total brightness of the binary will be reduced substantially, producing a so-called *primary minimum*. But when the bright star is in front of the dim star, the total brightness will be reduced only slightly, resulting in a *secondary minimum*. From examining the light curves of eclipsing binaries the astronomer can, in principle, determine a wide range of information, such as the sizes of the stars and the orientation of the orbit of the binary. If the eclipsing binary also happens to be a spectroscopic binary, by correlating information from the light curve with the speeds of the stars in their orbits, the astronomer can calculate the masses of each of the stars.

The important point is that it is possible to learn about the masses of stars from observing binaries. In fact, this is the *only* way astronomers can discover how much matter is in a star. If a star is just sitting alone by itself in space, the astronomer has no way of measuring the star's mass. But, if he sees two stars orbiting each other, he can use Newtonian mechanics to learn about their masses.

From studying binary systems, astronomers have been able to obtain fairly accurate measurements of the masses of a few dozen stars. In examining these stars, one very important fact has emerged: the masses of main sequence stars are directly related to their absolute magnitudes. Dim main sequence stars have low masses, whereas bright main sequence stars have high masses. Main sequence stars, therefore, follow this rule: the brighter the star, the bigger is its mass. This correlation between brightness and mass is called the *mass-luminosity relation*.

The mass-luminosity relation can be illustrated best with the aid of a graph, as shown in Figure 7–29. On the vertical axis of the graph the brightness of stars is measured (absolute magnitude or luminosity). On the horizontal axis the masses of stars are measured. For convenience, this axis is scaled in *solar masses*, rather than in kilograms or tons. Each dot on the graph represents a main sequence star whose mass and brightness have been measured. The data clearly fall along a narrow strip, ranging from stars of low mass and low luminosity to those of high mass and high luminosity. From such observations, astronomers find that the masses of stars range from about one-tenth to fifty times that of the sun. It is believed that stars that try to form with masses much less than 1/10-solar mass turn into planets. On the other hand, calculations reveal that, if a star tries to

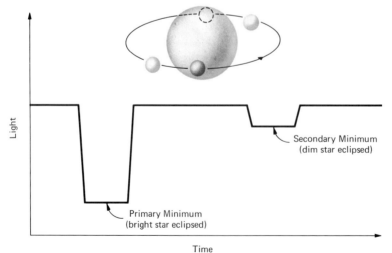

Figure 7–28. An Eclipsing Binary. An eclipsing binary is a double star in which the two stars alternately pass in front of each other. By measuring the total magnitude of an eclipsing binary for many nights, the astronomer can draw a light curve. From the precise shapes of the primary and secondary minima, many of the properties of the two stars can be deduced.

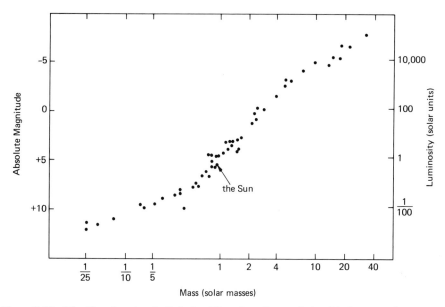

Figure 7–29. *The Mass-Luminosity Relation.* There is a direct relationship between the masses of stars and their luminosities. Low mass stars have low luminosities, while the most massive stars are also the most luminous.

form from a cloud of gas substantially more massive than 50 solar masses, the "star" is unstable and breaks up, perhaps to form several smaller stars.

These methods have enabled the astronomer to obtain some truly fundamental information about stars in general. He knows how massive stars are, he knows how bright they are, and he knows how hot their surfaces are. This knowledge of the masses, luminosities, and temperatures of stars, expecially as summarized in the H-R diagram and the mass-luminosity relation, has provided astrophysicists with the basic clues they needed to understand the life cycles of stars. As will be discussed in Chapter 8, this knowledge, along with the laws of physics, has provided mankind with the fascinating story of how stars are born, how they grow old, and what happens to them when they die.

7.6 Some Unusual Stars

The majority of stars in the sky—including our sun—are main sequence stars. The second most common type of star easily observed with the naked eye are red giants. Virtually every reddish-appearing star you can see in the nighttime sky is a red giant. Finally, there is a third large class of stars called white dwarfs. Although white dwarfs can be seen only with the aid of telescopes, astronomers have every reason to believe that they are very common. These three types of stars constitute the three major groupings of stars in the H-R diagram. They are found either singly, alone in space, or as members of binary or multiple star systems.

Although most stars observed by astronomers fall into one the these three common classifications, occasionally stars are found that are highly unusual. For example, there are a number of stars in the sky that are variable; their magnitudes vary over periods of time. There are roughly twenty thousand stars known to astronomers that exhibit such changes in their brightness. They are called *variable stars*. One reason why a star can vary its brightness is that it happens to be an unresolved eclipsing binary. About one fifth of all variable stars are eclipsing binaries. However, the remaining four fifths of all variable stars are doing something very unusual to change their light output.

A well-known and very important type of variable star recognized by astronomers since the late 1700s is the *Cepheid variable*. A Cepheid variable is identified by the characteristic way it changes its brightness. Figure 7–30 shows a light curve of a typical Cepheid variable to illustrate this characteristic behavior. These stars always brighten rapidly and dim slowly in a very regular fashion. Any time an astronomer notices a star changing its magnitude in this manner, he can be quite sure that he is observing a Cepheid variable.

Cepheid variables are intrinsically very bright stars. They have average absolute magnitudes in the range from −1.5 to −5. The *period* of variability (the time it takes the star to go through one cycle of maximum brightness to minimum brightness back to maximum brightness) depends on the particular Cepheid variable in question. The periods of Cepheid variables cover a range of from 3 to 50 days. Polaris, the North Star, is a Cepheid variable. Its apparent magnitude varies from 2.5 to 2.6 with a period of about 4 days.

Relatively speaking, Cepheid variables are quite rare. Fewer than a thousand such stars are known to astronomers. Yet, as will be discussed in Chapter 12, Cepheid variables play a very important role in the history of astronomy.

Profound insight into the nature of Cepheid variables comes from studying

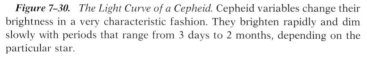

Figure 7–30. *The Light Curve of a Cepheid.* Cepheid variables change their brightness in a very characteristic fashion. They brighten rapidly and dim slowly with periods that range from 3 days to 2 months, depending on the particular star.

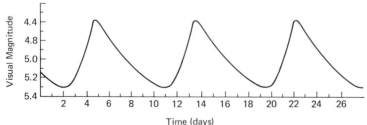

their spectra. First of all, the spectral class of a Cepheid variable changes slightly as the star brightens and dims. At maximum brightness, the spectrum of a Cepheid variable corresponds to a slightly hotter star than at minimum brightness. These stars are hotter when they are brighter, cooler when they are dimmer.

An even more striking discovery to come from the spectra of these stars is that their spectral lines exhibit Doppler shifts with exactly the same period as the light variations. When a Cepheid variable is brightening, its spectral lines are slightly blue-shifted from their average wavelengths, indicating that the surface of the star is coming toward us. When the Cepheid variable is dimming, the spectral lines are shifted slightly toward the red end of the spectrum, indicating that the star's surface is receding from us. From these observations, astronomers realize that Cepheid variables are actually pulsating in size. At maximum light, the star is big and hot. At minimum light, the star is smaller and cooler. The changes in temperature in a cycle are typically several hundred degrees, whereas the diameter of the star varies by a few per cent from its average size.

Cepheid variables are good examples of unstable stars. The physical processes in the atmospheres of ordinary stars are in balance. The atmosphere of one of these Cepheid variables, however, behaves somewhat like a spring. When the star is contracting, high pressures build up in it. These pressures get so high that they stop the contraction and start the star expanding. But, in expanding, the star goes too far; it gets too big. The pressures in it get so low that gravity stops the expansion and the star begins to contract again.

Similar kinds of "instabilities" are responsible for all *pulsating variable stars*. Something happens on or inside the star that causes it to expand or contract. As the star changes in size, its brightness changes. These pulsations can be highly periodic, such as in the case of Cepheid variables and *RR Lyrae variables*, or they can be semiregular, changing their brightness at unpredictable intervals.

RR Lyrae variables are the second most common variety of variable stars. Almost three thousand stars of this type have been identified and catalogued. RR Lyrae variables have much shorter periods than Cepheid variables. RR Lyrae stars, which are usually found in star clusters, have periods of less than 1 day. Many take only a few hours to go from maximum brightness down to minimum brightness and back up to maximum again.

By contrast, the most common variable stars are the *long-period variables*. These stars look like red giants, except that they brighten suddenly with somewhat unreliable periods. A good example is a star in the southern sky called Mira (o Ceti). Mira is usually so faint that it can be seen only with the aid of a telescope. However, about once a year, it brightens by half a dozen magnitudes so that it is readily apparent to the naked eye, sometimes becoming one of the brightest stars in the constellation of Cetus. There are about four thousand of these long-period variables known to astronomers.

There are many different types of pulsating variables. Long-period variables are the most numerous. RR Lyrae variables are the second most common type of pulsating variable. Cepheid variables, although considerably rarer, are

Figure 7-31. *Nova Herculis.* These two photographs show Nova Herculis as it appeared before and after its outburst in 1934. (*Yerkes Observatory*)

astronomically very important. In addition, there are several other classes of pulsating variables, such as RV Tauri stars, β Canis Majoris stars, and δ Scuti stars that astronomers have discovered. In all cases, the physical conditions in the stars are unstable, and the stars periodically change their sizes and temperatures, which causes the stars to change in brightness.

Although most variable stars are pulsating variables, there is a second broad category of stars that change their brightnesses due to some kind of explosive event. These stars are called *eruptive variables*. In general, eruptive variables can be thought of as stars that blow up rather than pulsate.

A common type of eruptive variable are the *novae*. A nova is a star that suddenly brightens and then gradually fades away. As its name suggests (*nova* is the Latin word meaning "new"), when an astronomer discovers a nova, it looks as though there is a new star in the sky. Actually, a very dim star, which has gone unnoticed for many years, has suddenly become very bright. When a star becomes a nova, it brightens very rapidly, usually in fewer than 24 hours. Typically, the star increases its brightness by about 10 magnitudes. It remains at maximum brightness for only a couple of days and then fades away during the next several months. Figure 7-31 shows two photographs of Nova Herculis, a nova that occurred in the constellation of Hercules in 1934, before and after the outburst. The star that caused this nova was originally so dim that it almost cannot be seen on the earlier photograph. In addition, Figure 7-32 shows the *light curve* of a typical nova, to illustrate the rapid rise to maximum brightness and the gradual decline.

During a nova outburst, a star is actually undergoing a major explosion. Spectroscopic studies of novae reveal that the outer atmosphere of the star is blown off to form an expanding shell of gas around the star. In doing so, the star is ejecting trillions of tons of gas into space at speeds up to 1,000 miles per second. Many years after the outburst it is often possible to see this gas around the star, as in the case of Nova Persei, shown in Figure 7–33. Typically, the total amount of ejected gas is only about a ten thousandth of the mass of the star.

Current astronomical data and theories strongly suggest that all novae are members of binary stars that are very close together. One star in the binary dumps matter onto the other star, which becomes overloaded, unstable, and finally blows up.

A far more spectacular type of eruptive variable are the so-called *supernovae*. When a star becomes a nova, its luminosity increases by a factor of about 10,000. However, if a star becomes a supernova, it flares up to almost a billion times its original brightness. A light curve of a supernova resembles that of an ordinary nova, except that a supernova becomes many times brighter. At maximum light, a supernova might reach an absolute magnitude of −20, which means that the exploding star is emitting as much light as 10 billion suns!

Supernovae are rare phenomena. A supernova was observed in A.D. 1054 by Chinese astronomers in the constellation of Taurus. This "guest star," as the Chinese astronomers called it, became so bright that it could easily be seen in broad daylight. In 1572, Tycho Brahe discovered a supernova in Cassiopeia, and in 1604 a supernova was seen in Serpeus and was described by both Kepler and Galileo. A decent, naked-eye supernova has not been seen since then, and most astronomers feel that we are probably overdue for a whopper. Nevertheless,

Figure 7–32. *The Light Curve of a Nova.* The light curve of a nova serves as a permanent record of the star's magnitude during its outburst. A rapid rise in brightness is always followed by a gradual decline.

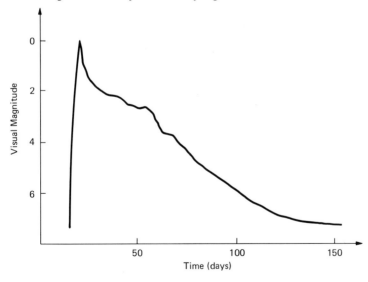

astronomers do discover supernovae in distant galaxies. Figure 7–34 shows such a supernova in a galaxy in the constellation of Coma Berenices. Sometimes a supernova can outshine all of the stars in a galaxy combined.

When astronomers turn their telescopes to the locations where supernovae have occurred, they usually find a large, expanding cloud of gas. Such an object is called a *supernova remnant* and stands in mute testimony to the fact that a star has blown itself up in a cataclysmic explosion. The famous Crab Nebula, about which we will have more to say in Chapter 9, is seen in Taurus at precisely the location of the guest star recorded by the Chinese. In the constellation of Cygnus, astronomers have found the Veil Nebula, which is the remnant of an ancient supernova that must have blazed forth in the sky during prehistoric times.

In comparison to novae, a supernova is truly a spectacular event. During a nova explosion, a star ejects only a ten thousandth of its mass at speeds of less than 1,000 miles per second. But, if a star becomes a supernova, it can blow off more than an entire solar mass with velocities up to 10,000 miles per second. As will be seen in the next chapter, the processes that produce a supernova are the death throes of a dying star. As its final act, a star bursts forth in brilliant glory—momentarily—outshining even an entire galaxy of stars, before it permanently fades from view.

Figure 7–33. *Nova Persei.* Nova Persei exploded in 1901. This photograph, taken in 1949 with the 200-inch telescope, shows expanding gases surrounding the star. (*Hale Observatories*)

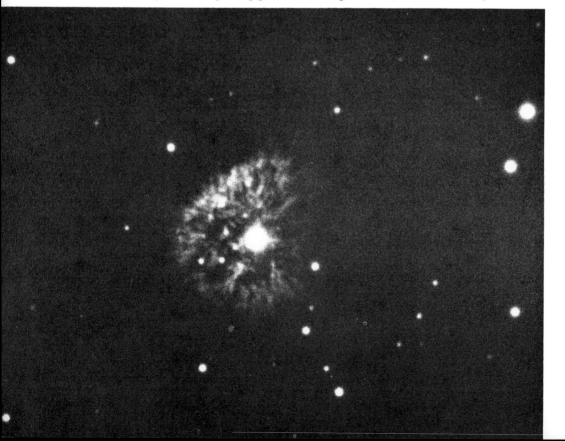

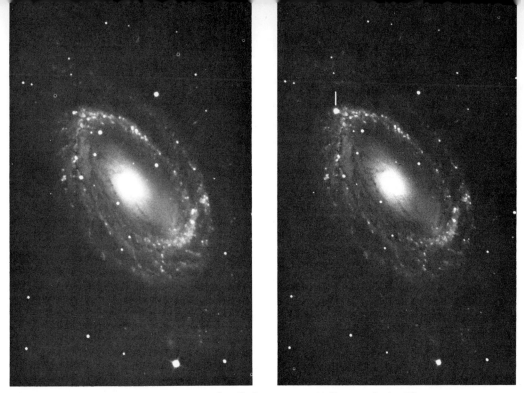

Figure 7-34. *A Supernova.* Astronomers often find supernovae in distant galaxies. The supernova shown here exploded in 1940 in the galaxy called NGC 4725 which is located in the constellation of Coma Berenices. (*Hale Observatories*)

Figure 7-35. *A Supernova Remnant.* The famous Veil Nebula in the constellation of Cygnus is the result of a star which became a supernova about 50,000 years ago. (*Hale Observatories*)

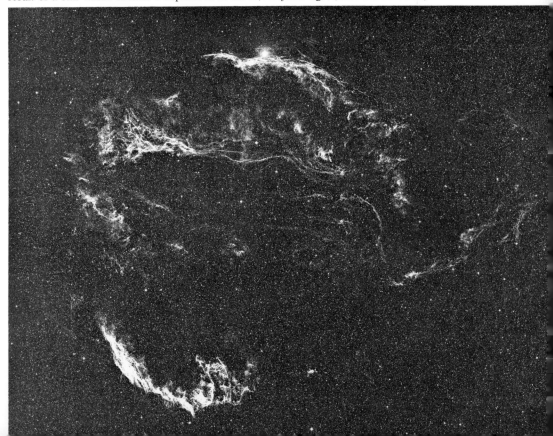

Questions and Exercises

1. What is meant by parallax, and how can it be used to measure the distances to stars?
2. Contrast and compare the apparent magnitudes of stars with the absolute magnitudes of stars. What do these terms mean? How do they differ?
3. What is meant by bolometric magnitude?
4. What, approximately, is the absolute magnitude of the brightest known stars? What, approximately, is the absolute magnitude of the dimmest known stars?
5. What is the difference between a neutral and an ionized atom?
6. Discuss how the surface temperature of a star can affect the appearance of its spectrum.
7. What is meant by the spectral class of stars?
8. To what do the letters O, B, A, F, G, K, M refer?
9. What is the spectral class and approximate surface temperature of a star whose spectrum is dominated by very strong hydrogen lines?
10. What is the spectral class and approximate surface temperature of a star whose spectrum is dominated by very strong bands of titanium oxide?
11. What is a Hertzsprung-Russell diagram?
12. Discuss the options of expressing the real brightness of stars either in terms of absolute magnitude or solar luminosity.
13. Discuss the relationship between stellar temperatures and color indices.
14. What is meant by the main sequence?
15. Compare and contrast the temperatures and sizes of red giant and white dwarf stars.
16. Briefly discuss how the size of a star can affect the appearance of its spectrum.
17. Describe the method of spectroscopic parallax.
18. Contrast and compare globular clusters and galactic clusters. How do their appearances differ? How do their color magnitude diagrams differ?
19. Briefly describe the method of main sequence fitting. How does this method allow the astronomer to determine the distance to a cluster?
20. What is meant by the photosphere, chromosphere, and corona of the sun? How is each most easily observed?
21. What is meant by the differential rotation of the sun?
22. What is the sunspot cycle?
23. Describe the appearance of sunspots.
24. What does a monochrometer do?
25. What is a prominence and how are prominences related to filaments?
26. What is an activity center?
27. What is thermonuclear fusion, and what role does it play in the sun?
28. Discuss the neutrino problem.

29. Contrast and compare visual binaries, spectroscopic binaries, and spectrum binaries.
30. What is meant by the light curve of an eclipsing binary?
31. What is the mass-luminosity relation, and why is it important?
32. How can an astronomer recognize a Cepheid variable from its light curve?
33. Briefly discuss the physical processes that occur in a pulsating variable.
34. What is a nova?
35. Contrast and compare novae and supernovae.

Stellar Evolution

8.1 The Birth of Stars

To ancient man the heavens must have seemed eternal, permanent, and unchanging. The stars and constellations that greet us when we walk outside on a clear night are virtually identical to what our most ancient ancestors saw as they gazed up into the star-filled sky. Yet, in spite of all appearances, man has realized that the heavens *must* be changing. Stars can be seen in the sky because they emit light. In order to give off light, stars must use up energy. And, as the sources of this energy are depleted, the stars must change; they must evolve. The reason why stars give the illusion of being permanent and unchanging is simply due to the fact that this evolution is very slow. Milestones marking major changes in the structure of a star during its life cycle are separated by hundreds of millions or even billions of years, far longer than the memory of men or the span of recorded history.

In the last chapter, a wealth of information about stars was discussed. It was learned what kinds of stars exist and how the astronomer obtains important data about these stars, such as their masses, temperatures, and luminosities. Also, from studying our sun, it was shown that a star is a large sphere of hot gases whose energy is produced by thermonuclear reactions at its core. The astrophysicist is now in a position to ask a variety of relevant questions: Where do stars come from? How are they born? How do they change as the nuclear "fuel" at their cores is used up? What happens to stars after all of this fuel is depleted? How do stars die and what becomes of them?

With the aid of modern computers to do the calculations, questions such as these can be answered. The method, in principle, is very simple. The astrophysicist takes known information about stars and combines this data with the laws of physics in a computer. The resulting computations tell him a great deal about the properties and processes in stars. In other words, the astrophysicist constructs *theoretical models* of stars in his computer. Since he cannot live long enough to see stars evolve in the sky, by pressing a button, he "evolves" his theoretical stars in his computer. Of course, at various stages in his computations, he compares the results of his calculations with data from astronomical observations to be sure he is on the right track. The final result is the fascinating story of stellar evolution.

When astronomers look into space with their telescopes, they frequently find large clouds of gas. These objects are usually called *nebulae*, from the Latin word meaning "cloud." The Orion nebula, which can just barely be seen with the naked eye as the middle "star" in Orion's "sword," is a good example. In examining these nebulae, some dark areas are frequently observed. These dark areas are not "holes" in the nebula, but rather are large clouds of cool gas that stand out in contrast to the brighter background. In the famous Horsehead nebula (see Plate 6) the outline of these cooler clouds strongly resembles a chess piece, and in the Trifid nebula (see Plate 7) dark clouds seem to divide the nebula into three parts.

Think for a moment about one of these cool clouds in space. As might be expected, such a cloud would not be perfectly homogeneous but rather would contain some lumps. In one of these lumps the gas is a little denser than in the neighboring regions of the cloud. Since such a lump contains more matter than surrounding regions, it will have a slightly stronger gravitational field. This gravitational field will attract some of the nearby gas and, therefore, the lump will begin to grow. As the mass of the lump increases, its gravitational field also becomes stronger, and it attracts still more gas, which continues to cause the lump to grow. This cycle repeats itself until pretty soon this lump has attracted trillions upon trillions of tons of gas into a sphere many times the size of our solar system. The embryo of a star has been created.

It should be noted that this process whereby an inhomogeneity (that is, a lump) grows by accretion will occur only if the temperature in the cloud is low, perhaps 10° above absolute zero (10°K) or even colder. If the temperature were high, then the speeds of the atoms in the gas would also be high. Due to the high velocities of the atoms, the process of condensation could never get started. The gas must be cool so that the speeds of the atoms are low, since only then will a lump be able to hold on to the gas that it attracts. For these reasons, astrophysicists believe that cool clouds of gas floating in space are the nurseries of the stars.

The astrophysicist then turns his attention to the nature of one of these large spheres of gas that has condensed from an inhomogeneity in the cool regions of a nebula. He finds that such a ball of gas is unstable against collapse. There is simply nothing to hold up all that gas, and the huge ball begins to contract. The shear weight of trillions upon trillions of tons of gas pressing inward under the force of gravity causes the ball to begin shrinking rapidly. As all this gas gets compressed into a smaller and smaller volume, the pressure, density, and temperature inside the sphere begin to rise dramatically. As the temperature rises, the contracting sphere begins radiating light and a *protostar* is formed.

The contraction of one of these protostars proceeds very rapidly. Eventually, since the gases are becoming more and more compressed, the temperatures at the center of a collapsing protostar reach several million degrees. At temperatures this high, the nuclei of hydrogen atoms are traveling so fast that they fuse together to become the nuclei of helium atoms. Thermonuclear reactions turn on. The tremendous outpouring of energy from these thermonuclear reactions gives rise to an outward pressure that finally stops the inward contraction. A star is born!

In examining the properties of contracting protostars with the aid of his computer, the astrophysicist finds that the laws of physics tell him several interesting things. First of all, if a star tries to form from a small ball of gas less massive than $1/_{10}$-solar mass, the temperatures inside such a protostar never get terribly high. Thermonuclear reactions never turn on. The object becomes a planet rather than a star. Similarly, if a star tries to form from a very large quantity of gas, more massive than about sixty-five suns, the protostar is unstable and breaks up. From these theoretical discoveries astronomers now understand why the masses of stars fall in the range from $1/_{10}$-solar mass to about 60 solar masses.

The astrophysicist can ask his computer what those newly formed stars that do have masses within the allowed range look like. He can ask about their surface temperatures or spectral class; he can ask the computer to calculate their luminosities or absolute magnitudes. He finds that *all* these newly created stars *must* be main sequence stars.

In a classic series of computations in the mid 1960s, the Japanese astrophysicist C. Hayashi was able to show theoretically how contracting protostars turn into main sequence stars. When a protostar first begins emitting light, it is still large and quite cool. On the H-R diagram, such a star would appear in the upper righthand corner along with the red giants, which also are large and cool. Contracting protostars are, however, *not* to be confused with red giants. Very rapidly, protostars decrease in size while maintaining nearly constant surface temperatures. If astronomers were lucky enough to find such a protostar in the sky, they could plot its path during its evolution on the H-R diagram. According to Dr. Hayashi's calculations, this path initially would go straight down on the diagram, as shown in Figure 8-1. This is because the surface temperature remains roughly constant, although the star is decreasing in size. As the star gets smaller it emits less light and, therefore, its brightness decreases. Eventually, the heat building up at the center of the star reaches the star's surface. When this happens, the path of the star on the H-R diagram veers sharply to the left and begins moving toward the main sequence. Finally, when thermonuclear reactions turn on, the star stops moving around on the H-R diagram and settles down on the main sequence.

It should be emphasized that this *Hayashi contraction* of a protostar occurs *very* rapidly. If the protostar is massive (say, around 10 to 40 solar masses), pressures and temperatures build up so fast that the protostar becomes a main sequence star in only a few thousand years. That's no time at all on the astronomical scale. For less massive stars such as our sun, it takes only a few million years for a protostar to evolve into a main sequence star. Since there is less matter, the pressures and temperatures build up more slowly and, therefore, it

Chapter 8 Stellar Evolution

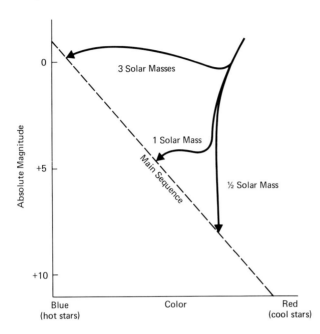

Figure 8–1. *Hayashi Contraction.* As new-born stars rapidly contract, they follow "Hayashi tracks" on the H-R diagram. Only after thermonuclear reactions are ignited at the cores of such stars do they settle down on the main sequence.

takes a longer time for the star's core to get hot enough to ignite thermonuclear reactions. Still, compared to the age of the universe, a few million years is only an instant in time.

Since the initial Hayashi contraction occurs so fast, and since astronomers have been carefully observing the stars for only a few centuries, there is virtually no hope of being lucky enough to "catch" a protostar in its very earliest stages of contraction. However, as the star gets closer to the main sequence, its evolution proceeds more slowly. Astronomers do have the chance to catch stars that are just about to settle down on the main sequence.

The obvious place to look for such stars is in the gas clouds where stars might be forming. The beautiful nebula called NGC 2264 in the constellation of Monoceros (see Figure 8–2) provides a good example of the missing link between protostars and main sequence stars. With the aid of a photoelectric photometer, the astronomer has no trouble measuring the visual (V) magnitudes and color indices (B-V) of the stars in NGC 2264. A graph showing the results of such observations in the form of a H-R diagram is given in Figure 8–3. For comparison, the line in the diagram is where the main sequence should be. Thus, most of the stars in NGC 2264 lie just above the main sequence. They haven't quite made it. They are still in the final stages of contraction as the thermonuclear reactions in their cores are turning on. Many of these stars are eruptive variables of the so-called *T Tauri* type. Such stars show rapid and erratic variations in brightness probably associated with the fact that they are contracting.

By applying the laws of physics to what is seen in the sky, astrophysicists have succeeded in understanding how stars are born. Due to the complexity and tediousness of the calculations, this level of understanding was totally impossible only a few short decades ago, before the development of huge computers. Many modern astronomers (this writer included) never look through telescopes. Rather, they spend long hours in front of flashing electronic lights, whirling magnetic tapes, and stacks of IBM cards doing in a fraction of a second that which nature takes millions or billions of years to accomplish in the sky.

Figure 8–2. Very Young Stars. Many new-born stars are found in this beautiful nebula called NGC 2264 in the constellation of Monoceros. The gas from which these young stars recently formed is easily seen. (*Hale Observatories*)

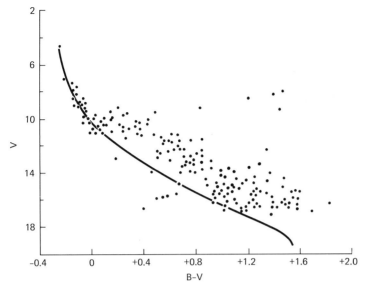

Figure 8-3. An H-R Diagram for NGC 2264. The Rosette Nebula (see Plate 8) contains many very young stars. The observed magnitudes and color indices of these stars places them slightly above the main sequence on a H-R diagram. (*Adapted from Dr. Walker*)

8.2 Maturity and Old Age

Stars are born in cold clouds of gas floating in space. Small inhomogeneities grow by accretion into large spheres of gas that contract under the force of gravity. As this contraction proceeds, the gases in these protostars heat up until finally thermonuclear reactions turn on. With the ignition of thermonuclear reactions, contraction stops and stars are born.

In order to understand the nature of young stars, the astrophysicist examines the theoretical properties of spheres of gas (that is, model stars) that have thermonuclear reactions occurring at their cores. He finds that such model stars can exist only if their mass lies in the range from $1/10$- to about 60 solar masses. An even more remarkable discovery arising from such calculations is made when the astrophysicist asks his computer about the appearance (surface temperature and luminosity) of his stable model stars. In all cases, the answers always correspond to main sequence stars. Finally, with the aid of computers, astronomers can begin to understand and appreciate the true significance of the main sequence on the H-R diagram.

The first thing to realize is that all main sequence stars have thermonuclear reactions occurring at their centers. In these reactions, hydrogen is converted into helium with the accompanying release of huge amounts of energy. It is this outpouring of energy that gives rise to the pressures that support stars and prevent any further collapse. This nuclear process is called *hydrogen burning*, although nothing is being burned in the usual sense. When you light a match and start a fire, chemical reactions occur that involve only the electrons in atoms. By contrast, hydrogen burning is a nuclear reaction: four hydrogen nuclei are fused together to produce one helium nucleus. This hydrogen burning may occur directly (the *proton-proton chain*), or it may involve carbon as a catalyst (the *CNO cycle*), depending on conditions in the star. The final result is, however, essentially the same: hydrogen is converted into helium. Astrophysicists have, therefore, come to the important realization that all main sequence stars are stars that are burning hydrogen at their cores.

Another important insight arising from calculations involving model stars deals with the fact that the precise location of a star on the main sequence depends on the star's mass. If a massive star is created (say, having a mass somewhat greater than 10 solar masses), the crushing inward pressure of the outer layers of the star easily heats up a large region at the star's core to temperatures of 10 million degrees. Thermonuclear reactions readily ignite and hydrogen burning proceeds at a furious rate. This rapid consumption of hydrogen in the large core of one of these massive stars produces so much energy that the star appears bright *and* hot. Such a star is an O or B star on the main sequence.

By contrast, if a star forms from a relatively small amount of gas (say, slightly less than 1 solar mass), the pressures and temperatures do not build up as easily or dramatically as in the case of a more massive star. A much smaller region at the core of a low mass star gradually heats up to 10 million degrees and the thermonuclear reactions turn on more slowly. Hydrogen burning proceeds at a leisurely rate and the star appears both cool *and* dim. Such a star is a K or M star on the main sequence.

These ideas explain the appearance of the mass-luminosity relation discussed in Section 7.5. In massive stars, hydrogen burning occurs at a very rapid rate, which causes the stars to be very bright. In low mass stars, hydrogen burning occurs at a much slower rate; since much less energy is produced, these stars are dim.

By way of summary, we see that the main sequence on the H-R diagram is a gradual progression of brightness, temperature, *and* mass. Newly formed high mass stars are hot and bright; they are located on the upper end of the main sequence. Young low mass stars are cool and dim; they appear on the lower end of the main sequence. Stars with intermediate masses (for example, our sun) have intermediate luminosities and surface temperatures; they fall in the middle section of the main sequence.

The obvious question that now comes to mind is "What happens next?" Hydrogen burning can't go on forever. Eventually, all the hydrogen at the core of a main sequence star gets used up. Eventually, the core of such a star will consist almost entirely of helium and the hydrogen burning must stop. For example, at

Chapter 8 Stellar Evolution

the center of our sun, 600 million tons of hydrogen are converted into helium each second. Fortunately, the sun is so huge and contains so much hydrogen that these thermonuclear processes can proceed at the current rate for a total of 10 billion years. But, the sun is $4\frac{1}{2}$ billion years old now, which means that it has about $5\frac{1}{2}$ billion years left as a main sequence star. What will the sun look like 6 billion years into the future when all the hydrogen fuel at its center is gone?

It is important to recall that the ignition of thermonuclear "fires" at the core of a star is the process that stopped the initial contraction of the protostar. As soon as all the hydrogen at a star's core is used up, hydrogen burning shuts off. The energy sources that gave rise to the pressure that supported the star are gone. Almost immediately the helium-rich core of the star begins to contract rapidly under the force of gravity. As the core contracts, the pressures, densities, and temperatures rise dramatically. Fairly soon, perhaps after only a few million years for a star slightly more massive than the sun, the region in the star surrounding the core becomes quite hot, as temperatures at the star's center approach 100 million degrees. But these regions above the core still contain fresh, unused hydrogen left over from when the star was formed. Therefore, when temperatures in the middle regions of the star finally reach 10 million degrees, thermonuclear reactions again turn on. This time, however, the hydrogen burning occurs in a *shell* surrounding the core.

These two processes, namely the contraction of the core and the ignition of a hydrogen burning shell, occur deep within the star. As seen from the outside, nothing really spectacular appears to be going on at first. During the first 10 million years or so after the cessation of hydrogen burning at the core, the star gets just a little brighter. But as soon as the energy from the recently ignited hydrogen-burning shell reaches the outer layers of the star, the star begins to expand. This new outpouring of energy from the hydrogen-burning shell pushes the atmosphere of the star millions of miles out into space in all directions. As the star's atmosphere expands, the gases cool. As seen from a great distance, the star appears to be getting bigger, cooler, and, therefore, redder. It is rapidly evolving away from the main sequence on the H-R diagram toward the region of the red giants.

Things are happening very rapidly inside a star at this critical stage in its life cycle. While the energy from the hydrogen burning shell is causing the star to expand to many thousands of times its original size, the core continues to contract under the immense force of its own weight. The helium-rich matter in the core becomes more and more compressed and the temperature rises to 100 million degrees. Finally, at this incredible temperature, a totally new and exotic thermonuclear reaction is ignited. Helium nuclei are fused together to form carbon nuclei in a process known as *helium burning*. When helium burns, three helium nuclei (each consisting of two protons and two neutrons) are fused into one carbon nucleus (which contains six protons and six neutrons). This nuclear reaction releases huge amounts of energy, and this new outpouring of energy gives rise to enough pressure to halt the contraction of the star's core.

To summarize events up to this point, two major new changes have occurred inside the star that have completely transformed its appearance. First of all, the hydrogen-burning shell has caused the star to expand to a diameter of several hundred million miles. In doing so, the star's atmosphere has cooled down to about 3,000 to 4,000°K and the star appears very red. Secondly, since the star is now very large and since helium burning at the star's core is producing huge amounts of energy, the star is very bright; its absolute magnitude lies somewhere between 0 and −5. The star has become a red giant!

A star that has become a red giant is literally in its declining years. Pretty soon, after perhaps only a couple of billion years, it will be over the hill and on its way to the grave. Helium burning at the core of a red giant is the major source of energy inside such a star, providing it with the heat and light that it radiates into the blackness of space.

In a classic series of calculations performed in the mid-1960s, Dr. Icko Iben at MIT succeeded in computing the *evolutionary tracks* of stars on the H-R diagram as they evolve off the main sequence to become red giants. Figure 8–4 shows

Figure 8–4. *Evolution to Red Giants.* After a star has used up all the hydrogen fuel in its core, it becomes a red giant. In becoming a red giant, stars follow the evolutionary tracks shown here on the H-R diagram. (*Adapted from Dr. Iben*)

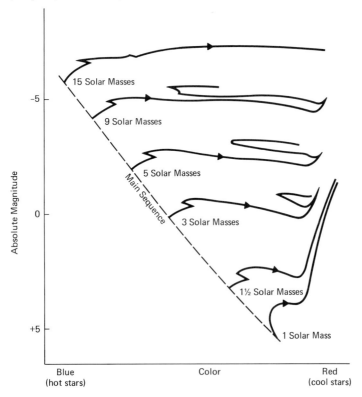

these paths. Notice that, although stars start off from very different locations on the main sequence due to their different masses, they all end up in roughly the same general location on the H-R diagram as red giants. Although stars spend many years on the main sequence and many years as red giants, only a short amount of time is spent in between. Finally, Dr. Iben's calculations reveal that red giants wander around in the upper righthand corner of the H-R diagram as they get older and older. Some of these stars can even become unstable and begin to pulsate as they near the end of their lives. Such stars, which periodically get bigger and smaller, hotter and cooler, are seen in the sky as Cepheid variables and RR Lyrae variables!

In about $5\frac{1}{2}$ billion years from now our own sun will become a red giant. Following the cessation of hydrogen burning at the sun's center, the atmosphere of our star will expand while its core collapses. Eventually, after perhaps only 1 billion years, the sun will have grown so large that its surface will reach almost out to the orbit of Mars. Our earth will be engulfed by the red giant sun. The oceans will boil and the continents will melt. The entire earth will be vaporized as it is swallowed by the sun. Of course, it would be foolish to try to speculate what "men" or other creatures inhabiting our planet during this far distant era will do to avoid the inevitable catastrophe. After all, there seems to be a high probability that through our own incredible stupidity we will have reduced this planet to a barren wasteland even before the end of the twentieth century. Yet, how many seeds fall to the ground in order to produce one tree? How many sperm die so that one egg is fertilized? How many races of intelligent creatures have appeared on the myriad of distant planets in the universe so that one, just one, is able to survive with the supreme gift of knowledge? Perhaps that's just the way nature works.

8.3 Star Clusters

Astronomers have been observing the skies for a very short period of time. In particular, detailed observations of stars have been made only within the last century. Yet, these stars are billions of years old. They evolve so slowly that scientists have no hope of actually seeing major changes in the structure of stars in the courses of their lives. The entire span of human history is incredibly brief compared to the time it takes for stars to evolve. Although a single human being can never live long enough to observe the process of stellar evolution, he can apply the laws of physics to what he sees and, thereby, understand the life cycle of stars. From this understanding he knows that the vastly different types of stars he finds in the sky are, in fact, intimately related. They look different primarily because they are at different stages in their evolution; they have different ages.

The situation facing the modern astronomer can be compared to that of a hypothetical intelligent insect: an ephemera—whose entire life span lasts only a few hours. When this hypothetical insect is born, he looks around the forest and sees trees, rotting logs, and green shoots sprouting from the ground. Since the bug will live for less than a day, the trees appear to be everlasting and unchanging objects in his world. Yet, if this insect is really intelligent and carefully examines what he sees in the forest and ponders the meaning of what he has observed, a fascinating theory might arise in his mind. Perhaps trees are not eternal. Perhaps the green shoots sprouting from the ground grow into huge trees that on their death fall to the ground and become the rotting logs to enrich the soil for future generations of trees. He then might realize that he will not live long enough to see any of these changes, but that the power of his intellect has given him deep insight into the life cycle that occurs in the forest.

Consider yet another analogy. Imagine that someone comes to you with a reel of motion picture film. At random he takes one frame from somewhere in the middle of the movie. You are allowed to examine this one picture in great detail. Then you are asked to figure out the plot of the movie. Astronomers have one such picture; it's called the H-R diagram. From studying the properties of the stars represented on the H-R diagram, astrophysicists have succeeded in figuring out the story of stellar evolution. Indeed, scientists now have a fairly good idea as to what the rest of the "movie" must look like.

When an astronomer draws a H-R diagram, as shown in Figure 7–6 (see Section 7.3), he includes as many stars as possible. Any star in the sky for which the absolute magnitude and spectral class are known gets plotted on the graph. Even though the resulting diagram illustrates the full range of stellar brightness and temperatures, there is a confusing hodgepodge of ages. Young stars, middle-aged stars, and old stars are *all* plotted together on the graph. For the purposes of understanding stellar evolution, it would be more advantageous to draw a series of H-R diagrams where each diagram contains only those stars that have roughly the *same* age. This is made possible by drawing H-R diagrams of the stars in individual star clusters.

As was discussed in Section 7.3, there are many clusters of stars in the sky. A cluster such as the Pleiades or M3 consists of a number of stars that formed in the same region of space, from the same clouds of gas, at roughly the same time. In other words, when the astronomer observes a star cluster in the sky, he can feel sure that all of the stars in the cluster have essentially the same ages. Therefore, by drawing separate H-R diagrams for the stars in individual clusters, the astronomer should be able to sort out the effects of age. These H-R diagrams will look like Figures 7–12 and 7–13.

The most convenient way of expressing the final results of the observations of star clusters is with a *composite* H-R diagram, as shown in Figure 8–5. In this figure, the H-R diagrams of several different clusters were plotted on top of each other after their main sequences had been lined up. Notice that some clusters, such as NGC 2362 and h and χ Persei have most of their stars on the main sequence. By contrast, in clusters such as M67 and NGC 188, only the low mass, cooler stars are on the main sequence, while a number of stars lie in the red giant region. Finally, clusters such as M11 and the Hyades seem to be in between these

Chapter 8 Stellar Evolution

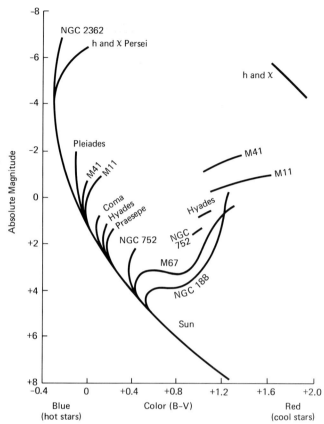

Figure 8–5. *A Composite H-R Diagram.* The H-R diagrams of several clusters have been plotted on top of each other on this graph. Clusters such as NGC 2362 have most of their stars still on the main sequence. Clusters like M67 have many stars on their way to becoming red giants. (*Adapted from Dr. Sandage*)

two extremes. They still have most of the main sequence left, but they also contain some red giants.

Perhaps the best explanation of these H-R diagrams for clusters comes from the work of the German astrophysicist, R. Kippenhahn. With the aid of a computer, Dr. Kippenhahn carefully examined the evolution of a large number of model stars having a variety of masses. In essence, he used the laws of physics to create a theoretical star cluster inside his computer. This theoretical cluster is affectionately known as M007. At various intervals, Dr. Kippenhahn stopped the computer to ask what the stars in M007 look like; he asked the computer to print out the surface temperatures and luminosities of all the stars in his cluster. The final results are illustrated in a series of H-R diagrams shown in Figure 8–6. They

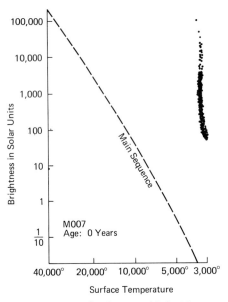

Figure 8–6a. The cluster at birth. All protostars have nearly the same temperature. The most massive stars are the brightest.

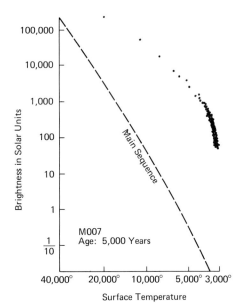

Figure 8–6b. The cluster is 5000 years old. The most massive stars have moved toward the main sequence.

Figure 8–6c. The cluster is 100,000 years old. The most massive stars have reached the main sequence. They are the O stars.

Figure 8–6d. The cluster is 3 million years old. Both A and B stars have now evolved onto the main sequence and are burning hydrogen at their cores.

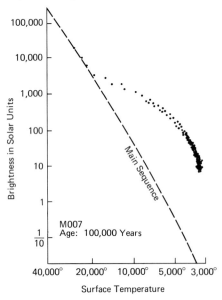

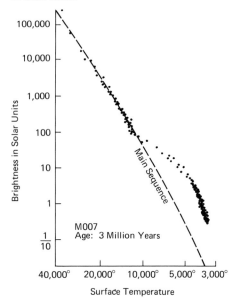

Chapter 8   Stellar Evolution

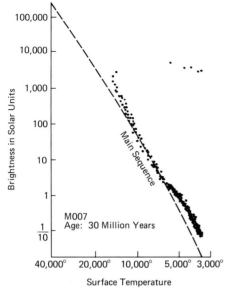

Figure 8–6e. The cluster is 30 million years old. The most massive stars (the O stars) have depleted all the hydrogen at their cores and evolved into red giants.

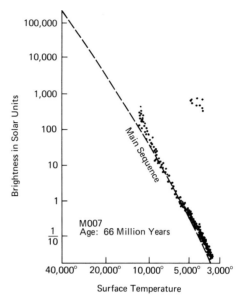

Figure 8–6f. The cluster is 66 million years old. The B stars have become red giants and the very low-mass stars (M stars) have almost arrived at the main sequence.

Figure 8–6g. The cluster is 100 million years old. The A stars have used up all the hydrogen at their cores and become red giants.

Figure 8–6h. The cluster is 4¼ billion years old (same as the age of the sun). The main sequence of the cluster has "burned down" to the F stars which are in the process of becoming red giants.

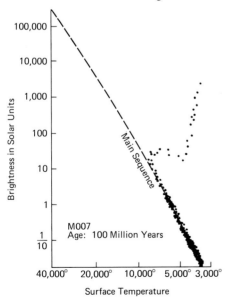

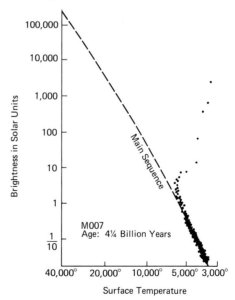

may be thought of as comprising representative pictures from the "movie" of stellar evolution.

When the stars in M007 first appear on the H-R diagram, they are located in a vertical strip in the red giant region, as shown in the first graph in Figure 8–6. Actually, they are not really stars at all at this time. They have just condensed out of a theoretical gas cloud and, as yet, thermonuclear reactions have not been ignited. They are protostars that are about to begin Hayashi contraction. It is important to note that the distribution of protostars in this verticle strip is according to mass. The most massive protostars are the brightest and lie near the top of the strip. The least massive stars are the dimmest and, therefore, are located near the bottom of the strip.

After 5,000 years, the most massive protostars are ready to ignite hydrogen burning in their cores. As shown in the second graph in Figure 8–6, these stars are moving rapidly toward the main sequence. All the rest of the stars are still undergoing Hayashi contraction.

When M007 is 100,000 years old (the third graph in Figure 8–6), the most massive stars finally will arrive on the main sequence. They are the hottest, brightest and bluest stars that can exist. They are the O stars. Due to their huge masses, the pressures and temperatures necessary to spark thermonuclear reactions at their cores were achieved easily. They are burning hydrogen at a furious rate. Meanwhile, moderately massive protostars are moving toward the main sequence. They soon will be ready to ignite their thermonuclear fires. But notice that the very low mass stars (less than 1 solar mass) do not seem to have done anything. They are all still undergoing Hayashi contraction and are just a little dimmer than they were 100,000 years earlier.

After 3 million years, as shown in the fourth graph of Figure 8–6, all the massive stars lie on the main sequence. B and A stars have joined the O stars in burning hydrogen at their cores. And even the less massive stars are beginning to move toward the main sequence. Temperatures at their cores are finally approaching 10 million degrees and hydrogen burning will soon turn on. Compare this theoretical H-R diagram with the observational H-R diagram of NGC 2264 shown in Figure 8–3. From this comparison we conclude that NGC 2264 is only a few million years old.

At an age of 30 million years (fifth graph in Figure 8–6), we notice that the O stars are gone! They burned hydrogen so fast that the fuel was soon depleted. Their cores contracted; their atmospheres expanded. They are now red giants consuming helium at their centers. Meanwhile, stars of intermediate mass have finally made it to the main sequence. F and G stars have joined the B and A stars burning hydrogen at their cores. In addition, the least massive stars are almost ready to settle down on the main sequence.

At this point it should be noted that Dr. Kippenhahn's computer unfortunately cannot handle stars after they cease to be red giants. When all the helium is burned up at the core of one of the model red giants, the computer literally throws the star out; it cannot do the calculations. The star disappears from the diagram in this stage in its evolution. The ultimate fate of these dying stars will be discussed in considerable detail in Chapters 9 and 11.

By the time M007 is 66 million years old, the B stars have evolved into red giants, as shown in the sixth graph in this series. All the remaining stars in the cluster now lie on the main sequence. Notice also that the main sequence is burning down like a candle. As the cluster gets older and older, the main sequence becomes shorter as the more massive stars evolve into red giants. This is especially apparent when examining the last two graphs in the series. At an age of 100 million years, the A stars in M007 have turned into red giants, and after $4\frac{1}{4}$ billion years even the F stars have left the main sequence.

From these calculations concerning M007, astronomers have come to the important realization that the appearance of the H-R diagram of a cluster tells its age. Returning to the composite H-R diagram in Figure 8–5, by comparison with the behavior of M007, it can be concluded that NGC 2362 and the double cluster called h and χ Persei are comparatively young. They are probably only about 10 million years old.

In the case of the Pleiades and M11, the main sequence has burned down to the extent that there are no O and B stars in these clusters. They must be almost 100 million years old.

In the case of the Hyades cluster, even the A stars have turned into red giants. This cluster of stars must have been formed about a billion years ago. Similarly, in M67, the F stars are becoming red giants and, therefore, this cluster must have an age of 5 billion years.

The cluster NGC 188 is the oldest cluster shown on the composite H-R diagram. The main sequence has burned down all the way to the location of the sun. Even the G stars have exhausted the supply of hydrogen fuel at their cores and are ready to evolve into red giants. The stars in NGC 188 must be 10 billion years old.

This gives a fairly good understanding of how stars are born and how they live out most of their lives. But what happens to a star after all the helium at its core is exhausted? Where do the red giants go? In the next chapter we will concentrate on some of the bizarre things that can happen when stars die.

8.4 How Stars Die

All the reddish-appearing stars seen in the nighttime sky (Aldebaran, Betelgeuse, Antares, Arcturus, and so on) are red giants whose light is produced primarily by helium burning at their cores. But after only a few billion years most of the helium in such stars will have been converted into carbon, and helium burning will cease at their centers. At this stage, events will occur very rapidly. The details of precisely what happens are not as yet known. In any case, astrophysicists know that the ultimate fate of such dying stars depends critically on their masses. Therefore, the death of stars must be discussed according to their masses.

The primary reason for the lack of detailed knowledge about the death of stars is due to the fact that computers have not been built that can handle the calculations. Up to this point in the life cycles of stars, changes have been occuring fairly slowly. As long as these changes occur slowly, the astrophysicist can recreate the processes of stellar evolution with present-day computer technology. However, when a star dies, violent events inside the star can take place during a period of days or even hours. Computers just do not exist that can handle such rapid and dramatic events. Perhaps in the 1980s the necessary computer technology will finally have been developed. Only then will astronomers be able to plot precise evolutionary tracks on their H-R diagrams to illustrate the post-red giant stages. Hopefully, a young astrophysicist will come along in the 1980s and do for post-red giant stages what scientists such as Hayashi, Iben, and Kippenhahn have done up to this point. Nevertheless, astronomers feel that they have a fairly good idea as to what these anticipated computations ultimately will reveal.

If a star has a very low mass, say less than about 1-solar mass, it probably dies a quiet death. After hydrogen burning shuts off at the star's center, the core begins to contract. Due to the low mass of the star, the core of the star may *never* get hot enough for helium burning to proceed at a reasonable rate. Since very little energy is being produced inside the star, the *entire* star will contract. As seen from the outside, the star gets very small and, therefore, becomes very dim. The resulting compression heats the gases at the star's surface to fairly high temperatures, somewhere between 10 and 40 thousand degrees. The star is hot and dim. It has become a white dwarf!

A white dwarf is a dead star. No thermonuclear fires are burning inside such stars. Either all the fuels have been used up *or*, as in the case of stars with very low mass, the pressures and temperatures never get high enough to ignite the fuels in the first place.

If a dying star is a little more massive than in the previous case, its death is somewhat more interesting. Stars with masses approximately the same as the sun do indeed develop pressures and temperatures at their cores high enough to ignite substantial helium burning as they grow into red giants. But after several billion years, the helium at their centers has been converted into carbon and helium burning shuts off due to the lack of fuel. Just as before, the cores of such stars again contract and heat up. But now the temperatures surrounding the cores get high enough to ignite a *helium-burning shell*. As the helium in the middle regions of such stars begins to burn, this new outpouring of energy causes the stars to expand to even greater extents than before. In fact, this expansion can be so great that the outer layers of such stars actually *separate* from the inner parts of the stars.

A star of more than 1 solar mass will, therefore, blow off its atmosphere into space as it begins to die. As much as 10 to 20 per cent of the entire mass of the star may be ejected in this fashion. After all the fuel in the helium-burning shell

Figure 8-7. A White Dwarf. The brightest-appearing star in the sky, Sirius, is a double star and has a white dwarf companion. This white dwarf is shown here along side Sirius. (*Courtesy of R. B. Minton*)

is exhausted, thermonuclear fires are forever extinguished and what remains of the star turns into a white dwarf.

As astronomers look into space with their telescopes, they often find expanding shells of gas surrounding small, white-hot stars. These shells of gas are called *planetary nebulae*. This unfortunate name arises from a historical accident; through the low-quality telescopes used by astronomers more than 100 years ago, these objects looked like fuzzy planets. Planetary nebulae are among the most beautiful objects in the sky. Plates 9 and 10 are very fine examples of shells of expanding gas surrounding dying stars. After only 30,000 to 40,000 years from now, these shells of gas will have expanded so far and become so dim that they will no longer be visible. A planetary nebula is, therefore, a *very* temporary phenomenon in the sky. For purposes of illustration, the probable evolutionary track of a 1.2 solar mass star over the entire course of its life is shown in Figure 8–8. Future calculations may confirm that this is the path followed by a star on the H-R diagram that produces a planetary nebula.

A star that is substantially more massive than the sun experiences an even more dramatic fate at the end of its life. In such a star the compression of its

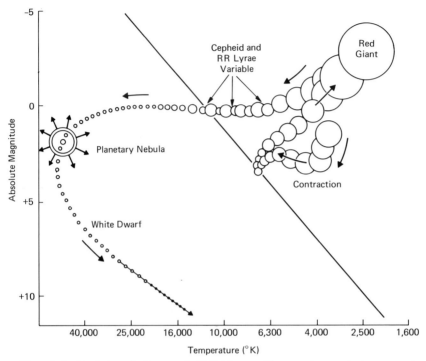

Figure 8-8. *Evolution of a Star on the H-R Diagram.* The complete evolutionary track of a star like the sun is shown on this H-R diagram. The star spends a long time (billions of years) as a main sequence star, as a red giant, and finally as a white dwarf. Other regions of the H-R diagram are traversed very quickly.

carbon-rich core, after the termination of helium burning, heats its center to between 600 million and 700 million degrees. At these temperatures *carbon burning* ignites. Carbon nuclei are fused together to produce magnesium, sodium, neon, and—finally—oxygen. After the carbon has been used up, carbon burning stops, and the core of the star again contracts. When temperatures of 1 billion degrees are achieved, *oxygen burning* commences. Oxygen nuclei are fused together to produce sulfur and phosphorus. Meanwhile, a hydrogen-burning shell, a helium-burning shell, and perhaps even a carbon-burning shell have been moving outward in the star in search for fresh fuel. It is even possible that *silicon burning* might be ignited at the star's center at temperatures in excess of 1 billion degrees. At 3 billion degrees, a wide range of thermonuclear reactions is occurring inside the star, whose matter is becoming enriched with heavy elements, especially iron.

A star that has reached this stage is ready to die. When the core of the star is very rich in iron, the nuclear reactions turn off for the last time. Nuclear physicists have shown that thermonuclear fusion cannot occur with nuclei heavier than iron. The star's core implodes violently and heats up the gases in the outer shells to extreme temperatures when nuclear reactions are proceeding at a furious rate. In a matter of seconds *all* the known chemical elements are produced. Finally, these cataclysmic events give rise to a *shock wave* that blows the star apart. The star ends its life in one incredibly brilliant explosion. The star has become a supernova!

As we have learned (see Section 7.6), astronomers have discovered many supernova remnants in the sky. Some of these objects can be observed with ordinary telescopes, such as the Crab nebula (Plate 11) and the Veil nebula (Plate 12). Other supernova remnants can be seen only in radio light. Tycho Brahe's supernova of 1572 is an example of this second type. The radio source known as 3C10 is all that is left of the star that blazed forth more than 400 years ago. Figure 8-9 shows a map of 3C10 as seen by radio astronomers.

During a supernova explosion, a star can eject a substantial fraction of its mass into space. Due to the nuclear reactions that occurred inside the star just before its death, this gas, which is cast off during the explosion, is very rich in heavy elements. Astronomers, therefore, say that supernovae *enrich* the interstellar medium. Indeed, processes such as those that occur in supernova are the *only* source of heavy elements. Astronomers have very strong evidence that only hydrogen and helium existed after the creation of the universe. As we will see in Chapter 14, no other elements could have survived in the primordial fireball from which the universe was born. Yet, as we look into space we see heavy elements. The spectra of stars and galaxies in the sky show spectral lines of virtually *all* the well-known elements. Indeed, the earth is composed largely of iron, and our own bodies are made up of carbon, oxygen, and nitrogen. Since these elements were not present at the creation of the universe, we must come to the realization that every atom in our bodies was created billions of years ago deep inside some long-since dead star.

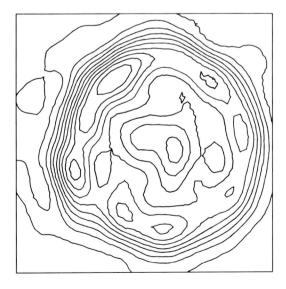

Figure 8-9. *Tycho Brahe's Supernova.* Radio astronomers sometimes find supernova remnants that are invisible through optical telescopes. The remnant of a supernova observed by Tycho Brahe (called 3C 10 by radio astronomers) is a good example. (*Adapted from Drs. Weiler and Seielstad*)

Questions and Exercises

1. Briefly describe the conditions and processes in which stars are formed.
2. What is a protostar?
3. Why are there no stars with masses much less than $1/10$-solar mass?
4. What is meant by Hayashi contraction?
5. What physical processes occur inside a star to stop its contraction and cause it to become a main sequence star?
6. In what kinds of objects are astronomers most likely to find young stars?
7. What is meant by hydrogen burning?
8. What group of stars on the H-R diagram have hydrogen burning occurring at their centers?
9. How is the mass of a star related to the rate at which thermonuclear reactions occur?
10. What happens to a star that has used up all the hydrogen "fuel" at its core?
11. What is going on inside a star that is in the process of becoming a red giant?
12. What is meant by helium burning?
13. When will the sun become a red giant?
14. What is meant by a composite H-R diagram?
15. How do the H-R diagrams of open clusters differ from those of globular clusters? Why?
16. What is a white dwarf?
17. What is a planetary nebula?
18. Sketch the complete evolution of a star such as our sun on an H-R diagram. Briefly describe the processes occurring inside the star as it moves from one place to another on the diagram.
19. How are supernova remnants formed?
20. Briefly describe the kinds of processes that occur in stars to create the "heavy" elements.
21. What is meant by the statement that supernovae "enrich" the interstellar medium?

White Dwarfs and Neutron Stars

9.1 White Dwarfs As Dead Stars

Calculating theoretical models of stars is a complicated business. The astrophysicist cannot just sit down and dream up what might happen. Rather, he must use the laws of physics as *tools* to make painstaking calculations of what happens to a ball of hot gases containing trillions upon trillions of tons of matter. No wonder astrophysicists take advantage of computer technology to assist them in their computations.

The construction of model stars—a task that ultimately gives the story of stellar evolution—makes use of *five* basic equations that express the necessary laws of physics. These are the *equations of stellar structure*. First of all, since gravity holds a star together, there is an equation based on Newton's law of gravitation to relate the pressure to the mass and density of matter inside the star. This relationship is the *equation of hydrostatic equilibrium*. Secondly, there is a simple equation that tells the astrophysicist how the mass and density of matter inside the star are connected, the *mass equation*. Since stars are hot and, therefore, radiate light, there must be an equation that tells the astrophysicist how the temperatures inside the star are related to the outward flow of light. This third equation is the *equation of radiative transport*. In addition, nuclear reactions deep inside stars produce the energy by which they shine. There is, therefore, an *equation of energy generation*. And finally, the *equation of state* contains information about the properties and behavior of gases in a star; it relates the pressure of a gas to its temperature, density, and chemical composition.

The whole idea behind these complicated-sounding details is that the astrophysicist must apply laws of nature in order to construct model stars. These laws are contained in the five equations of stellar structure that the astrophysicist plugs into his computer. His computer then grinds out answers such as the absolute magnitude and spectral class of the model stars. If these answers agree with what astronomers observe in the sky, scientists joyfully conclude that they understand the properties of these stars. If the computer comes up with weird nonsense, the astrophysicist must go back to the drawing board to figure out where he went wrong in applying the laws of physics.

An interesting example of astrophysicists spending a lot of time at the drawing board deals with the structure and properties of dead stars. Recall that when a protostar first forms out of condensations in cool clouds of interstellar gas, the young protostar simply contracts under the force of gravity. There is, as yet, nothing to hold it up; it is unstable and cannot support the crushing weight of the trillions upon trillions of tons of gas from which it was created. But soon the resulting compression heats up the gases at the protostar's center, thereby igniting thermonuclear fires. When hydrogen burning is ignited, the tremendous outpouring of energy produces pressures that can support the outer layers of the star. Contraction then stops and the star becomes stable.

But what happens after a star has used up *all* of its thermonuclear fuel? When no more sources of energy remain, pressures that once supported the star are eliminated. The star must collapse in on itself. Pressures inside the star will get higher and higher, and the density inside the star will get higher and higher as the atoms in the star are crushed together. Is there anything in nature to stop this collapse? Are there perhaps exotic and unusual forces that might build up inside the dying star sufficiently powerful to halt the contraction? Even today these fascinating questions continue to plague the minds of astronomers and astrophysicists alike. The answers require close scrutinization of some of the most bizarre laws of physics.

Think for a moment about what must be happening to matter inside a dying and collapsing star of approximately 1 solar mass. As the star collapses, the atoms inside the star are squeezed closer and closer together. As the collapse proceeds, they collide violently with each other, stripping themselves of their electrons. Soon, all the atoms are almost completely ionized and the matter inside the star consists of bare nuclei floating in a "sea" of electrons.

Under "ordinary" conditions, neutral atoms are mostly empty space. Electrons orbit the nuclei at specific "allowed" distances. However, if the atoms in a gas are completely ionized, the electrons and the nuclei can be squeezed into a volume much smaller than if the atoms were still in tact.

The compression cannot go on indefinitely. Long before the matter inside a collapsing star reaches the point where the electrons and nuclei are "touching" each other, certain laws of physics come into play. In particular, an important law of quantum mechanics, the *Pauli exclusion principle,* explains that no two identical electrons can occupy the same small volume. In other words, there is a minimum volume into which a given number of electrons can be squeezed. Just as the laws of quantum mechanics require that electrons can orbit the nuclei of atoms only at allowed distances, these same laws dictate that there is a certain minimum allowed volume into which a given number of electrons can be confined. If we were to try to squeeze more electrons into this minimum volume than is permitted by the Pauli exclusion principle, we would find our efforts meeting with tremendous resistance.

The application of the Pauli exclusion principle and quantum mechanics to highly compressed matter has profound implications for dying stars. As a star such as our sun contracts after all its nuclear fuel is exhausted, the atoms inside it soon become highly ionized. The pressures and densities in the star rapidly rise until the electrons finally are squeezed into the smallest volumes permitted to them under the laws of quantum mechanics. At this stage, the matter in the star is said to be *degenerate*, and the electrons comprise a *degenerate gas*. This degenerate sea of electrons produces an enormous pressure that vigorously resists further compression. Each and every small volume inside the dead star contains the maximum allowed number of electrons; to try to squeeze any more electrons into these volumes would violate the laws of physics. The resulting pressure stops the contraction of the star. The star has become a white dwarf!

In the early 1930s, the brilliant astrophysicist S. Chandrasekher, now at the University of Chicago, worked out the theoretical details of the structure and properties of white dwarfs. He discovered that when all of the matter of a dying star of about 1 solar mass collapses down to a sphere about 10,000 miles in diameter, the degenerate electron pressure is strong enough to prevent further contraction. White dwarfs are stable because the degenerate electrons provide the pressure to hold up the star. From Dr. Chandrasekher's calculations, it is evident that a typical white dwarf has roughly the same mass as our sun, except all that matter is contained in a ball about the same size as our earth! Since the star is so highly compressed, one tablespoon of matter from a white dwarf weighs 1,000 tons!

One of the most surprising results to come from Dr. Chandrasekher's theoretical investigations of white dwarfs is that there is a unique relationship between their masses and diameters. In other words, if the mass of a white dwarf is known—say from a binary star containing a white dwarf—using Dr. Chandrasekher's formulas astronomers can calculate what the diameter of the star *must* be. Figure 9–1 shows this mass-diameter relationship in the form of a

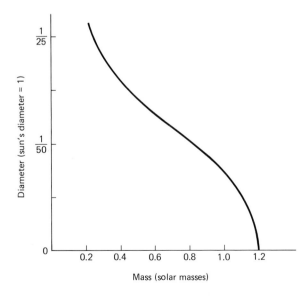

Figure 9–1. *Mass-Diameter Relation for White Dwarfs.* The greater the mass of a white dwarf, the smaller is its size. If the mass of a "dead" star exceeds $1\frac{1}{4}$ solar masses, it cannot be a white dwarf.

Figure 9-2. *A White Dwarf.* The companion to Sirius is a white dwarf. Its mass must be less than the Chandrasekhar limit of $1\frac{1}{4}$ solar masses.

graph. The important point to realize from this graph is that there is an upper limit to the mass of a white dwarf. White dwarfs cannot exist with masses greater than about $1\frac{1}{4}$ solar masses. This is the *Chandrasekher limit*. All white dwarf stars must have masses less than this critical limit of $1\frac{1}{4}$ suns.

The existence of the Chandrasekher limit came as a shock to astronomers in the 1930s. From the study of binary stars, it was obvious that many of the brightest stars in the sky have masses much greater than $1\frac{1}{4}$ times that of the sun. So, astronomers concluded that, in order for a massive dying star to become a white dwarf, it *must* eject a large fraction of its matter into space as it dies. It was popular to believe, for example, that when a 40 solar mass star exhausts all of its nuclear fuel, it must blow off at least $38\frac{3}{4}$ solar masses of matter in a supernova explosion so that what remains will be below the Chandrasekher limit.

This conclusion seemed to be supported by observations. The masses of all known white dwarfs lie in the range predicted from Dr. Chandrasekher's calculations. Observations of binary stars containing white dwarfs shows that their masses lie between $\frac{1}{10}$- and $1\frac{1}{4}$ solar masses. In addition, white dwarfs seem to be very common. For example, the brightest star seen in the nighttime sky, Sirius, is actually a double star containing a white dwarf. Figure 9-2 shows a photograph of Sirius and its white dwarf companion. When astronomers *esti-*

mate how many massive stars have probably died, and compare this number with how many white dwarfs probably exist in space, the two numbers are nearly equal. In other words, it appears as though there are enough white dwarf corpses around to account for all the stars that have used up their nuclear fuel and died. Although—strictly speaking—this conclusion is *wrong,* it does indeed seem that most stars in fact become white dwarfs at the end of their life.

Let us recapitulate the properties of white dwarfs. Astronomers find that their masses are typically about the same as our sun, but *always* less than $1\frac{1}{4}$ solar masses. They are roughly about the same size as the earth, with diameters of about 10,000 miles. They are supported by the degenerate pressure arising from closely packed electrons, and one cubic inch of matter from a white dwarf has a mass of 1,000 tons.

Due to their small size, white dwarfs are very dim. They are also very hot, with surface temperatures that range from about 5,000° to 50,000°K. No nuclear reactions are going on inside white dwarfs. As they radiate light into space they cool off, just as a hot cup of coffee cools off. This cooling process is very slow; it takes a white dwarf more than 10 billion years to cool down to 3,000°K. Gradu-

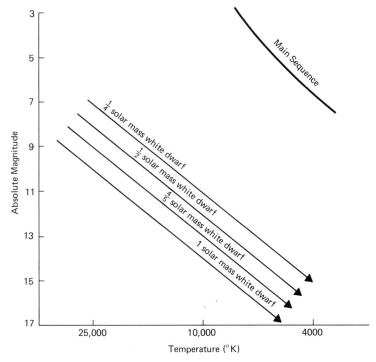

Figure 9–3. Cooling Curves. White dwarfs simply cool off with age. As they cool off, their luminosity gradually decreases. This means that they slowly move to the lower right on a H-R diagram.

Figure 9–4. The Conservation of Angular Momentum. The conservation of angular momentum is illustrated by an ice skater doing a pirouette on the ice. As she pulls in her arms, her rate of rotation speeds up.

ally they become dimmer and colder. Their evolutionary tracks on the H-R diagram simply follow *cooling curves,* as shown in Figure 9–3.

In addition to being small, dim, and hot, white dwarfs are probably rapidly rotating. To see why this is so, imagine an ice skater doing a pirouette, as shown in Figure 9–4. As she pulls in her arms, her rate of rotation speeds up. This is a direct consequence of a law of physics called the *conservation of angular momentum.* Similarly, if a star such as the sun (which rotates about once a month) were to collapse down to a white dwarf, its rate of rotation would also speed up. Indeed, when our sun becomes a white dwarf billions of years from now, it probably will be rotating once every few minutes.

Finally, it is reasonable to expect white dwarfs to have strong magnetic fields. As we learned in Section 7.4, our sun possesses a weak magnetic field. The strength of the sun's magnetic field is about the same as the strength of the earth's geomagnetic field. But, when the sun collapses to become a white dwarf, its magnetic field will be compressed. The sun's magnetic field—now spread out over a quadrillion square miles of its surface—will be confined to an area no bigger than the surface of the earth. As the sun's magnetic field is compressed, its strength will increase dramatically until it is about one hundred thousand times its present value.

During the early 1970s, astronomers succeeded in detecting these intense magnetic fields associated with white dwarfs. By measuring the Zeeman splitting of spectral lines in the light from white dwarfs, the astronomer Dr. J. R. P. Angel at Columbia University has discovered magnetic fields between one million and 100 million times stronger than the magnetic field of the sun.

At this point in the discussion of dying stars, some nagging questions should exist in the minds of careful readers. Astronomers know that stars have masses up to about sixty times that of the sun. They also know that massive stars, although they contain great reserves of nuclear fuel, evolve very rapidly. Finally, white dwarfs have masses that *must* be less than $1\frac{1}{4}$ solar masses. Is it really possible that *every* dying star becomes a white dwarf? By what absurd act of clairvoyance is a massive star smart enough to eject just enough of its mass so that its dead core contains less than the critical $1\frac{1}{4}$ solar masses? It would be absurd to presume that a star near the end of its life cycle has some way of saying to itself: "Well, now that I am about to croak, I must be sure to blow off enough gas so that what remains of me is less than Dr. Chandrasekher's limit." Suppose that a dying 40 solar mass star ejects only 35 solar mass as it becomes a supernova. What happens to the remaining 5 solar mass object? It is clearly too massive to become a white dwarf.

In the late 1930s, the Soviet physicist L. Landau and the physicist J. Robert Oppenheimer, "the father of the atom bomb," theoretically predicted the existence of two new types of dead stars: the *neutron star* and the *black hole*. Most astronomers dismissed these two possibilities as fantasies. After all, there seemed to be enough white dwarfs around to account for all the massive stars that already have died. Actually, astronomers just did not know what to look for!

9.2 The Discovery of Pulsars

In the summer of 1967, a team of radio astronomers at Cambridge University headed by Dr. Antony Hewish were making a series of observations with their radio telescope. Suddenly they realized that they were picking up some very unusual signals. These signals consisted of regular bursts of radio noise at perfectly spaced intervals of about one second, like the ticking of a clock. The idea immediately came to mind that perhaps they had detected a man-made spacecraft. After all, the Russians usually launched their satellites in secret and perhaps they were just hearing the transmissions from one of those Soviet vehicles. After a few days of observation, however, the idea had to be discarded. A spacecraft, even if it is on a mission to a distant planet, should appear to change its location with respect to the stars as it orbits the sun; this pulsating radio source remained fixed in the sky. They had to be observing something very far away from our solar system. The object was called a pulsating radio star or *pulsar*, for short. This first pulsar to be discovered was named CP 1919, which stands for "Cambridge Pulsar at right ascension 19^h19^m."

Within a few months, three more pulsars were discovered by the Cambridge radio astronomers. These new sources were called CP 0834, CP 1133, and CP 0950, all named in the same fashion as the first pulsar. During detailed observations of these pulsars late in 1967, a remarkable discovery was made. Looking at a chart record of the signals of any one of these pulsars, such as that shown in Figure 9–5, we notice that some signals are weak whereas others are strong. Yet, careful scrutiny reveals that each of these bursts of radio noise arrives at the earth with incredible regularity. It soon became apparent that the ticking of these pulsars is more regular and precise than that of any other object that has ever been discovered in the universe! For example, radio bursts from CP 1919 arrive at the earth every 1.3373 seconds. Only the very best man-made atomic clocks can rival pulsars in their precision and accuracy.

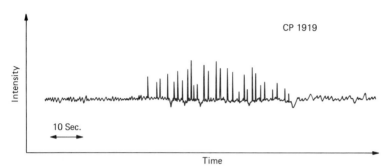

Figure 9-5. *A Pulsar Recording.* The pulses of radio noise from a typical pulsar are shown in this chart recording from a radio telescope. Some pulses are weak and some are strong, but the timing of the pulses are extremely precise.

This discovery of pulsars was kept secret for several months while the Cambridge astronomers checked and rechecked their observations. It isn't often that an astronomer just happens to stumble across an amazing new type of object in the sky; the Cambridge team wanted to make sure they really knew what they were talking about before the word got out.

Most professional astronomers will remember that day in February 1968 when they went to their mailboxes and opened their copies of the journal *Nature*. Astronomers were already very familiar with all sorts of pulsating stars in the sky. There were Cepheid variables, RR Lyrae variables, and Mira-type variables. But *never* had they seen or heard of anything like this! Immediately astronomers and astrophysicists around the world became very busy trying to explain the mechanism that might be responsible for the incredibly precise bursts of radio noise from each of these four pulsars. Everyone began dreaming up all kinds of theories. The scientific journals were flooded with papers as astronomers frantically raced to be "first" in explaining pulsars. Not before or since has more pure nonsense been produced in the name of science in such a short period of time.

Take the *LGM theory*, for example. It tells us that there are a lot of stars in our galaxy and that, in all probability, many of these stars have planets going around them. Of these billions of unseen planets, there must be many that are habitable. If life is a common phenomenon in the universe, surely intelligent creatures have evolved on a few of these planets. Also, according to the LGM theory, some of these creatures may be further advanced then the human race. Perhaps these creatures have even mastered the art of interstellar spaceflight! Obviously, these creatures would find it to their advantage to place navigational beacons around our galaxy to aid in their travels among the stars. Thus, according to the LGM theory, the pulsars are powerful radio transmitters made by a race of highly advanced creatures. Needless to say, the letters "LGM" stand for "little green men."

We will not belabor the multitude of reasons that demonstrate that the LGM theory is an absurd hypothesis. Suffice it to say that observations soon indicated that no intelligent creature would ever construct radio transmitters that have the properties of pulsars. We can expect a great deal more "class" in the design of radio transmitters made by a race of advanced creatures. Astronomers, therefore, turned their attention to examining the properties of pulsars in the hope of arriving at a natural explanation of these incredibly precise signals.

When astronomers discover a new class of objects in the sky, one of the first things they want to know is the distances to these objects. Are they fairly nearby, like the stars in the sky? Or are they very remote, far beyond the stars in our own galaxy? To decide this issue, astronomers find it very useful to make a map of the sky. If a map of the visible sky is drawn with the Milky Way across the center, as shown in Figure 9–6, most of the stars are concentrated strongly across the middle of the diagram, along the Milky Way. If the astronomer draws a similar map of the sky showing the locations of pulsars, examining how the pulsars are distributed should enable him to make a good guess as to their location. For example, if the pulsars are distributed randomly over the sky, then they are probably very distant objects in the universe, lying far beyond the stars in our galaxy. However, if a map of the pulsars in the sky shows that they are concentrated in the regions occupied by the Milky Way, then the pulsars must be fairly nearby.

By the early 1970s, enough pulsars had been discovered so that a map of their locations revealed something about their distribution in the sky. Figure 9–7 shows such a map for the first sixty-five known pulsars. The dotted line runs down the middle of the Milky Way. Clearly, the pulsars are strongly concentrated toward the Milky Way. From comparing the map of the stars (Figure 9–6) and the map of the pulsars (Figure 9–7), astronomers concluded that pulsars must be located among the stars in our galaxy. This is one important clue suggesting that pulsars somehow may be associated with stars.

During the late 1960s and early 1970s, observations of pulsars proceeded at a furious rate. At every major radio astronomy observatory around the world, programs were set up to search for new pulsars. A particularly successful observing program was initiated at the Molonglo Radio Observatory in Australia where astonomers discovered about thirty pulsars. Today, more than one hundred pulsars are known.

As more and more pulsars were discovered, astronomers found that they could make some general statements about their properties. First of all, the *periods* of pulsation (that is, the time between successive bursts) range from about 0.03- to 4 seconds. The most rapid pulsar is NP 0532, whose period is 0.0330911-second; the slowest pulsar is NP 0527, with a period of 3.745491 seconds. (The letters *NP* stand for National Radio Astronomy Observatory pulsar, thereby giving credit to the observatory where the pulsars were first discovered.) Most pulsars have periods of around 1 second.

Most of the time, a pulsar is "off"; no radiation is being emitted. However, at incredibly precise intervals, a strong burst of radio noise is observed. This burst typically lasts less than one twenty-fifth of a second. Careful observations of the

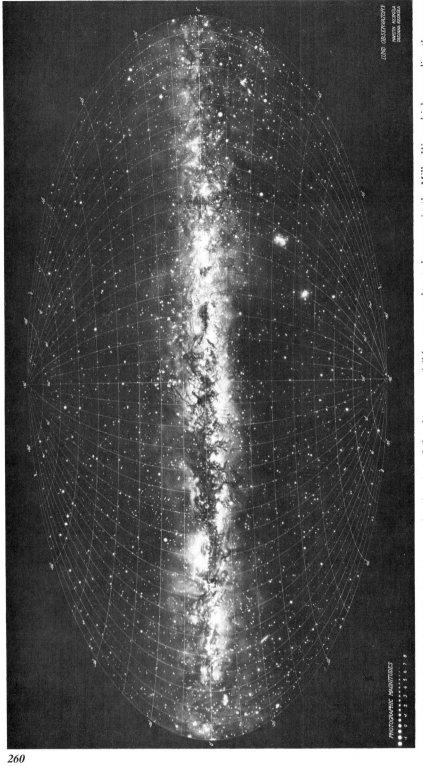

Figure 9–6. The *Visible Stars.* As shown in this drawing of the sky, most visible stars are located near or in the Milky Way which runs directly across the center of the map. (*Lund Observatory*)

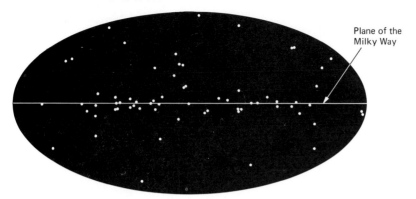

Figure 9–7. The Locations of Pulsars. The locations of 65 pulsars are shown in this map which is drawn in the same fashion as the map of the visible stars (Figure 9–6). Since the locations of the pulsars mimic the locations of stars, it is reasonable to suppose that pulsars are scattered among the stars in the sky. (*Adapted from Dr. Manchester*)

intensity of the radio noise received during the few hundreths of a second when a pulsar is "on" reveal that these bursts have detailed "structure." The actual *pulse shape*, or *pulse profile*, of a burst of radio noise depends on which pulsar the astronomer is observing. Figure 9–8 shows average pulse profiles of four different pulsars, all drawn to the same scale. They all look different. Typically, for example, CP 0950 has a very sharply peaked pulse profile, whereas the pulses from CP 1133 show two humps. These highly magnified views of average pulses must be trying to tell us something about the details of the properties of pulsars.

Whatever pulsars actually are, astronomers have been quick to realize that pulsars can be used as tools to probe the nature of interstellar space. When an astronomer observes a pulsar, he can tune his radio telescope to different frequencies, or wavelengths, much in the same way that you tune your car radio to different stations just by turning a knob. In doing this, astronomers have discovered that the *same* pulse from a pulsar arrives at the earth at *different* times, depending on the wavelength to which a telescope is tuned. A given pulse arrives at the earth later as the astronomer looks at longer wavelengths. This phenomenon is called *velocity dispersion* and is caused by electrons in space between earth and the pulsar. In other words, when a pulsar emits a burst of radio noise, it gives off this noise at *all* wavelengths at exactly the same instant. But as this burst of noise travels toward earth, the electrons in space slow down the radio waves. These electrons slow down the long wavelength part of the burst much more easily than the short wavelength part of the burst. The long wavelength part of the signal is delayed more than the short wavelength part. The size of the resulting delay depends on precisely how many electrons there are between earth and the pulsar. By measuring how much a particular pulse is delayed at longer and longer wavelengths, the radio astronomer can calculate

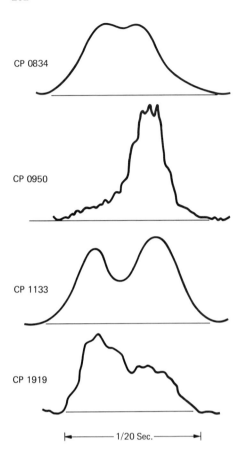

Figure 9–8. *Pulse Profiles.* The detailed shape or profile of a pulse of radio noise from a pulsar varies widely from one pulsar to the next. Four typical pulse profiles are shown here.

the number of electrons in space. From such observations, astronomers conclude that there are about two electrons per cubic inch in interstellar space.

Similar observations of pulses at different wavelengths also can be used to measure the interstellar magnetic field. Due to an effect called *Faraday rotation*, detailed properties of radio waves are altered as they pass through a magnetic field in space. From such observations, astronomers conclude that there is a general magnetic field in the galaxy with an average strength of roughly one millionth the strength of the earth's geomagnetic field.

The important point to realize in these two complicated-sounding types of observations is that pulsars can be used as tools. The stuff in space affects how radio waves move through space. Most notably, the electrons and magnetic fields in interstellar space have profoundly affected the properties of the pulses from pulsars by the time they arrive at the earth. By detailed observations of these pulses, astronomers are now able to probe the properties of the near-perfect vacuum between the stars.

As observations of pulsars proceeded, contrary to initial indications, it was

Chapter 9 White Dwarfs and Neutron Stars

noticed that *all* pulsars are slowing down. Although the periods of pulsars are extremely precise, after many months of careful observation, astronomers noticed that these periods were very gradually getting longer and longer. For example, the first known pulsar, CP 1919, was initially shown to have a period of 1.3373 seconds. In about 3,000 years from now, its period will have increased to 1.3374 seconds. This is typical of the rate at which pulsars are slowing down. Pulsars lose about one billionth of a second each month. Although this may seem to be an unimportant observation, it does have a profound effect on the theoretical work of astrophysicists. Some theories were being talked about that predicted a gradual speeding up of pulsars. These theories were promptly and permanently discarded.

In connection with pulsar periods, a remarkable event was observed in February 1969. Radio astronomers had been examining the pulsar PSR 0833-45 (PSR stands for "pulsar"; the remaining numbers give its approximate location). This is the second fastest known pulsar; its period is about 0.0892-second. Some time during the last week in February (when no one was looking), it suddenly speeded up! On February 24, 1969, the period was 0.0892092-second. When astronomers again observed this pulsar on March 3, 1969, its period was 0.0892090-second. For some unknown reason, it was pulsing one-fourth of a millionth of a second faster than during the week before. This type of a sudden speed-up in the period of a pulsar is called a *glitch*. Glitches also have been observed in the period of NP 0532, the fastest known pulsar.

As more and more data about pulsars accumulated, several interesting facts began to emerge: pulsars have very regular periods; pulsars slow down; and

Figure 9–9. *A Glitch.* Pulsars very gradually slow down. On rare occasions, a sudden speed-up in the period of a pulsar is observed. As shown in this graph, around March 1, 1969, the pulsar PSR 0833-45 suddenly sped up by $\frac{1}{4}$ millionth of a second.

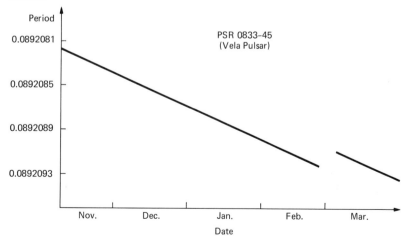

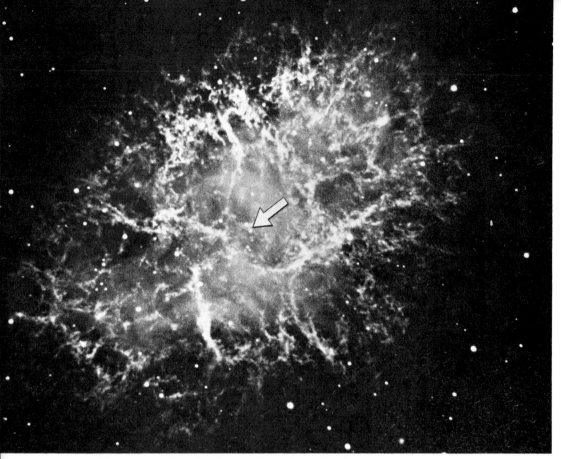

Figure 9–10. *The Crab Pulsar.* A pulsar is located at the center of the Crab nebula (see Plate 11 for a color photograph). This pulsar, NP 0532, has the shortest period of all known pulsars. Its position is indicated by the arrow. (*Lick Observatory*)

Figure 9–11. *The Gum Nebula.* A pulsar is located very near the center of this huge supernova remnant. This pulsar (PSR 0833-45 = the Vela pulsar) has the second shortest period of all known pulsars. (*Cerro Tololo Observatory, courtesy of Dr. Bok*)

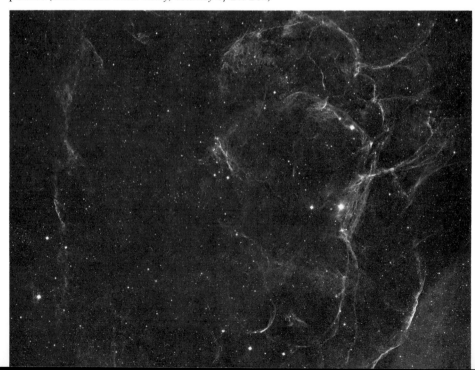

Chapter 9 White Dwarfs and Neutron Stars

their periods very gradually get longer. Therefore, perhaps pulsars with very short periods are young, and pulsars with very long periods are old. There are two pulsars with periods of less than a tenth of a second: NP 0532 in Taurus and PSR 0833-45 in Vela. NP 0532 is located in the middle of the Crab nebula (which is known to be a supernova remnant) and is, therefore, called the *Crab pulsar*. PSR 0833-45, the *Vela pulsar*, is located in the middle of the Gum nebula (named after Dr. C. S. Gum, who discovered the nebula in the early 1950s), which is also a supernova remnant. In other words, the two youngest, most recently created pulsars are associated conclusively with the remains of supernovae. Could it be that a pulsar is a dead star? Is it possible that dying stars that are too massive to become white dwarfs somehow turn into pulsars? In order to answer these fascinating questions, astrophysicists had to examine the theoretical properties of a massive dying star in greater detail than ever before.

9.3 Pulsars and Neutron Stars

Up until the late 1960s, most astronomers believed that white dwarfs were the *only* possible final state for a star. There seemed to be enough white dwarfs in the sky to account for all the stars that have died. Although Dr. Chandrasekher had proved that the masses of white dwarfs *must* be less than $1\frac{1}{4}$ solar masses, it was generally assumed that somehow all stars manage to get below this critical Chandrasekher limit. The general conclusion was that, during the process of dying, all stars—even the most massive—somehow succeeded in ejecting enough material into space in the form of a planetary nebula or supernova so that their dead cores would be less than $1\frac{1}{4}$ solar masses. But with the discovery of pulsars, some astronomers began to have grave doubts.

Think for a moment about a massive dying star. All of its nuclear fuel has been used up, and it begins to contract under the force of gravity. Suppose, furthermore, that during a supernova explosion this dying star succeeds in ejecting all but 2 solar masses of its matter. In other words, the collapsing core is *above* the Chandrasekher limit. What happens?

As this 2 solar mass dead star contracts under its own weight, the degenerate electron pressure, which would support a white dwarf, is *not* strong enough to stop the contraction of the star. The star simply gets smaller and smaller. The huge pressures that arise from the degeneracy of the electrons cannot support the crushing weight of the outer layers of the star. As the star continues to shrink, the pressures and densities inside it become so enormous that a remarkable process begins to occur. When the compression of the material inside the star reaches the point at which each cubic inch contains a billion tons of matter, atomic particles begin getting squeezed inside each other. In particular, the electrons in the star are squeezed into the nuclei of the atoms. When this happens, the negatively charged electrons combine with the positively charged protons to produce neutrons. Fairly soon after this metamorphosis begins, almost the entire star is transformed into neutrons. The entire star has become one gigantic atomic nucleus. It has become a *neutron star!*

In thinking about the properties of matter inside a neutron star, scientists are quick to realize that the same basic physical laws that applied to white dwarfs should still be relevant. In particular, the Pauli exclusion principle tells us that no two identical neutrons can exist in the same space. This law of physics states that there is a minimum volume into which a given number of neutrons can be squeezed. In other words, the star again becomes *degenerate,* but this time it is the neutrons that produce the degeneracy. When the star is about 90 per cent neutrons, these neutrons comprise a *degenerate gas.* This peculiar gas gives rise to an incredibly enormous pressure that vigorously resists any further compression. Each and every small volume inside the star contains the maximum allowed number of neutrons; to try to squeeze any more neutrons into these volumes would violate the laws of physics. The resulting pressure stops the contraction of the massive dying stellar core. A neutron star is born!

Theoretical models based on this scenario tell us that a neutron star is about 15 miles in diameter. When all the matter of a 2 solar mass star has collapsed down to a sphere 15 miles in diameter, the pressure from the degenerate neutrons is sufficient to support the star. Matter inside such a neutron star is incredibly dense. One heaping tablespoonful of material from a neutron star would weigh 3 billion tons here on the earth!

The initial ideas concerning neutron stars date back to the 1930s. In 1934, W. Baade and F. Zwicky at Caltech proposed the concept of a neutron star; the first detailed calculations were done by Oppenheimer and Volkoff in 1939. These works were largely ignored as interesting fantasies until the discovery of pulsars in the late 1960s. In 1968, primarily as a result of suggestions made by Dr. T. Gold at Cornell University, it was realized that neutron stars have many attractive features that are related to pulsars. This is especially apparent when we consider the formation of a neutron star in greater detail.

It seems quite reasonable to suppose that all stars rotate and have magnetic fields. For example, our sun is a typical, garden-variety star. It rotates about once a month and has a weak magnetic field. The strength of the sun's magnetic field is roughly of the same intensity as the earth's magnetic field. So, consider carefully what happens to the magnetic field and rotation rate of a star such as the sun as it collapses to become a neutron star.

In connection with the discussion of white dwarfs, a law of physics called the conservation of angular momentum has been discussed. This law explains how fast things rotate—whether they speed up or slow down—as they change shape. For example, a direct consequence of this law is that an ice skater doing a pirouette on the ice will rotate more rapidly when she pulls in her arms (see Figure 9–4). For precisely the same reason that the ice skater speeds up when she pulls in her arms, the rotation rate of a star will increase as it shrinks in size. As a collapsing star becomes more and more compressed, it will rotate faster and faster. Indeed, if a star such as the sun were to collapse down to a neutron star

Chapter 9 White Dwarfs and Neutron Stars

only 15 miles in diameter, it would be rotating faster than once a second. This seems to be a very enticing clue to the mystery of pulsars. Young pulsars (for example, NP 0532, the Crab pulsar) emit bursts of radio noise about a hundred times a second. When a slowly rotating star such as our sun collapses down to a neutron star, it should be rotating about a hundred times a second.

Just as the rotation rate of a collapsing star would increase, the strength of its magnetic field also should rise dramatically. In the discussion of white dwarfs, remember that when a star such as our sun contracts to form a white dwarf, the strength of its magnetic field will increase by approximately a hundred thousand times. The star's magnetic field, which initially was spread out over a sphere 1 million miles in diameter, becomes compressed and is then confined to the surface of a white dwarf whose diameter is less than 10,000 miles. If, however, the stellar collapse proceeds to the neutron star stage, the magnetic field is then squeezed down to the surface of a star whose diameter is only 15 miles. The resulting magnetic field will be more than a billion times stronger than it was before collapse. Therefore, astronomers speculate that neutron stars have magnetic fields that are a trillion times stronger than the sun's magnetic field.

To summarize the expected properties of neutron stars, we have found that

1. A neutron star is small. Its diameter is about 15 miles.
2. A neutron star is dense. One cubic inch of material from a neutron star contains about three billion tons of matter.
3. A neutron star is rotating rapidly. Just after formation (before it begins to slow down appreciably) it probably rotates at least one hundred times per second.
4. A neutron star should have a very intense magnetic field. The magnetic field of a neutron star should be between a billion and a trillion times stronger than the average present-day strength of the sun's magnetic field.

Now, think for a moment about the earth. The earth is rotating and the earth has a magnetic field. *But* the earth's axis of rotation is *not* pointed in exactly the same direction as the axis of the earth's magnetic field. A compass does not point to "true north." A compass points to the magnetic North Pole of the earth, which is located in northern Canada. By analogy with the earth, it would be surprising if the magnetic axis of the neutron star were exactly parallel to its rotational axis. Rather, we would expect the magnetic axis to be tilted with respect to the axis of rotation. The resulting configuration is shown in Figure 9–12.

To complete this story of neutron stars, it is necessary to discuss what is happening on the surface of the star. The matter at the surface of a neutron star is not nearly as dense as the matter inside the star. There are still many protons and electrons that have not been squeezed inside each other to form neutrons in the outermost few inches of the star's surface. According to electromagnetic theory, these charged particles (protons and electrons) will be accelerated outward at the star's north and south magnetic poles. As this happens, the charged particles emit electromagnetic radiation. We, therefore, have arrived at

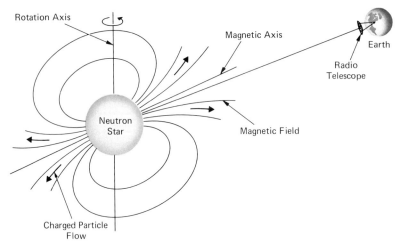

Figure 9–12. *A Neutron Star.* Pulsars can be explained as rapidly rotating neutron stars with intense magnetic fields. Every time the beams of radiation emanating from the magnetic poles of the neutron star shine toward the earth, astronomers observe a pulse of radiation.

the very important realization that a beam of light should be coming out of the north pole of the star and a second beam of light should be coming out of the south pole!

As the neutron star rotates, these two beams of radiation will sweep across the sky. *If* the earth is located in the right direction, every time the star rotates we will have the opportunity to look straight down the north pole or south pole of the star. For a fraction of a second we will see a burst of radiation, as the beam from the neutron star sweeps past our line of sight. We will see a pulsar!

All of the known properties of pulsars can be fully explained by this so-called *oblique-rotator model.* The rotating neutron star acts like a lighthouse beacon. Every time the beam of radiation sweeps past the earth, a pulse can be seen. If the neutron star rotates once a second, we see pulses once a second. In addition, since the neutron star is putting out all this light into space, it must be using up energy. As it uses up energy, it must gradually slow down. Indeed, pulsars are slowing down just enough (one billionth of a second each month) to account for the energy output.

The oblique-rotator model of pulsars has been so successful in explaining the mysterious pulses first observed in 1967, that few astronomers have any doubts about the existence of neutron stars. Consequently, many astrophysicists are currently engaged in detailed calculations that might help explain many of the fine points of pulsars. For example, it now seems that the *crust* of a neutron star, the outermost layer of which is about 2 miles thick, should be crystalline and very brittle. So it is reasonable that cracks and fissures could develop in this brittle crust and result in *starquakes.* During a starquake, the crust of the star suddenly would readjust itself due to the stresses that have been building up from the slowing of the pulsar's rate of rotation. This is the most probable explanation of the glitches. The most powerful and devastating earthquakes that occur here on the earth measure $8\frac{1}{2}$ on the *Richter scale*. By comparison, the starquakes that produce the glitches in the Crab pulsar would measure $18\frac{1}{2}$ on the Richter scale. The starquakes that produce the glitches in the Vela pulsar probably occur much deeper inside the neutron star; they would measure as high as 23 on the Richter scale!

All this raises some interesting questions. If this oblique-rotator model is so great, why are only radio pulses seen from pulsars? Why shouldn't pulses in visible light be seen? Why shouldn't X-ray, γ-ray, and infrared pulses be detected? Astronomers began to wonder about these possibilities back in the late 1960s.

In January 1969, a team of astronomers at the Steward Observatory in Arizona pointed their 36-inch reflecting telescope at the center of the Crab nebula. Using a photomultiplier, they found that a star in the middle of the Crab nebula was actually blinking on and off thirty times a second. Its period was exactly the same as the radio pulsar NP 0532. A few months later, another team of astronomers at Kitt Peak succeeded in photographing the Crab pulsar in visible light as it blinks. This was accomplished by using a TV camera and a rotating disc, as shown in Figure 9-13. Slits in the rotating disc alternately block out or let

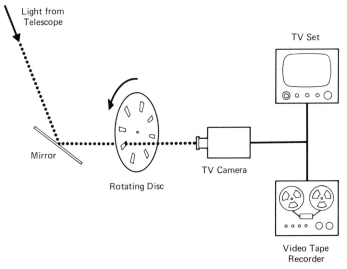

Figure 9-13. *TV Observations of Pulsars.* Using this apparatus, astronomers at the University of Arizona first detected visible pulsations of the Crab pulsar (NP 0532). Light from the Crab pulsar is "chopped" by a rotating disc and the results are viewed on a television screen.

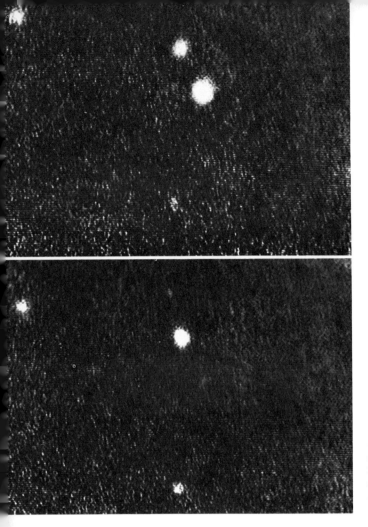

Figure 9–14. TV Photographs of the Crab Pulsar. Two television photographs of the stars at the center of the Crab nebula are shown here. The Crab pulsar is clearly seen to be blinking on and off 30 times each second. (*Lick Observatory*)

through light from the pulsar. By rotating the disc at different speeds, they were able to catch the pulsar at various parts of its cycle. A series of photographs showing the variation in brightness of the pulsar during one cycle is shown in Figure 9–15. It is interesting to notice that in addition to the main pulse there is a secondary pulse. This is probably because NP 0532 is oriented in such a way that we are able to see *both* the north and south poles of the rotating neutron star. Therefore, during one period of 0.331-second we observe *two* bursts of energy. Based on these observations, Dr. Erika Bohm-Vitense at the University of Washington was able to calculate the relative orientations of the rotational axis, magnetic axis, and direction to the earth for the Crab pulsar. Figure 9–16 shows the configuration of the rotating neutron star based on her work.

In 1969, it was discovered that observable pulses of X rays, γ rays and infrared light are also coming from the Crab pulsar. In fact, back as far as June 1967, a team of astronomers at Rice University flew a balloon carrying an X-ray telescope. They observed the Crab nebula. Two years later they reexamined their data and found that they had indeed observed X-ray pulses from the center of the nebula, even before pulsars had been discovered! No one had expected rapid

Chapter 9 White Dwarfs and Neutron Stars

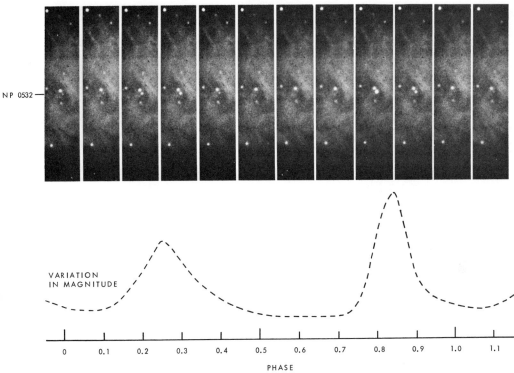

Figure 9-15. *The Crab Pulsar.* This series of photographs shows the visual appearance of the Crab pulsar over one cycle. For comparison, the "light curve" is shown below the photographs. (*Kitt Peak Observatory*)

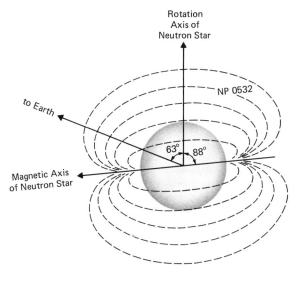

Figure 9-16. *A Model of the Crab Pulsar.* The Crab pulsar can be explained in terms of a rapidly rotating neutron star with an intense magnetic field oriented as shown in this diagram. (*Adapted from Bohm-Vitense*)

bursts of X rays from the Crab nebula and, therefore, up until 1969, no one had looked for any.

Perhaps even more surprising is the fact that, back in 1942, Dr. R. Minkowski, at Berkeley, proposed that an unusual star with no spectral lines at the center of the Crab nebula might be the core of the dead star that produced the nebula. He was looking right at the Crab pulsar, but no one noticed that it was blinking on and off thirty times a second.

The Crab pulsar, NP 0532, is the *only* pulsar that has been observed over a wide range of wavelengths. Figure 9–17 shows how the pulses in X rays, visible light, and radio waves are correlated. Searches for optical and X-ray pulses from other pulsars have been going on since the late 1960s. For example, a great deal of effort has been made to try to observe visible pulses from the Vela pulsar. However, all such searches have met with failure. The Crab pulsar is, therefore, one of the most important pulsars known to astronomers.

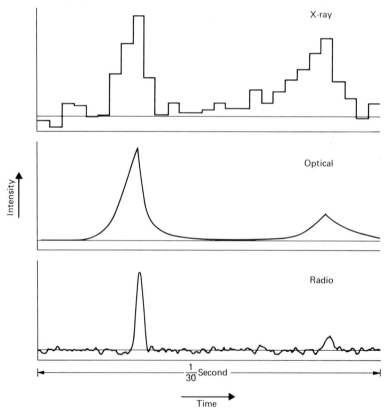

Figure 9–17. X-Ray, Optical, and Radio Observations. Pulses of radiation at all wavelengths are observed coming from the Crab pulsar. These graphs show how the X-ray, optical, and radio pulses are correlated over one cycle.

The Crab pulsar is the youngest of all known pulsars. All other pulsars have periods longer than that of NP 0532 and, therefore, are considerably older. Perhaps as pulsars get old and slow down, they are no longer powerful enough to emit substantial bursts of visible or X-ray radiation. They can only be observed with radio telescopes. Nevertheless, the fact that pulsars are dead stars created during supernovae explosions is now crystal clear. Many pulsars do not seem to be located near supernova remnants, but that is probably because the nebula has dissipated and vanished from view. Also, there are many supernova remnants that do not seem to contain pulsars. This is probably because either the dead star failed to turn into a neutron star (it could have become a white dwarf or—as will be discussed in later chapters—a black hole) *or* the beam of radiation from the rotating neutron star never shines toward the earth.

The discovery of pulsars is unquestionably one of the most important events in the history of astronomy. In particular, the Crab pulsar has supplied astronomers with strong evidence that pulsars are rapidly rotating neutron stars. Astronomers have been observing the neutron star at the center of the Crab nebula for many years. However, until very recently, no one knew how to observe this star in order to discover its true nature. Indeed, the actual formation of this neutron star was seen on July 4, 1054, as recorded in Part 9, Chapter 56 of *Sung Shih*, or the *History of the Sung Dynasty*. In this document, the Chinese historian Toktaga tells us that "on a chi-chhou day in the fifth month of the first year of Chih-Ho Reign-Period a 'guest star' appeared at the south-east of Thien-Kuan, measuring several inches. After more than a year, it faded away."

9.4 Black Holes in Space?

In the early 1970s, astronomers concluded that there are at least *two* possible final states in stellar evolution. A star could become a white dwarf, in which case it is supported by the pressure supplied by degenerate electrons. Or a star could become a neutron star, in which case it is supported by the pressure supplied by degenerate neutrons. The existence of white dwarfs had been known to astronomers for many years. With the discovery of pulsars, convincing evidence for the existence of rapidly rotating neutron stars with intense magnetic fields was rapidly accumulated.

The discovery of neutron stars prompted many astrophysicists to construct detailed theoretical models of dead stars. Using methods very similar to those described at the beginning of this chapter, astrophysicists found that they could calculate a variety of interesting things about what goes on inside white dwarfs and neutron stars. For example, there is now a field of theoretical study affectionately known as *neutron star geology*. We referred briefly to some of the results of calculations by the geologists who study neutron stars when we spoke about starquakes as the cause of glitches in pulsars. By 1975, the geological picture of a neutron star consisted of a solid crust and a solid core separated by a region of *superfluid* matter at a temperature of almost one billion degrees. In the laboratory, *superfluidity* is only observed in liquid helium at temperatures very close to absolute zero. Superfluid helium exhibits a number of very weird properties. For example, superfluid helium in a glass or jar will creep up the sides of the container in defiance of gravity.

One of the most fundamental theoretical discoveries to come from the recent work of astrophysicists is that there is an upper limit to the mass of a neutron star. We recall that from the work of Dr. Chandrasekher, the upper limit to the mass of a white dwarf is $1\frac{1}{4}$ solar masses. In a dead star whose mass is above the Chandrasekher limit, the degenerate electron pressure is not sufficient to halt gravitational contraction and support the star. Similarly, there is an upper limit to the mass of a neutron star, above which the degenerate neutron pressure is not sufficient to halt gravitational contraction and support the star. In view of the implications and importance of this result, in this section we will examine how astrophysicists arrive at this conclusion.

In order to construct models of dead stars, the astrophysicist takes the laws of nature, particularly those involving nuclear physics and degenerate matter, and programs them into his computer. He then applies these laws to imaginary massive spheres of matter containing no nuclear fuels and asks the computer to calculate the properties of such spheres. He can ask his computer about sizes, temperatures, pressures, densities—all sorts of things—of dead stars. For example, he can ask: "For a dead star of such-and-such solar masses, what is the density (tons per cubic inch) at the star's center?" The computer, if properly programmed, will produce a strange-looking wiggly line on a graph, as shown in Figure 9–18.

Anyone who has ever used a computer knows that you should never really trust them. More precisely, in doing stellar structure calculations, the careful astrophysicist will occasionally check the computer's answers by laboriously grinding out the computations by hand with pencil and paper. In this way, the astrophysicist feels confident that he is not being misled by the computer's results, since the computer does not have the physical intuition to evaluate itself and insure that its answers are always physically reasonable. Therefore, we must examine the wiggly curve given in Figure 9–18 to see what portions are physically meaningful and what portions should be discarded as nonsense.

Imagine standing on an ordinary bathroom scale, measuring how much you weigh. Suppose that while you are standing on the scale, someone hands you a brick. It is obvious that when you take the brick, your weight, as measured by the scale, will increase. You and the brick together weigh more than just you alone.

Now examine Figure 9–18. This graph tells us the density of matter (measured in tons per cubic inch) for dead stars of various masses. Imagine adding matter to a dead star—for example, by laying bricks onto its surface. By analogy with the experiment with the bathroom scale, the density at the center of the star must go up. In other words, increase the star's mass, and its central density also must increase. This amounts to nothing more than a little common sense. Therefore,

Chapter 9 White Dwarfs and Neutron Stars

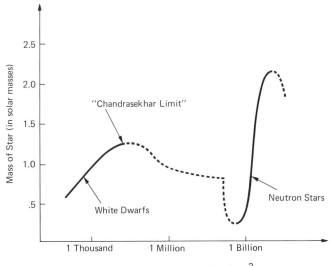

Figure 9–18. *Mass vs. Density of "Dead" Stars.* Calculations concerning the structure of dead stars show two stable configurations. If degenerate electron pressure is sufficient to support the star, a white dwarf results. If degenerate neutron pressure is sufficient to support the star, a neutron star is formed.

only those portions of the wiggly curve that correspond to increasing density with increasing mass will be considered physically reasonable. Only those portions of the curve that slope toward the upper righthand corner of the graph can represent stars that could actually exist in nature. These portions of the curve are drawn with a solid line. Those portions of the curve drawn with a dotted line cannot be related to real stars. The dotted portion of the curve represent stars that would have the absurd property of increasing central density with decreasing mass. If you go on a diet and lose ten pounds, but your bathroom scale tells you that you weigh *more* than before, obviously your scale is broken. You should discard your old scale and purchase a new one that will give correct answers. Similarly, astronomers discard models of dead stars that exhibit increasing density with decreasing mass; they cannot exist in nature.

There are only two portions of the curve in Figure 9–18 that can represent real stars. At densities of around 1,000 tons per cubic inch, degenerate electrons are supporting the star; it is a white dwarf. At densities around 10 billion tons per cubic inch, degenerate neutrons are supporting the star; it is a neutron star. The familiar Chandrasekher limit stands out on the white dwarf portion of the curve. Up to a density of about 20,000 tons per cubic inch, degenerate electron pressure is powerful enough to resist compression. At higher densities, when the matter in the star is squeezed even tighter, this is no longer the case. The curve on the graph reaches a maximum at $1\frac{1}{4}$ solar masses and then turns over.

Behavior similar to the curve in Figure 9–18 is exhibited in the case of neutron stars. Degenerate neutron pressure can support the star up to densities of about 20 billion tons per cubic inch. The curve on the graph reaches a maximum at $2\frac{1}{2}$ solar masses. This is the upper limit for the mass of a neutron star. Neutron stars cannot exist with masses greater than $2\frac{1}{2}$ times that of the sun!

A second important discovery to come from the calculations that are Figure 9–18 is that there are *only* two stable types of dead stars. At densities higher than 20 billion tons per cubic inch, the curve *always* slopes downward, corresponding to increasing density with decreasing mass. In other words, there are *no* stable dead stars with densities higher than 20 billion tons per cubic inch.

Another question the astrophysicist could ask of his computer goes like this: "Given a dead star of such-and-such a mass, how big is it?" Figure 9–19 shows the answers provided by the computer, again in graphic form. In this diagram two curves are plotted. The solid line is for dead stars that are *not* rotating. However, if a star is rotating, the atoms in the star will feel an outward *centrifugal force*. Therefore, we suspect that a rotating dead star might be able to contain more matter than one that is not rotating. Perhaps these forces due to rotation would assist the degenerate pressures in supporting the star. Indeed, our suspicions are confirmed by calculations. The dashed curve in Figure 9–19 corre-

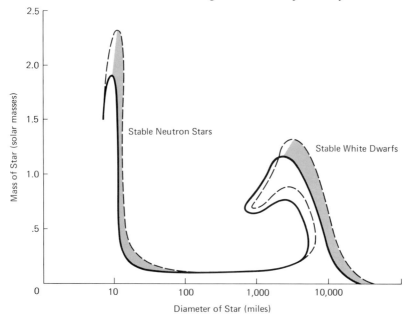

Figure 9–19. *Mass vs. Size of a "Dead" Star.* Calculations concerning the diameters of dead stars show that white dwarfs are about the same size as the earth. Neutron stars are only 10 miles in diameter. The solid curve is for non-rotating stars while the dashed curve is for stars rotating at the "break up" velocity.

sponds to rapidly rotating dead stars, stars that are rotating at their breakup velocity. Since the solid curve is for nonrotating stars, and the dashed curve is for stars rotating as fast as possible, models of real dead stars *must* be located *in between* these two extreme curves on the graph. In addition, since real dead stars can only be white dwarfs or neutron stars, there are only two small regions on the graph in Figure 9–19 that correspond to what can exist in nature. These regions are the shaded areas labeled stable white dwarfs and stable neutron stars.

If rotation is included to aid in supporting a dead star, the maximum mass of a white dwarf is about $1\frac{1}{2}$ solar masses. Such a star has a diameter of about 5,000 miles. Similarly, the maximum mass of a rapidly rotating neutron star is about 3 solar masses. Its diameter is about 10 miles. These calculations are extremely critical because they tell us that there are no dead stars with masses greater than three times that of the sun under any conditions whatsoever.

But what happens to a dying star that does not eject all but 3 solar masses of its matter during a supernova? Suppose a huge star has 5 or 6 solar masses of material left over after it becomes a supernova. What happens to the remains of a dead star whose mass exceeds the upper limit for a neutron star?

As indicated, a dead star whose mass is greater than three times the sun cannot become a white dwarf. Degenerate electron pressure is far too weak to hold up the enormous crushing weight of the star's outer layers. Similarly, it cannot become a neutron star. The degenerate neutron pressure is also too weak to support the star. In fact, there are *no* physical forces whatsoever that are strong enough to hold up the star. The star simply gets smaller and smaller and smaller.

Back in Chapter 2, we discussed Newton's law of gravitation. To illustrate how gravity works, we considered the hypothetical problem of how much a man standing on the earth would weigh if the earth were squeezed down to a small size. We found that if we could squeeze all the matter of the earth down into a small ball, the man's weight would become enormous. Therefore, as all the matter of a very massive dying star collapses in on itself, the intensity of the gravitational field above the star can become enormous. Soon the intensity of this incredible gravitational field is so strong that Newton's law of gravity no longer properly describes what is going on. Instead, astronomers must turn to a new theory of gravity, the *general theory of relativity*. According to general relativity, when the collapsing massive star is only a couple of miles in diameter, the intensity of the gravitational field is so great that the very fabric of space and time fold in over themselves and the star disappears from the universe! What is left is called a *black hole*.

We, therefore, conclude that there are *three* possible final states for stellar evolution. Stars less massive than $1\frac{1}{2}$ suns (if rotation is included) can become *white dwarfs*. Stars with masses up to 3 suns can become *neutron stars*. But dead stars with masses greater than 3 solar masses must become *black holes*.

The possibility of black holes existing in the universe has been known since 1939 as a result of calculations performed by Drs. Oppenheimer and Snyder.

Astronomers, most of whom tend to be extremely conservative, scoffed at these ideas as pure fantasies. However, in the very recent past it has become entirely feasible that black holes do exist, and that the theory of relativity plays a critical role in the nature of the universe. Indeed, it seems quite impossible to understand the universe at all without thinking in four dimensions, without speaking about the warping of space and time. We will, therefore, turn our attention to one of the most beautiful, elegant, and profound creations of the human mind, the special and general theories of relativity.

Questions and Exercises

1. In your own words, briefly discuss how an astrophysicist "constructs" a theoretical model of a star.
2. What is the Pauli exclusion principle?
3. What is meant by a degenerate electron gas, and in what kinds of stars is it important?
4. What is the Chandrasekhar limit?
5. Roughly, what is the diameter of a typical white dwarf?
6. As white dwarfs cool off, how do they move on the H-R diagram?
7. Why is it reasonable to suppose that many white dwarf stars are rotating rapidly?
8. What evidence is there for magnetic fields associated with white dwarfs?
9. When was the first pulsar discovered?
10. About how many pulsars are known?
11. What is the range in periods of pulsars?
12. Briefly discuss two ways in which pulsars can be used as tools to probe interstellar space.
13. In what kinds of objects are the Crab and Vela pulsars located?
14. Under what conditions can the matter in a star be converted almost completely into neutrons?
15. What is meant by a degenerate neutron gas, and in what kinds of stars is it important?
16. Compare and contrast the sizes, masses, and densities of neutron stars and white dwarfs.
17. Briefly describe the oblique-rotator model of a neutron star.
18. How does the oblique-rotator model account for pulsars?
19. What are starquakes, and how might they be related to glitches?
20. How are visible pulses of the Crab pulsar observed?
21. Give two reasons why pulsars have not been discovered in every supernova remnant.

22. What is the maximum mass of a white dwarf?
23. What is the maximum mass of a neutron star?
24. Name the three possible final states of stellar evolution.
25. What might happen to a dead star whose mass is greater than 3 solar masses?

The Special Theory of Relativity

10.1 The Crisis in Classical Physics

Some of the most important advances in man's understanding of physical reality came about as a direct result of the work of James Clerk Maxwell. Up until the early 1800s, the phenomena of electricity and magnetism were considered to be separate and unrelated. Electricity had something to do with Lyden jars, galvanic cells (batteries), and Ben Franklin's kite. Magnetism involved strange iron rocks that always pointed toward the north when suspended from a string. However, the experimental work of great physicists such as M. Faraday and H. C. Oersted demonstrated that there is an intimate and profound relationship between electric and magnetic fields. It was not until Maxwell succeeded in formulating his *electromagnetic theory* that the full implications of these relationships could be appreciated. The four *Maxwell equations* at the heart of this electromagnetic theory form the complete foundation of modern understanding of all electrical and magnetic phenomena. Why light bulbs, TV sets, radios, or almost anything else works can be explained completely by applications of Maxwell's equations.

One of the most remarkable discoveries to come from Maxwell's work is that his four equations can be combined mathematically in such a way as to give *wave equations* that can be thought of as actually "predicting" the existence of light. As explained in Chapter 5, this discovery led scientists to realize that radiowaves, X rays, γ rays, and visible, infrared, and ultraviolet radiation are all really the same thing: electromagnetic radiation.

In mathematically deriving the wave equations that describe electromagnetic radiation, something rather unexpected occurs. In the middle of these fundamental equations, a number pops up that has the precise value of 186,000 miles per second. This number is a speed; it tells how fast electromagnetic waves move through empty space. This is the first time in physics that a velocity appears in equations at the most basic and fundamental level. The implications are truly profound.

In discussing electromagnetic waves in Chapter 5, an analogy was drawn with

Chapter 10 The Special Theory of Relativity

water waves on the surface of the ocean. Many of the properties of light can be appreciated and understood if we think about how water waves behave. Yet, water waves move across the surface of the ocean because there is water in the ocean. Obviously, if the oceans dried up, water waves would not exist. Physicists during the nineteenth century believed that, just as water is needed to propagate waves on the ocean's surface, some *medium* must exist in space that generates electromagnetic waves. If empty space were *really* empty, how could light waves move? Wouldn't this be like trying to have water waves where there isn't any water?

As a result of this line of thought, physicists in the mid-1800s decided that something must exist in a vacuum and in empty space. Light obviously moves through vacuums and empty space, so there must be something that supports light waves. This "something" is called the *ether*. The ether is everywhere. It is the ether that wiggles as electromagnetic waves move through empty space, just as the surface of the ocean moves up and down as it is traversed by water waves. Furthermore, this velocity that appears in Maxwell's equations—186,000 miles per second—must be the speed with which electromagnetic waves move through this mysterious, hypothetical ether. As will be discussed shortly, this bizarre concept of the ether is total nonsense.

So, more than 100 years ago, physicists believed in the ether. Empty space was permeated by this ether, and the speed of electromagnetic radiation through the ether was thought to be 186,000 miles per second. But some scientists soon began to have serious doubts about the existence of this mysterious stuff that is supposed to exist everywhere.

In the 1880s, the American physicist A. A. Michelson decided to try to detect the existence of the ether. He realized that if the ether exists everywhere, then the earth must be moving through the ether as it orbits the sun. In fact, since the speed of the earth about the sun is $18\frac{1}{2}$ miles per second, the ether must be flowing past the earth. Indeed, we could think of the earth as an island in the middle of a river of ether that is streaming past us.

Imagine sitting with some friends on an island in the middle of a large river. Suppose that you have two identical boats equipped with identical outboard motors. In still water, both of these boats move at exactly the same speed, say 10 miles per hour. But the water flowing past the island is not still; the river is flowing downstream at 5 miles per hour. Now perform an experiment. One of your friends jumps into a boat and travels 1,000 feet toward the bank of the river and then turns around and comes back. He is traveling *perpendicular* to the direction in which the river is flowing. Meanwhile, another friend jumps into the second boat and travels 1,000 feet downstream and then turns around and comes back. He is traveling *parallel* to the direction in which the river is flowing. Let's suppose that two posts have been planted in the riverbed 1,000 feet from the island so that your friends know exactly where to turn around. The setup of your experiment is shown in Figure 10–1.

If the river were not moving (if you were on a lake rather than a river) each of your friends would take exactly the same time to make his journey. They have

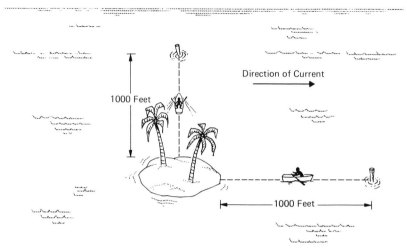

Figure 10–1. *Boats in a River.* Two people use identical boats and traverse identical distances in a flowing river. One person travels perpendicular to the river current while the other person travels parallel to the river current. The person who travels perpendicular to the current always will complete the round trip in a shorter period of time.

identical boats that travel identical distances. But the river *is* flowing. Therefore, the real speeds of the boats in this moving water will *not* be the same. For example, the fellow who travels parallel to the river flow is helped by the current in going downstream but must fight the current on the return trip upstream. His real speed going downstream is 15 miles per hour (10 miles per hour for the boat *plus* 5 miles per hour for the current), but his real speed going upstream is 5 miles per hour (10 miles per hour for the boat *minus* 5 miles per hour for the current). The whole point of this experiment is that your two friends, traveling in identical boats over identical distances, will take *different* times to complete their journeys. The fellow who travels perpendicular to the stream *always* completes the journey in less time than the fellow who travels parallel to the flow of the river. The mathematically inclined reader can verify that, with the numbers given in this experiment, the boat that travels perpendicular to the stream completes the journey in $2\frac{1}{2}$ minutes, as opposed to 3 minutes for the round-trip time for the second boat traveling parallel to the stream.

The situation of the island in the middle of a river is analogous to the ether flowing past the earth. From this analogy, Michelson, later joined by E. W. Morley, proceeded to devise an experiment to detect the ether streaming. The apparatus they used is called a *Michelson interferometer* and is shown schematically in Figure 10–2.

In the Michelson interferometer, a beam of light is emitted by a lamp that travels toward a beam splitter at the center of the apparatus. The beam splitter allows about half of the light to pass through to a mirror at M_1 while the rest of the light is reflected at right angles to a mirror at M_2. The mirrors at M_1 and M_2 return these beams to the center of the apparatus. These beams travel identical distances and, after they come together, they are observed by the scientist, as shown in the diagram. When these beams are reunited, they will *interfere* with each other and produce *interference fringes*. The scientist will, therefore, see a pattern of dark and light bands as he looks into the apparatus.

It is important to note that the Michelson interferometer is attached to the earth, and the hypothetical ether is streaming past the earth just as the water in the river was streaming past the island. Therefore, the beam of light that travels perpendicular to the motion of the earth will complete the round-trip to and from the beam splitter in less time than the beam that travels parallel to the earth's motion. After Michelson and Morley set up their experiment, all they had to do was sit and wait because as the earth rotates, the two arms of the apparatus will interchange roles. If at 6:00 A.M. one arm is perpendicular to the earth's motion while the other is parallel, by 12:00 noon the earth will have turned through 90° and the first arm will be parallel to the earth's motion while the second arm is perpendicular. As the two arms of the apparatus interchange roles, the travel times of light also will be exchanged. This, in turn, means that the interference fringes should *shift* back and forth as the earth rotates.

In spite of long and careful observation, Michelson and Morley *failed* to detect any fringe shifts at all. No matter in which direction the arms of the interferometer were pointing, the travel time of light did not change. There are, therefore, only two logical conclusions to come from this so-called *Michelson-Morley experiment*. Either the earth is not moving (which is absurd!), or there is something *very* wrong with physics (which is frightening!).

The idea that something really weird was going on in physics was also apparent from theoretical considerations involving Maxwell's equations. All the difficulties centered about the appearance of this number, 186,000 miles per second, in the wave equations for light. By the end of the nineteenth century, therefore, a severe crisis had developed in physics. Physical reality was not behaving according to the standards of common sense!

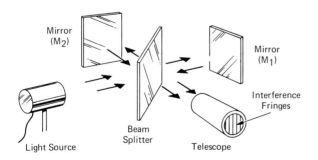

Figure 10–2. The Michelson Interferometer. The Michelson interferometer splits a beam of light in two. One of the resulting beams travels perpendicular to the earth's motion while the other beam travels parallel to the earth's motion.

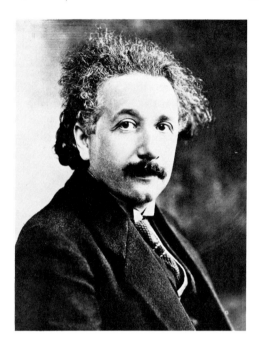

Figure 10-3. Albert Einstein, 1879–1955. (*Yerkes Observatory*)

In 1905, a young physicist named Albert Einstein finally succeeded in resolving the dilemma of the physics of light. For many years he worked with the theoretical aspects of Maxwell's equations, focusing his attention on the meaning of the appearance of the velocity of light in the mathematics. He considered that Maxwell's equations were trying to say something more far-reaching and profound than ever before imagined. He attributed unique and special properties to the speed of 186,000 miles per second, usually symbolized by the letter c. Dr. Einstein found that he could solve all the difficulties with light if he assumed that *the speed of light is an absolute constant.* He supposed that the speed of light is *always* 186,000 miles per second, and the problems of physical reality disappeared. In fact, with this assumption Maxwell's equations take on an even deeper level of meaning.

Thus, with the beginning of the twentieth century, everything known about nature was inexorably pointing to the idea that the speed of light is an absolute constant. This means that whoever measures the speed of light, however he measures the speed of light, regardless of how he is moving, regardless of how the source of light is moving, he will *always* come up with exactly the same answer: 186,000 miles per second. It is an *absolute* constant.

Although all the problems and inconsistencies with light vanish with this important assumption, new difficulties appear. This assumption is "unnatural"; it is directly opposed to common sense. To see why this is so, imagine someone standing on a train that is traveling down the tracks at a speed of 50 miles per hour. Suppose he throws a rock in front of the train with a speed of 10 miles per hour. As seen by someone standing alongside the train tracks, the resulting speed of the rock will be 60 miles per hour (50 miles per hour for the train *plus* 10 miles per hour for the rock), as shown in Figure 10–4. This is straightforward common sense. Now suppose that the fellow on the train turns around and throws another rock with the same speed (10 miles per hour) toward the caboose. As seen by a person standing alongside the train tracks, the speed of this second rock is 40 miles per hour (50 miles per hour for the train *minus* 10 miles per hour for the rock). This also is common sense. But now suppose that the fellow standing on the train turns on a light. According to Dr. Einstein's assumption, the speed of light is the same in *all* directions. Toward the front of the train, toward the back of the train, straight up or straight down, the person alongside the train tracks will always measure the same speed: 186,000 miles per second. In fact, if someone were flying past the train in a rocket ship going 99 per cent of the speed of light, he also would measure the speed of light to be exactly 186,000 miles per second. This is diametrically opposed to physical intuition and common sense.

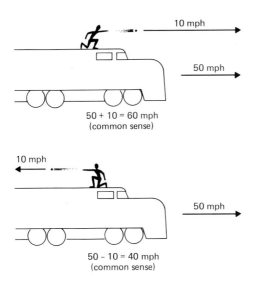

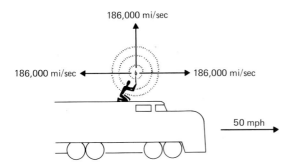

Figure 10–4. Common Sense and The Speed of Light. If a person standing on a moving train throws a rock, simple arithmetic is used to calculate the speed of the rock relative to the ground. However, if a person shines a beam of light from the moving train, anyone who measures the speed of this light will always discover that it is moving at 186,000 miles per second, regardless of the speed of the train.

Yet, everything known about physical reality points to the inescapable conclusion that the speed of light is an absolute constant.

Once we are willing to accept the absolute constancy of the speed of light, the most trusted concepts of time, space, and mass must be altered. These concepts must now be approached from a totally new and exciting viewpoint, the viewpoint of the special theory of relativity.

10.2 The Birth of Special Relativity

Astronomy has the dual distinction of being both the oldest science and one of the most fascinating fields of modern research. It is truly remarkable that many of the most important concepts in astronomy are only a few decades old, in spite of the fact that man has been studying the heavens for thousands upon thousands of years. Indeed, up until a century ago, man did not even know where the stars were. Only after the techniques of stellar parallax were developed, could astronomers actually measure the distances to the stars seen in the nighttime sky. As discussed in Chapter 7, the distances to the stars are immense and new yardsticks, or units of measure, were invented to express these distances more conveniently. One very useful yardstick turns out to be the *light year*, how far light travels during one year. A light year is very nearly equal to 6 trillion miles.

Expressing the distances to the stars in terms of light years is extremely convenient. Sirius, the brightest appearing star in the sky, is 9 light years away. In the constellation of Orion, Betelgeuse is 490 light years away, and the distance to Rigel is 810 light years.

In thinking about these distances, it appears that perhaps nature is trying to say something very profound. For example, Aldebaran, the bright red giant star in the constellation of Taurus is 68 light years away. What does this really mean? It means that the light that strikes our eyes when we look up at Aldebaran on a cold winter's night actually left the star 68 years ago. We are not seeing what the star looks like now. We are seeing how the star appeared 68 years ago, before the beginning of the World War I. Similarly, if an astronomer photographs a galaxy whose distance is 300 million light years (there are a large number of galaxies seen in the constellation of Coma Berenices at this distance), the light that is exposing his photographic plate is 300 million years old. He is actually seeing how the galaxy looked 300 million years ago during the Carboniferous period, when reptiles first appeared on the face of our planet. In other words, the inescapable conclusion is that when we look out into space, we are actually looking backward in time! Just by looking up into the nighttime sky and being truly aware of what we are seeing, we must realize that space and time are intimately interwoven. We cannot speak about one without also talking about the other.

Chapter 10   The Special Theory of Relativity

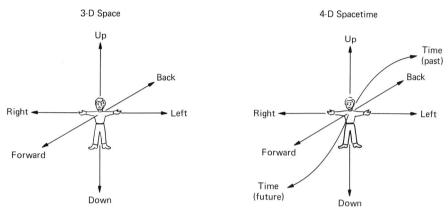

Figure 10–5. *Three and Four Dimensions.* Although we have freedom to move in the three dimensions of space, we realize that we are really moving through the four dimensions of space-time.

We are three-dimensional creatures. We have freedom to move only in the three *spatial* dimensions: up-down, forward-back, left-right. But from just looking up at the stars, we realize that we must consider time as a fourth dimension. As we are born, grow old, and die, we are three-dimensional creatures moving through a four dimensional *space-time*, inexorably forced to go from the past into the future.

But what is meant by space-time? To illustrate space-time, imagine taking an airplane flight from Los Angeles to Seattle with a 1-hour stopover at San Francisco. Draw a graph of your trip as shown in Figure 10–6. On the vertical axis measure the time on your wrist watch while on the horizontal axis, and plot how far you have traveled. Flying from Los Angeles to San Francisco and on to Seattle, you can draw a line on this graph, as shown in Figure 10–6, based on when and where you are. This is a two-dimensional *space-time diagram.* Your path, or *trajectory*, in this two-dimensional space-time is simply the line that you have drawn.

As another example of space-time, suppose you go into a room, walk over to a lamp and then to a chair, as shown in Figure 10–7. On the left side of Figure 10–7, the path in ordinary space is shown. But to see what your path looks like in space-time, a three-dimensional graph must be drawn. On one axis of the graph measure how far you have moved in the east-west direction. On a second axis measure how far you have moved in the north-south direction. Finally, on the third axis measure what time it is. Your path in this three dimensional space-time is shown on the right side of Figure 10–7. Going from the door to the lamp to the chair, your path rises higher and higher in the diagram because you are also moving through time, into the future.

We will not bother to draw a four-dimensional space-time diagram (three spatial axes and one temporal axis) because the artwork is hard to do, and also because it is really unnecessary. If the reader honestly understands the two-dimensional example, he knows all there is to know. This illustrates an important point. Very few astrophysicists working with relativity theory actually can visualize in four dimensions. But, to a large degree, this ability is unnecessary. If one really understands what is going on in two-dimensional space-time, any number of dimensions may be added. Work can be "generalized" to four dimensions with no trouble just by adding two more spatial axes. Therefore, the following discussions of relativity will be confined to two-dimensional space-time diagrams since two more dimensions can be added later on, if desired. The final conclusions always remain unchanged.

Although the diagrams illustrating two- and three-dimensional space-time are very picturesque, Figures 10–6 and 10–7 are not exactly the way physicists like to draw space-time. The conventional way physicists draw two-dimensional space-time is shown in Figure 10–8. In this diagram, the axes of the graph have been scaled in a very particular fashion. If 1 inch on the vertical axis represents 1 second of time, then we insist that 1 inch on the horizontal axis represent 186,000 miles. As a result, light rays travel along 45° lines, since for every second which passes, they cover 186,000 miles.

When space-time diagrams are drawn in this fashion, we realize that all of space-time divides up into three types of regions. In particular, suppose we say that the center of the diagram represents "here" and "now." As will be shown in

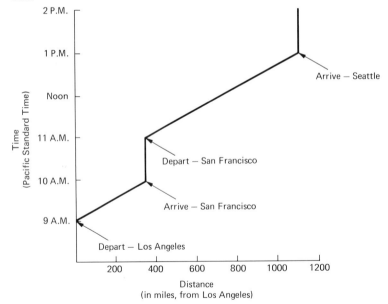

Figure 10–6. A Two-Dimensional Space-Time. This graph showing a trip from Los Angeles to Seattle is an example of a drawing of two-dimensional space-time.

Chapter 10 The Special Theory of Relativity

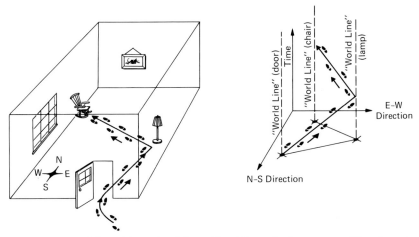

Figure 10–7. *A Three-Dimensional Space-Time.* The path of a person walking through a room (in ordinary space) is shown on the left. To draw a space-time diagram, the time on the person's wristwatch must be plotted on a third perpendicular axis of a graph, as shown on the right.

the next few pages, it is impossible to travel faster than the speed of light. Therefore, you can*not* travel from here and now to just anywhere in space-time. Since you can travel only at speeds less than the speed of light, man is forever restricted to the region labeled "future" in the diagram. This diagram has been drawn so that light rays move along 45° lines. If your speed is to be less than 186,000 miles per second, your path in this space-time diagram must be inclined

Figure 10–8. *Space-Time.* The conventional way of drawing a space-time diagram involves scaling the axes of the graph so that light rays travel along 45° lines.

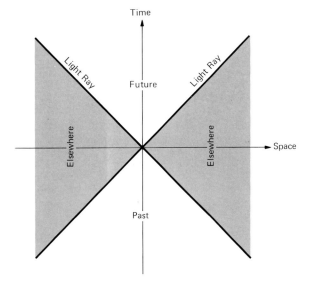

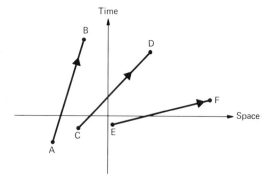

Figure 10–9. *Trips in Space-Time.* Since it is impossible to travel faster than the speed of light, only "time-like trips" are allowed. "Space-like trips" are forbidden.

at an angle of less than 45° from the vertical. Therefore, you always end up in the "future." You can never get from here and now into the region labeled "elsewhere" because, to do so, your path would have to be inclined at an angle greater than 45° from the vertical; your speed would have to be faster than the speed of light. Similarly, in order to get to here and now, you must have come from the region labeled "past."

From this example, it is possible to talk about three kinds of "trips" in space-time, as shown in Figure 10–9. You can easily go from point *A* to point *B* in this diagram. Your path always makes an angle of less than 45° with respect to the vertical. Lots of time passes by and not much distance is covered. The speed is less than the speed of light. Such a trajectory is called a *timelike trip* since it is dominated by the passage of time. The path connecting points *C* and *D* is inclined by exactly 45° with respect to the vertical. Due to the way the axes of the graph have been scaled, for every second of time that passes, 186,000 miles must be traversed. The speed of something traveling on such a path must be equal to the speed of light. This trajectory is, therefore, called a *lightlike trip*. Finally, it is impossible to get from point *E* to point *F.* Along this path, which makes an angle greater than 45° with respect to the vertical, a huge distance is covered in a very short period of time. The speed must be greater than the speed of light. Since space dominates such a trajectory, it is called a *spacelike trip.* Ordinary stuff (people, rocket ships, basketballs, and so on) can travel only along timelike paths in the universe. Light travels along lightlike paths and, so far as we know, nothing can travel along a spacelike path.

As we learned in the previous section, during the first decade of the twentieth century, Albert Einstein discovered that he could resolve all the difficulties surrounding Maxwell's equations if he assumed that the speed of light is an absolute constant. In addition, he found that he could achieve his goals in theoretical physics most easily if he treated time as a fourth dimension along with the usual three spatial dimensions. Einstein, therefore, found it most convenient to reformulate electromagnetic theory in the mathematical language of four-dimensional space-time.

As was also learned in the previous section, the idea that the speed of light is an absolute constant is directly opposed to everyday common sense. Ordinary things (motorboats in a river; rocks thrown from a moving train) just don't work that way. The very important point to realize is that if a physical theory is based on an assumption that is opposed to common sense, many of the predictions and conclusions of the theory will quite naturally also be in opposition to common sense. In resolving the problems with Maxwell's equations, Dr. Einstein found that he had to revise the most trusted and time-honored ideas about space, time, and mass. A whole new approach to the most fundamental nature of physical reality had to be developed. This new approach is contained in the *special theory of relativity,* which reveals some very unusual things about how the universe works.

In 1905, Dr. Einstein published his special theory of relativity, which has revised almost everything known about reality. The special theory of relativity tells what reality is like from the viewpoint of four-dimensional space-time, in which the speed of light is an absolute constant. Specifically, this theory tells what happens at velocities near the speed of light.

At the heart of the special theory of relativity are a set of equations called the *Lorentz transformations.* The Lorentz transformations tell what reality is like as seen by different "observers" who are moving relative to each other at velocities near the speed of light. If the relative speeds between different observers is very low, then ordinary common sense can be used. In the experiment with throwing rocks from a moving train (see Figure 10–4 in Section 10.1) all the speeds are very small compared to the speed of light. In calculating the speeds of the rocks as measured by a person standing alongside the train tracks, common sense gave the correct answers. But, if the train were moving at 90 per cent of the speed of light, or if the rocks were thrown at 95 per cent of the speed of light, "common sense" would give crazy answers. More precisely, these answers could not be reconciled with any reasonable, consistent, coherent picture of reality.

To appreciate the meaning of the Lorentz transformations, consider an experiment. Suppose two scientists (that is, observers) are moving relative to each other at some very high speed. Perhaps, for example, one observer is here on the earth while the other is flying through the solar system in his rocket ship. These two observers are in communication with each other using radios and they ask each other questions. The first thing we notice is that something very strange has happened to time. Each observer will say that the other fellow's clocks are going slow. The scientist on the earth will conclude that all the clocks in the moving rocket ship have slowed down. The scientist in the rocket ship will say that all the clocks on earth are going slow. They have an argument. The scientist in the rocket ship checks his clocks carefully and finds nothing wrong. But, according to the observer on the earth, *everything* in the rocket ship has slowed down, even the metabolism, heartbeat, and thinking processes of the scientist in the ship; that's why he doesn't find anything wrong. Meanwhile, the scientist in the rocket ship comes to the same conclusions about the clocks of the fellow on the earth. We, therefore, arrive at the very important conclusion that *as seen by a stationary observer, moving clocks go slowly.*

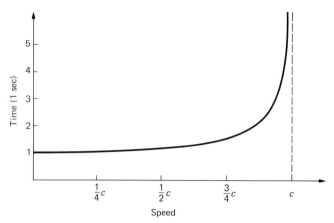

Figure 10-10. *The Dilation of Time.* According to a stationary observer, the clocks of a moving observer appear to slow down. This graph shows how long one second would appear to last for clocks moving at various speeds.

This slowing down of time is expressed very precisely by the Lorentz transformations. The Lorentz transformation for time is shown graphically in Figure 10–10. In particular, this graph answers the question "How long does 1 second on a moving clock seem to last as measured by a stationary clock?" As seen from this graph, if the relative speed between the stationary and moving clocks is low (less than about one half of the speed of light, or $\frac{1}{2}c$), the two clocks agree. One second on the stationary clock is the same as 1 second on the moving clock. But, as the speed of light is approached, this slowing down of time becomes quite noticeable. For example, at about 85 per cent of the speed of light, 1 second on the moving clock will appear to last about 2 seconds on the stationary clock. Indeed, at the speed of light, the slowing down of time has become so severe that *time stops completely!*

Of course, who is to say which clock is "moving" and which clock is "stationary"! The scientist carrying the moving clock in his spaceship checks the clock very carefully. He compares it with other clocks in his spaceship, with his heartbeat, and with the rate of decay of radioactive chemicals. He concludes that his clocks are just fine; the clocks of the scientist on the earth are slowing down. This effect of the slowing down of clocks is often called the *dilation of time.*

Just as the dilation of time is built into the special theory of relativity, moving observers also will be unable to agree about measurements of space. Not only do the scientist on the earth and the scientist in the rocket ship argue about their clocks, they also find themselves arguing about their rulers. Each observer says that his rulers are fine, but the rulers of the moving observer *shrink* when held in the direction of motion. This shrinking of rulers is called the *Fitzgerald contraction* and is also a direct result of the Lorentz transformations. The Lorentz transformation for distances is shown in Figure 10–11. This graph answers the question "how long is a 1-foot ruler held parallel to the direction of motion of a moving observer, compared to the ruler of a stationary observer?" According to the stationary observer, the moving ruler will be shorter than 1 foot. For example, from the graph, a 1-foot ruler moving at 80 per cent of the speed of light will appear to be only about 7 inches long. Indeed, at the speed of light, the ruler will have shrunk down to nothing; it will have zero length!

This Fitzgerald contraction applies *only* to rulers held *parallel* to the direction of motion. Distances measured perpendicular to the direction remain unchanged. Thus, if a basketball were moving very near the speed of light, it would actually behave like a pancake. The diameter of the basketball parallel to the direction of motion would have shrunk, but any diameter measured perpendicular to the direction of motion would be unchanged. (Actually, if you watched a basketball moving near the speed of light, it would not *look* like a pancake. How things *look* near the speed of light is more complicated than this and will be treated in the next section.) Suffice it to say that distances measured parallel to the direction of motion depend on how the observer is moving. In order to have a consistent, coherent description of reality, a stationary observer must conclude that moving rulers held parallel to the direction of motion are compressed by precisely the amount given by the Lorentz transformations.

Finally, according to the special theory of relativity, two different observers moving with respect to each other cannot agree on the masses of objects. At rest, in a laboratory, for example, there is no possibility for debate. Scientists can stand around and measure the masses of such things as atoms or bricks, or whatever. When someone measures the mass of an object that is not moving, the answer he gets is called, appropriately, the *rest mass*. But if he measures the

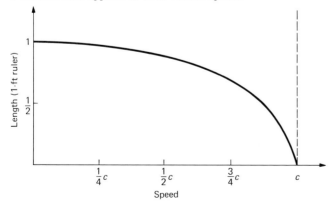

Figure 10–11. *The Fitzgerald Contraction.* According to a stationary observer, the rulers of a moving observer appear to shrink when held parallel to the direction of motion. This graph shows how long a one-foot ruler appears to be at various speeds.

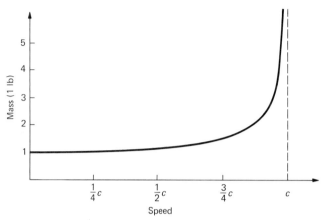

Figure 10–12. *The Increase in Mass.* According to a stationary observer, the masses of moving objects appear to increase with increasing speed. This graph shows how the mass of a 1 pound object appears to increase with speed.

mass of something that is moving, he gets an answer that is bigger than the rest mass of that object. The masses of objects increase with increasing speed, as prescribed by the Lorentz transformations. The Lorentz transformation for mass is shown graphically in Figure 10–12. This graph answers the question "Suppose I have an object with a rest mass of 1 pound; what would be the object's mass if it were moving at a high speed?" As in the case of the dilation of time, and the shrinking of rulers, noticeable effects do not occur until we start talking about speeds higher than about half the speed of light. But if the speed of an object is very high, that object will behave as though it has a mass much greater than its rest mass. In fact, at the speed of light, the masses of objects become infinite!

So, we have seen that from the special theory of relativity, the measurement of time, length, and mass depends on the velocity of the observer. With increasing speed, clocks slow down, rulers shrink, and mass increases. However, these effects are really noticeable only at speeds near the speed of light.

It might seem that the effects predicted by special relativity are restricted to the realm of science fiction because even when NASA sends its most powerful rockets to the moon, the resulting speeds are always very small compared to the speed of light. But this is not so. There are many machines that can make objects travel close to the speed of light. They are the atom smashers used by nuclear physicists in their experiments. Velocities of 99 per cent of the speed of light are not uncommon in such experiments. From the work of nuclear physicists, the predictions of special relativity have been confirmed to a very high degree of accuracy. In fact, the Lorentz transformations are a very important theoretical tool for the nuclear physicist. He finds that it is impossible to understand the results of his experiments without taking into account the dilation of time, the Fitzgerald contraction, and the increase in mass. Thus, there is absolutely no

Chapter 10 The Special Theory of Relativity

question about the validity of the special theory of relativity. The theory works. It predicts the correct answers in any situation just as accurately and precisely as the physics used by engineers to build bridges, cars, or TV sets.

At this point we can appreciate why it is impossible to go faster than the speed of light. Imagine a rocket ship with very powerful engines and a very large reserve of fuel. Astronauts board the rocket ship and blast off. The rockets burn continuously day after day, week after week, month after month. The rocket ship experiences a continuous acceleration that makes it go faster and faster. Meanwhile, scientists back at Mission Control in Houston are monitoring the flight. Since the rocket engines are burning continuously, scientists on earth notice that the speed of the rocket ship is constantly increasing. However, when the speed of the rocket reaches about $\frac{1}{4}$-billion miles per hour (half the speed of light expressed in miles per hour), the scientists at mission control notice that something unusual is starting to happen. The dilation of time is beginning to make itself felt. In particular, the earth-based scientist notices that the rocket engines seem to be burning the fuel slower than usual. The astronauts check their rocket ship and report that everything is A-OK; the engines are consuming the fuel at the standard rate. But, as the speed of the rocket continues to increase, this effect becomes even more pronounced. At 85 per cent of the speed of light, according to the earth-based scientist, the rockets are burning fuel at only half the rate they should be. It takes 2 minutes to consume the same amount of fuel that should have been consumed in 1 minute. Since the fuel is burning at a slower rate, the rocket ship is no longer accelerating as rapidly as it was. We can think of the rate at which chemical (or nuclear) reactions occur in the rocket engines as a clock. As time slows down, the rate at which fuel burns also slows down. As the rocket ship gets closer and closer to the speed of light, according to earth-based scientists the rocket engines provide less and less thrust. Each additional mile per hour of speed comes with increasing difficulty. In fact, at the speed of light, the rocket engines shut off completely. The final boost of energy needed to push the rocket up to a speed of 186,000 miles per second never comes. The slowing down of time is so effective that anyone trying to "break the light barrier" has to wait an infinite number of years for that final boost. Although this example focuses on the flight of a rocket ship, the same basic ideas apply to any vehicle or any process that might be used to reach the speed of light. Although it is possible to go 99.99 per cent of the speed of light, it would take an infinitely long time to reach 100 per cent.

The clever reader might conceive of some possible experiments that might "fool Mother Nature." Perhaps, as shown in Figure 10–13, one rocket could leave the earth at 95 per cent of the speed of light. Meanwhile, a second rocket could blast off from the earth in exactly the *opposite* direction, also at 95 per cent of the speed of light. Suppose an astronaut in the first rocket were to measure the speed of the second rocket. Wouldn't the answer be greater than the speed of light? The Lorentz transformations are ideally suited to answer a question like this. Detailed calculations show that the relative speed between the two rockets is 99.87 per cent of the speed of light. In other words, the barrier at the speed of

Figure 10–13. *You Can't Fool Mother Nature.* Two rocketships leave the earth going in opposite directions at 95% the speed of light, as measured by people on the earth. When the two astronauts communicate with each other, they find that they are moving apart at 99.9% the speed of light.

light is so powerful and pervasive that even the most commonsense proposals to exceed this limiting speed will always fail.

Since the Lorentz transformations are so very important, it is useful to illustrate graphically what they do to space-time. Consider an ordinary space-time diagram of a scientist on earth (he is the stationary observer), just like the diagram shown in Figure 10–8. This scientist compares notes with an astronaut traveling past the earth at a very high speed. The effects of the Lorentz transformations can be illustrated by drawing a space-time diagram of the moving astronaut on top of the space-time diagram of the stationary scientist, as shown in Figure 10–14. The space and time axes of the stationary observer are perpendicular to each other. However, according to this stationary observer, the space and time axes of the moving observer are *tilted* toward the 45°-light-ray line. The greater the speed of the astronaut, the larger is the degree of tilting. At the speed of light, the space and time axes of the moving observer meet at the 45°-light-ray line. As a result of this tilting, the two observers will not agree on the intervals of time or distances in space between events in the universe. If an *event* occurs at time t and location x, as seen by the stationary observer, according to the moving observer, this same event occurs at time T and location X, as shown in Figure 10–14.

From this graphic representation of the Lorentz transformations, we see that there is no such thing as *simultaneity* in the universe. Consider two events that, according to a stationary observer, occur simultaneously. Call them event A and event B. Since they occur simultaneously, they happen at the same time in the space-time diagram of the stationary observer. The time of event A, t_A, equals the time of event B, t_B, as shown in Figure 10–15. However, now examine how these two events look to an observer who is moving with respect to the stationary observer. The Lorentz transformations cause the axes of the moving observer to be tilted toward the 45°-light-ray line. Therefore, as shown in Figure 10–15, these

Chapter 10   The Special Theory of Relativity

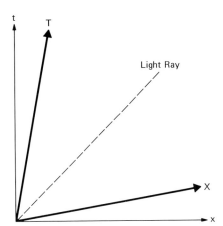

Figure 10–14. *The Effects of the Lorentz Transformations.* The Lorentz transformations can be displayed graphically in terms of what they do to space-time diagrams. According to a stationary observer, the space and time axes of the moving observer are tilted toward the light ray line.

two events occur at *different times* according to the moving observer. The time of event A, T_A, occurs later than the time of event B, T_B. In fact, from examining Figure 10–15, we see that the more distant event *always* appears to occur earlier, before the nearby event.

The formulation of the special theory of relativity has had a profound effect on the course of physics. We have seen that the absolute constancy of the speed of light has required that we reexamine all our traditional concepts of space, time, and matter. The most frequent applications of special relativity are found in nuclear physics since this subject often deals with particles moving near the speed of light. There are, however, some intriguing astronomical applications that will be examined in the next section.

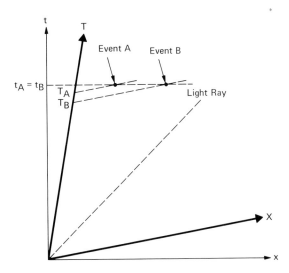

Figure 10–15. *"Simultaneity" Is Nonsense.* Two events (Event A and Event B) appear to be simultaneous to the stationary observer. A moving observer, however, will conclude that the two events occurred at different times.

10.3 Traveling Near the Speed of Light

The dilation of time is one of the very intriguing results to come from the special theory of relativity. If someone is moving at a high speed, in order to have a rational picture of physical reality (which of necessity includes Maxwell's equations and the Lorentz transformations), it *must* be concluded that his clocks are going slowly. This effect is often encountered in laboratory experiments. For example, if a nuclear physicist accelerates radioactive particles up to velocities near the speed of light, he finds that the so-called *half-life* of these particles becomes very long. Particles such as *mesons* last for only short periods of time when at rest in the laboratory; they radioactively decay very quickly into new atomic particles. But if these mesons are coming from a cyclotron and are traveling at very high speeds, they last much longer before turning into new particles. The slowing down of time affects everything, even the half lives of radioactive material.

This dilation of time has many paradoxical aspects. Imagine two astronauts, each in his own spaceship, traveling past each other near the speed of light. Each astronaut carefully checks the clocks in his spaceship and finds them in perfect working order. As the two astronauts communicate with each other, each one says that his clocks are fine but that the other person's clocks are going slowly. This, in itself, is contrary to common sense. If someone comes up to you on the street and tells you that your wrist watch is going slowly, you can explain the discrepancy between your clocks by saying that your wrist watch is fine, but his is going too fast. Not so in relativity. As the two people pass each other at a high speed, each one says that the other's clock has slowed down.

This matter of two people comparing clocks has given rise to a problem known as the *twin paradox*, which has been hotly debated for many decades. Suppose two people who have the same age (that is, they could be twins) each decide to go on a space flight, each in his own rocket. After traveling around the galaxy at velocities very close to the speed of light, what will they find when they get back together? Will they still be the same age? Will one be older than the other? Which one will be older? For purposes of illustration, we will examine a version of the twin paradox in which the confusion has been minimized.

Consider Harry and Tom. They are both 20 years old and they own an extraordinary rocket ship, as shown in Figure 10–16. Since the rocket ship can carry only one astronaut, they flip a coin to see who is going to make the round-trip journey to a star 25 light years away. Harry wins the toss and climbs aboard the rocket ship. This rocket ship travels at 98 per cent of the speed of light, and Harry makes the round-trip, covering a total distance of 50 light years (25 light years out to the star and 25 light years back). What do Tom and Harry look like when they get back together?

First of all, everyone agrees that the star is 25 light years away and that the rocket ship travels at 98 per cent of the speed of light. Tom, who stays behind on

Chapter 10 The Special Theory of Relativity 299

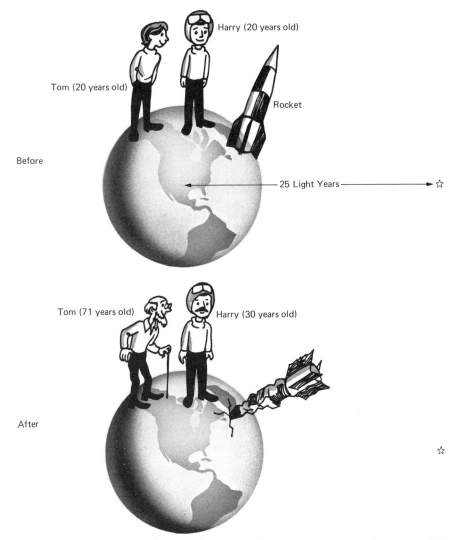

Figure 10-16. Harry, Tom, and the "Twin Paradox." Harry makes a round-trip to a star 25 light years away. He travels at 98% the speed of light. During this trip Harry ages by only 10 years. Tom, who stays behind on the earth, ages by 51 years while Harry is gone.

the earth realizes that it would take light 50 years to make the round-trip. Since it is only possible to travel *slower* than the speed of light, it must take *more* than 50 years for Harry to complete the trip. Actually, traveling at 98 per cent of the speed of light, it takes 51 years to travel out to the star and back. As measured by clocks on the earth, 51 years must elapse. Tom is 71 years old when he again sees Harry.

But, at 98 per cent of the speed of light, time is dilated by a factor of 5. For

Figure 10–17. *Walking in the Rain.* Even though the rain is falling straight down (i.e., no wind), a person walking in the rain must tilt his umbrella in front of him if he is not to get wet.

every year that passes on board the rocket ship, 5 years elapse according to clocks on the earth. Therefore, during the entire duration of the trip, Harry ages by only a little more than 10 years. Harry is 30 years old when he again sees Tom.

From this example, it seems that it would be possible to journey great distances to the stars and galaxies in the universe in reasonable periods of time *if* spacecrafts could travel at speeds *very* close to the speed of light. However, when adventurous astronauts returned to earth after completing their lengthy odyssey, they would find that their friends, colleagues, and relatives had long since passed away. After a lengthy journey at speeds like 99.999 per cent of the speed of light, several generations would have come and gone here on earth while the astronauts would be only middle-aged. The astronauts would probably be treated like curiosities stepping out of the past since the new inhabitants of the planet probably would have forgotten why their ancestors undertook such a journey in the first place.

It is interesting to contemplate what the sky would look like to astronauts traveling near the speed of light. To gain insight into this problem, imagine someone walking down the street in the rain holding an umbrella, as shown in Figure 10–17. Let us suppose that there is no wind, so that the raindrops are falling straight down. If this person is walking fast, he *must* tilt the umbrella forward in front of him in order to stay dry. The raindrops will seem to be coming toward this moving person, trying to strike him in the face. This effect is well known to anyone who has driven a car in the rain. Driving down a freeway

Chapter 10 The Special Theory of Relativity

at 55 miles per hour in a rainstorm, many raindrops pelt the windshield, while the back window of the car remains almost dry. Even though, in reality, the raindrops are falling straight down, it looks as if they are coming directly at the driver.

This same effect also occurs with light. The earth is moving around the sun with a speed of $18\frac{1}{2}$ miles per second. Just as the person walking in the rain must tilt his umbrella forward, the astronomer must tilt his telescope forward (ever so slightly) to insure that the light rays from the stars "fall down" his telescope tube. This effect is called the *aberration of light*. Since the speed of the earth is very small compared to the speed of light, this effect goes almost unnoticed. However, if future astronauts succeed in achieving speeds close to that of light, this effect will be very pronounced.

If a car is parked or is moving very slowly in a rainstorm, equal numbers of raindrops will fall on the front and back windows. If a car is moving very fast, almost all the raindrops will fall on the front windshield, while very few raindrops will strike the back window. It will seem like the raindrops are coming toward the car, striking the windshield head on. Similarly, for a spacecraft traveling near the speed of light, it will seem as though the light rays from the

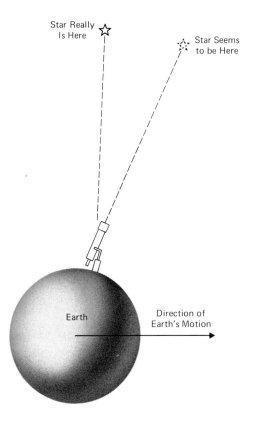

Figure 10–18. *Aberration of Starlight.* Since the earth is moving, telescopes focused on distant stars must be tilted slightly in the forward direction.

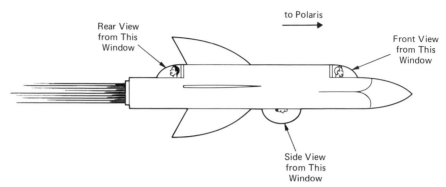

Figure 10-19. *A Spaceship.* This hypothetical spaceship is capable of travelling at speeds close to the speed of light. The spaceship has three big windows from each of which astronauts can see one half of the entire sky. The spaceship is traveling toward Polaris.

Figure 10-20. *Views from the Spaceship.* The hypothetical spaceship is travelling toward Polaris. Views from the three windows are shown for various speeds. As the velocity of the spacecraft approaches the speed of light, all the stars appear to converge on Polaris. (*Adapted from G. D. Scott and H. J. Driel*)

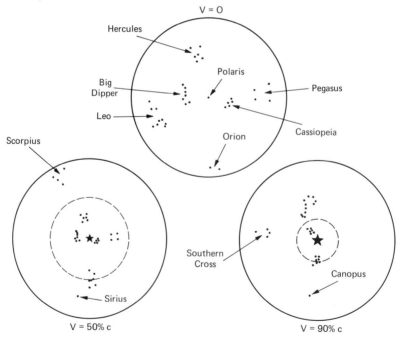

Front View

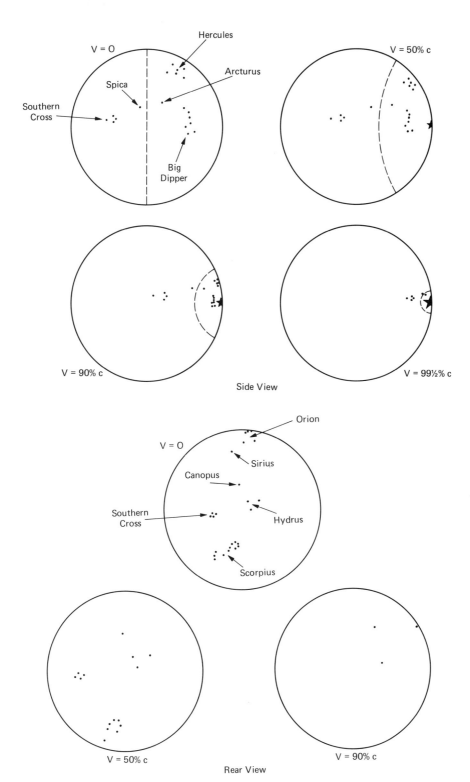

stars are coming toward the spacecraft, striking the window at the front of the vehicle head on. In other words, it will look as though all the stars in the sky are *in front* of the spacecraft!

Using the Lorentz transformations, it is possible to calculate how the appearance of the sky is distorted for astronauts traveling at various speeds. For purposes of illustration, suppose the spacecraft in question has three big windows, as shown in Figure 10–19. One window is in front, one in back, and one on the side. Each window allows the astronauts inside to see exactly one half of the sky. The spacecraft is aimed directly at Polaris, the North Star. The rocket engines are ignited and the spacecraft begins to move faster and faster. As the speed of the spacecraft increases to sizable fractions of the speed of light, the aberration of starlight becomes more and more noticeable. Stars that were behind the spacecraft now appear in front of it. Stars that were seen out of the rear window are now seen out of the front window. As the spacecraft approaches the speed of light, the stars in the sky seem to drift toward Polaris, toward the direction in which the astronauts are moving. In fact, at the speed of light, every star in the sky, every galaxy in the sky appears directly *in front* of the spacecraft! There is then only one brilliant, blinding light in the sky toward which the astronauts plunge; the rest of the sky is totally black!

On the previous pages there are a series of star charts that illustrate these intriguing effects. In particular, there are three series of charts: one series for the front window, one series for the side window, and one series for the rear window. Remember that half of the sky is visible through each window. The views labeled $v = 0$ are for zero speed; the spacecraft has not started moving. This is the undistorted view of the sky, just as we see from earth. To make the diagrams as simple as possible, only the most familiar constellations are shown. The remaining views are for velocities at various fractions of the speed of light. Notice how all the star images converge toward Polaris, toward the direction in which the spacecraft is moving.

This distortion of the appearance of the sky is partly within and partly outside of everyday experience and common sense. It is within our common experience in that the central ideas can be appreciated fully by thinking about driving a car down a freeway at a high speed in a rainstorm. Yet, in a fascinating way, it is far removed from common experience. According to common experience, if you move away from something, it seems to grow smaller. If you jump in a car and drive away from your house, the house will appear to get small and smaller and eventually will fade away into the distance. This absolutely is *not* true if your car travels near the speed of light. The aberration of light at extreme velocities tells us that images converge on the point toward which you are moving. At the speed of light, your house would appear in front of you.

From this example, we can try to imagine what would happen if astronauts were to leave the solar system in a rocket ship at speeds greater than 95 per cent of the speed of light. If they were to go directly away from the sun, as seen from the rear window the sun would get bigger and bigger as the spacecraft went faster and faster. At about 99 per cent of the speed of light, the sun would fill the entire view from the rear window and begin appearing around the edges of the front window. Even closer to the speed of light, the surface of the sun would fill almost the entire sky as it tried to converge from all sides on the point toward which the astronauts were headed. At these extreme speeds, most of the sun's surface would be concentrated in a ring closing in on this convergent point. And at the speed of light, this ring would collapse to a point directly ahead of the spacecraft.

Although it might seem that in this section we have stepped into an entertaining fantasyland, everything presented here is a direct result of the best description of physical reality known to the human mind: the theory of relativity. But if the reader has found things strange and bizarre, he should remember the words of Al Jolson: "you ain't heard nothin' yet!"

10.4 Do Tachyons Exist?

In 1928, the brilliant physicist P. A. M. Dirac reformulated quantum mechanics to include relativistic effects. One of the results of his work is the so-called *Dirac equation*, which fully describes the properties of subatomic particles. In particular, when Dirac applied his equation to electrons, he found that the equation had *two* types of solutions, one for "positive energy states" and one for "negative energy states." The first solution gave a detailed description of electrons, which, as previously discussed, are negatively charged particles of small mass usually found in orbit about the nuclei of atoms. Electrons correspond to the positive energy state solutions to the Dirac equation. But the other solution, corresponding to negative energy states, described mysterious particles that were unknown to physicists at that time. These peculiar particles are virtually identical to electrons, except that they have a positive electric charge. They are called *positrons* or antielectrons.

It is fairly common that solutions to complicated equations give several answers, some of which are physically reasonable, whereas others must be discarded as nonsense. The initial temptation was, therefore, to discard the second type of particles predicted by Dirac's equation. Perhaps they exist only on paper because of shortcomings in the mathematics of Dirac's theory. All the particles that were known to physicists up to the time of Dirac's work corresponded to positive energy state solutions. Furthermore, negative energy particles may be thought of as positive energy particles moving backward in time. A positron is an electron moving from the future into the past! Particles having these peculiar properties are referred to as *antimatter*.

In 1932, the American physicist C. D. Anderson was examining particles produced by *cosmic rays*. A cosmic ray is nothing more than a particle from outer space traveling at a velocity very close to the speed of light. Most cosmic rays consist of high-speed electrons or high-speed protons. They are produced by processes such as flares on the sun or supernovae explosions. When one of these cosmic rays enters the earth's atmosphere, it eventually collides with an atom in the air, breaking it into many pieces. These pieces usually consist of protons, neutrons, and particles called pi-mesons, all of which, in turn, strike additional nuclei lower in the earth's atmosphere, thereby producing still more particles. The final result is a *cosmic ray shower* of all types of subatomic particles that rain down on the earth. Figure 10–21 schematically shows such a cosmic ray shower. By examining the properties of the particles produced by

Figure 10–21. *A Cosmic Ray Shower.* High speed particles (usually protons or electrons) from outerspace are called cosmic "rays." When one of these particles hits an atom high in the earth's atmosphere, the atom is broken into many fragments. The resulting particles "rain" down on the earth, producing a so-called cosmic ray shower. (*Adapted from Drs. Goble and Baker*)

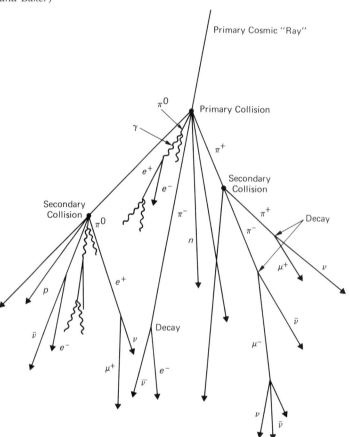

Chapter 10 The Special Theory of Relativity

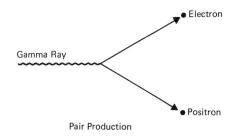

Pair Production

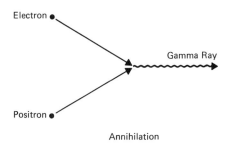

Annihilation

Figure 10–22. Pair Production and Annihilation. A γ ray can spontaneously turn into two particles, one of which is made of ordinary matter while the other is made of antimatter. Similarly, if a particle and an antiparticle collide, they annihilate each other and produce a γ ray.

cosmic rays, Anderson found that positively charged electrons were quite common. He had discovered the positrons predicted by Dirac's equation. Antimatter does indeed exist.

Today the existence of antimatter virtually is taken for granted. For every particle there must be an antiparticle. Since protons exist, antiprotons must also exist. Since neutrons exist, antineutrons must also exist. Whenever a nuclear physicist discovers a new particle in his experiments, everyone realizes that a new antiparticle also must exist. But antimatter does not last for very long. Whenever an antiparticle meets a real particle, they *annihilate* each other. Both particle and antiparticle disappear and produce a high energy γ ray. Similarly, pairs of particles and antiparticles can be created from gamma rays. This process is known as *pair production*. For example, under appropriate conditions a γ ray can spontaneously turn into an electron and a positron, or a proton and an antiproton. The processes of annihilation and pair production are shown schematically in Figure 10–22.

The discovery of antimatter stands in mute testimony to the predictive power of theoretical physics. Antimatter was predicted with nothing more than paper and pencil. The theoretical physicist needed only a little faith in his equations; he just had to wait for the experimental physicist to discover antiparticles in the laboratory.

One of the questions that often faces the theoretical physicist has to do with how much faith he should have in his equations. For example, just as Dirac's work allows for the existence of matter and antimatter, Einstein's work allows for the existence of slower-than-light particles as well as faster-than-light particles. This, of course, does not mean that faster-than-light particles *must* exist. Instead, one can fiddle mathematically with the equations of special relativity in such a way that they will describe particles that travel at velocities greater than the speed of light.

The people who talk about faster-than-light physics like to divide everything into three categories. Anything that travels slower than light is called a *tardyon*. People, atoms, positrons, and basketballs all travel at speeds less than the speed of light. They are all tardyons. Tardyons always travel along timelike paths in space-time.

There are things in the universe such as light rays that travel at exactly the speed of light. They are called *luxons*. Photons and neutrinos are examples of luxons. Luxons travel along lightlike paths in space-time.

Finally, hypothetical particles that travel at speeds greater than the speed of light are called *tachyons*, from the Greek word meaning "swift." Tachyons always travel along spacelike paths in space-time.

No one has ever seen a tachyon. In fact, as will be discussed shortly, there may be good reasons why they cannot exist. Also note that antimatter should not be confused with tachyons. All antiparticles travel at speeds slower than light; they are tardyons.

To appreciate how the mathematics of special relativity can be fiddled with to include tachyons, suppose someone were to ask you: "What is the square root of 9?" Obviously, $\sqrt{9} = 3$ because $3 \times 3 = 9$. But also $\sqrt{9} = -3$ since $(-3) \times (-3) = 9$. In other words, the question has two answers: both $+3$ and -3 are the square root of 9. Both answers are equally good, just as both electrons and positrons are equally good solutions to the Dirac equation.

But suppose someone were to ask: "What is the square root of -9?" The answer is *not* 3, since $3 \times 3 = +9$. Also, the answer is *not* -3, since $(-3) \times (-3) = +9$. In other words, there are no "real" numbers that can answer this question. To cope with the situation, mathematicians have invented "imaginary" numbers. The mathematician then says that the square root of -9 is equal to an imaginary 3, which usually is written $3i$. By definition, $(3i) \times (3i) = -9$.

As explained earlier, when measuring the mass of an object in the laboratory, the answer is called the *rest mass*, or *proper mass*. The answer is always a "real" number. Thus, for example, we could conclude that the rest mass of a brick is 3 pounds, from simply measuring how much a brick weighs. The brick will obey all the laws of physics, including the special theory of relativity. No matter what, it will always travel at speeds less than the speed of light. It is a tardyon brick.

If a mathematician takes a long, hard look at the equations of special relativity, he soon realizes that the theory also applies to objects that have imaginary proper mass, provided that these objects *always* travel faster than the speed of light. This is where the concept of tachyons comes from. An object can travel faster than light *if* it has an imaginary proper mass. While tardyons have real proper mass and always travel at speeds less than the speed of light, tachyons

Chapter 10 The Special Theory of Relativity

have imaginary proper mass and always travel at speeds faster than the speed of light. A typical tachyon brick, therefore, would have a proper mass of 3i pounds.

In this introduction to tachyons, the reader has not been spared some gruesome mathematical details. Suffice it to say that ordinary objects in the universe have masses that can be expressed with ordinary numbers. These ordinary objects (tardyons) must always travel at speeds less than the speed of light. But the special theory of relativity allows for the possibility of unusual objects that always travel at speeds greater than the speed of light. These unusual objects (tachyons) must have proper masses, which cannot be expressed with ordinary (real) numbers.

In many respects, tachyons have properties that are exactly the opposite of the properties of tardyons. For example, tachyons always travel faster than light, whereas tardyons always travel slower than light. Tachyons always travel along spacelike paths, whereas tardyons always travel along timelike paths. It is very difficult to speed up tardyons to velocities near the speed of light. Such a project requires powerful rockets or cyclotrons. Similarly, it is very difficult to slow down tachyons to velocities near the speed of light. Tachyons like to travel with infinite speed, and a great deal of effort is required to slow them down.

The bizarre properties of tachyons really become apparent when considering how tachyons look to different observers traveling at different speeds. Suppose you have a gun that shoots tachyon bullets and you shoot a tachyon bullet at a target. Figure 10–23 shows a space-time diagram of this process. The tachyon bullet leaves the gun at "here-and-now" in the center of the diagram and strikes

Figure 10–23. *A Tachyon's Path in Space-Time.* Tachyons, if they exist, travel along space-like paths in space-time. The tachyon in this diagram is created at here-and-now and is destroyed at a later time *t*.

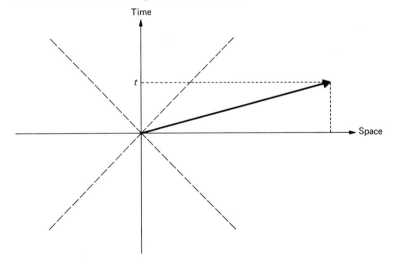

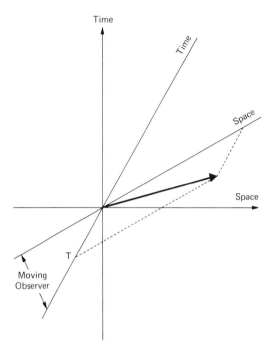

Figure 10–24. *Tachyons Violate Causality.* As seen by a moving observer, the tachyon (whose path is the same as in Figure 10–23) is created at here-and-now but is destroyed at time T. According to the moving observer, the tachyon is destroyed *before* it is created.

the target t seconds later. In traveling from here and now to the target, the tachyon bullet moves along a spacelike path inclined at an angle greater than 45° from the vertical. What does this simple experiment look like to someone moving past you at some high speed? To answer this question, draw the space and time axes of this second observer directly on top of the original space-time diagram. According to the Lorentz transformations, this second set of axes will be tilted toward the 45° light-ray line, as in Figure 10–14. The final result is shown in Figure 10–24, which is the same as the previous diagram except that the space-time diagram of the moving observer is drawn on top of the stationary space-time diagram. We now notice something very weird. Both you (the stationary observer holding the gun) and the moving observer agree that the tachyon bullet was fired from the gun at here-and-now. But, according to the moving observer, the tachyon bullet hit the target at time T, which is *earlier* than "now." In other words, as seen by the moving observer, the tachyon bullet hit the target *before* you fired the gun!

This simple example shows the *tachyons violate causality*. Any process involving tachyons can be viewed in such a way that effects happen *before* the causes. Tachyons can arrive at their destinations before they leave their starting places. The violation of the *law of causality* is the primary argument against the existence of tachyons. From everything we know, the physical universe *must* obey causality. This is the only rational possibility open to the human mind. If we see an event occur, we logically conclude that something must have happened earlier to have made that event occur. Things just don't happen for no reason at all. But if tachyons do exist, then events can occur before their causes and the universe is irrational at a very fundamental level.

Although most physicists dismiss the existence of tachyons on the basis of the causality argument, some scientists have spent many hours experimentally searching for tachyon particles. Such experiments usually involve the techniques of nuclear physics and all have met with negative results—with only one possible exception.

Think about a cosmic ray (for example, a high-speed proton or electron) striking the earth's upper atmosphere. On the average, the incoming particle hits an atom in the air at an altitude of about 14 miles. This is the first collision, and it starts a cosmic ray shower. However, it takes light about 60 millionths of a second to cover 14 miles. This means that no one on the earth's surface can know about the incoming cosmic ray until at least 60 millionths of a second *after* the first collision 14 miles up in the air. We must wait at least 60 millionths of a second for the particles (which travel slower than the speed of light) from the cosmic ray shower to reach detecting equipment on the ground.

But suppose a tachyon particle is created during the initial collision between the incoming cosmic ray and an atom in the upper atmosphere. This tachyon

Figure 10–25. *A Possible Observation of Tachyons?* If a tachyon is produced in a cosmic ray shower, it will reach the earth *before* all the other (ordinary) particles. Such a tachyon might produce a small shower just before the onset of the main shower caused by the rest of the (ordinary) particles.

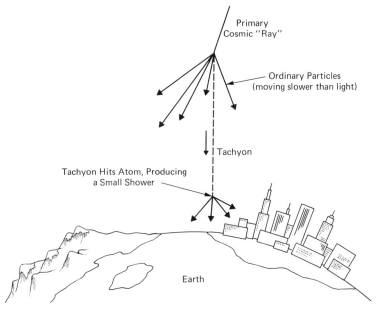

could reach the earth's surface in much less than 60 millionths of a second. As the tachyon enters the denser atmosphere, it might cause a small shower of particles beginning at an altitude of only a few thousand feet. In other words, *if* tachyons are produced by cosmic rays, we might expect to see a small shower of ordinary particles just before the arrival of the main shower of particles. This possibility is shown schematically in Figure 10–25.

Late in 1973, a team of physicists at the University of Adelaide in Australia carefully examined data from 1,307 cosmic ray showers. In particular, they looked carefully at what was going on up to a ten thousandth of a second *before* the arrival of the shower. They claim to have observed nonrandom events preceding the arrival of an extensive air shower. Indeed, their data does seem to indicate that they are detecting particles arriving roughly 60 millionths of a second before the shower.

This is the *only* experiment so far that even remotely suggests that tachyons may exist. All other experiments have produced totally negative results. Although the Adelaide experiment might just be plain wrong (for example, their equipment might not have been operating properly), the results are certainly tantalizing. If mankind ever conclusively discovers tachyons, if tachyons do exist in nature, then scientists must revise some of the most basic concepts of physical reality. *If* tachyons exist, causality is violated. Effects would occur prior to their causes. Physical reality would be fundamentally irrational!

Questions and Exercises

1. Briefly discuss the significance of Maxwell's equations.
2. What, presumably, is the ether?
3. How did A. A. Michelson propose to detect the motion of the earth through the ether?
4. What is the Michelson interferometer?
5. Briefly discuss the results and significance of the Michelson-Morley experiment.
6. Briefly discuss why the assumption of the absolute constancy of the speed of light is opposed to common sense.
7. What is a light year?
8. Briefly discuss the motivation for thinking of time as a fourth dimension.
9. Draw a two-dimensional space-time diagram in which light rays travel in 45° lines. Pick a point and label it "here-and-now." Draw those areas that correspond to the future, the past, and "elsewhere" for an observer at here-and-now.
10. Compare and contrast timelike, spacelike, and lightlike trips.

Chapter 10 The Special Theory of Relativity   313

11. What is meant by the Lorentz transformations?
12. Briefly discuss how the clocks, rulers, and masses of a moving observer change with speed, as seen by a stationary observer.
13. What happens to prevent a rocket ship (even if it has an infinite supply of fuel) from attaining speeds greater than the speed of light?
14. With the aid of a diagram, discuss how the axes of the space-time graphs of a moving observer are distorted, according to a stationary observer.
15. With the aid of a diagram, show that two events that appear to be simultaneous to one observer are not necessarily simultaneous to another observer.
16. Describe how it would be possible to travel huge distances through the universe (perhaps even *millions* of light years) during your own lifetime. When you returned to earth after such a trip, how might things be different compared to when you left?
17. What is meant by the aberration of starlight?
18. Briefly describe the appearance of the sky as seen by an observer who is moving at speeds close to the speed of light.
19. Compare and contrast matter and antimatter.
20. What is meant by pair production?
21. Compare and contrast tardyons, luxons, and tachyons. What kinds of paths do each follow in space-time?
22. With the aid of a diagram, show that tachyons violate causality.
23. Briefly describe how observations of cosmic ray showers might be used to discover tachyons.
24. Briefly discuss how some of mankind's ideas about reality might be affected by the conclusive discovery of tachyons.

The General Theory of Relativity

11.1 Gravitation and Warped Space-Time

According to legend, about 400 years ago a clever Italian physicist named Galileo Galilei performed an interesting experiment. He and his assistants climbed to the top of a building in the town of Pisa. From the top of this building, which happened to be tilted to one side due to poor engineering, they dropped several objects. In particular, they made the remarkable discovery that when two very different objects, such as metal and wooden weights, are released simultaneously from the top of the building, they strike the ground many stories below at precisely the same instant! In other words, everything falls at the same rate (that is, with the same acceleration) in a gravitational field. Indeed, if Galileo's experiment were performed in a vacuum (to eliminate air friction) a feather and a cannon ball released simultaneously from the top of the famous Leaning Tower would hit the ground at exactly the same time.

During the centuries following the work of Galileo, a lot of propaganda was developed that incorporated the uniform acceleration of freely falling bodies. The standard party line claims that gravity is a *force* having the property that all objects falling under the influence of gravity experience the same acceleration. How very bizarre! Surely no other forces in the universe work in this way. If you push on a cannon ball, it moves in a particular fashion. If you push on a feather, it moves in a very different manner. Surely there must be a better way to describe gravity than to think of it as a force.

As discussed in the previous chapter, it is both obvious and natural to incorporate time as a fourth dimension along with the three usual spatial dimensions. After all, it is flatly impossible to gaze up at the stars in the night sky without realizing that you are looking out into space *and* backward in time. Thinking of our own lives, we become aware that as we are born, grow old, and finally die, we are three-dimensional creatures moving relentlessly through four dimensions of space-time.

As we gaze out into the star-filled nighttime sky and ponder carefully what we see, in addition to conceiving of space-time, we also realize that gravity is the dominant interaction between everything in the universe. It is gravity that keeps

Chapter 11   The General Theory of Relativity

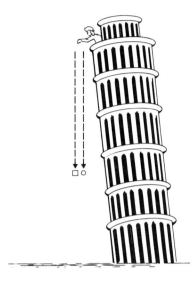

Figure 11-1. *Galileo and the Leaning Tower.* According to legend, Galileo discovered that all objects fall to the ground at the same rate, regardless of the detailed properties of the objects such as size, shape, color, chemical composition, etc.

the moon in orbit about the earth. It is gravity that holds the solar system together. It is gravity that determines the paths followed by the stars and galaxies. Indeed, gravity may even dictate the entire past and future of the universe as a whole! The existence of space-time and the significance of gravitation are two of the most profound realizations in astronomy that can occur to the human mind.

It is now possible to appreciate the true meaning of Galileo's legendary experiment from a much deeper level of insight into the nature of reality. Of all the possible paths from the top of the Leaning Tower to the ground, falling objects travel along one and only one path. Galileo could have stood at the top of the Leaning Tower and dropped cannon balls, bricks, shoes, and flowerpots over the edge. They *all* would have followed the same path to the ground. They *all* would have struck the ground at exactly the same place. At any point between the top of the Leaning Tower and the ground, they *all* would have had exactly the same speed. Yet, everything Galileo might have dropped from the Leaning Tower moves through space *and* time. A cannon ball would have been a little older when it hit the ground than it was when it left Galileo's hands. In other words, what Galileo's experiment was really saying is that out of the infinite possible paths in space-time connecting the top of the Leaning Tower of Pisa to the ground below, objects moving solely under the influence of gravity will consistently choose to follow one and only one path. Galileo's experiment cries out to us that truly to understand gravitation we should *not* talk about "forces," but rather we should examine the geometry of paths in space-time. Surely it would be a monumental triumph of the human intellect to devise a theory of gravitation that speaks of gravity in terms of the geometry of space and time.

In 1916, one of the greatest physicists of all time, Albert Einstein, published his famous *General Theory of Relativity*. He had succeeded in "generalizing" the mathematical tools of space-time physics of special relativity to include gravitation. The general theory of relativity is all about gravity; it tells us how gravity works. The central idea behind the entire general theory of relativity is this: *gravity warps space-time*. The gravitational field of an object, whether it be the earth, the sun, or a galaxy, manifests itself by giving a *curvature* to the very fabric of space and time. Objects such as planets or cannon balls move along the paths in that curved space-time. In other words, matter tells space-time how to curve and curved space-time tells matter how to move! No longer do scientists need to think of gravity as a force. Gravitation is expressed entirely in terms of the geometry of space and time!

At the heart of the general theory of relativity there is a set of equations. These so-called *field equations* tell precisely how space-time is warped by the gravitational field of an object such as the earth, or the sun, or a galaxy. If there is no matter, then there are no sources of gravity and space-time is *flat*. It is flat in the same sense that a table top or the floor is flat. But if there is matter somewhere in space-time, then in the vicinity of this matter, space-time is *curved*. The stronger the gravitational field, the greater is the curvature. Space-time is curved in the same sense that the surface of a football or basketball is curved. The field equations of general relativity tell precisely how much and in what directions space-time is curved by the presence of the gravitational field of an object.

To get a better feeling for what the field equations mean, we can draw a picture of curved space near an object such as the sun. As is the case with *all* material objects, the sun has a gravitational field; that's what keeps the solar system together. In order to draw a picture of the warped space-time around the sun, for convenience, we will "suppress" two of the four dimensions. The resulting picture is shown in Figure 11–2. *If* space-time were two dimensional, this is what curved space would look like. This is called an *embedding diagram*. In order to visualize what warped space-time looks like, simply ignore two of the four dimensions. Notice that far from the sun, space is flat. Far from any object, the gravitational field is so weak that it is virtually unnoticeable. But as you move closer to the sun, space is more highly curved; you feel a stronger gravitational field.

Figure 11–2. *Warped Space Near the Sun.* The effects of gravity on the curvature of space-time can be visualized with the aid of an "embedding diagram." Such a diagram shows how space is curved *if* space were two-dimensional. (*Adapted with permission from* Gravitation *by Charles W. Misner, Kip S. Thorne and John A. Wheeler. W. H. Freeman and Company. Copyright © 1973*)

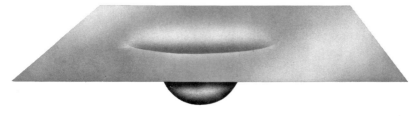

Chapter 11 The General Theory of Relativity

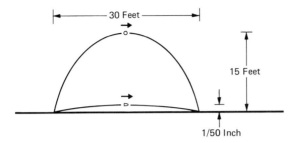

Figure 11–3. *The Ball and the Bullet.* Although the ball and the bullet start from the same place and end up in the same place, their paths look very different in space. (*Adapted with permission from* Gravitation *by Charles W. Misner, Kip S. Thorne and John A. Wheeler. W. H. Freeman and Company. Copyright © 1973*)

But it is not enough just to know how space-time is warped by gravity. It is necessary to know how things such as cannon balls and planets move in the warped space-time surrounding a massive object such as the earth or the sun. To do this we make the very simple and beautiful statement that "Nature operates in the most efficient fashion possible." More precisely, it is assumed that of all the infinite possible paths connecting two events in space-time, a freely falling body will choose to travel along the *shortest* path. If space-time were flat (no gravity present), then the trajectories followed by objects would be ordinary straight lines. After all, a straight line is the shortest distance between two points. But where space-time is curved, it is impossible to have perfectly straight lines, just as it is impossible to draw a perfectly straight line on the surface of a football. Nevertheless, even on a curved surface or in curved space-time there must be one curve connecting two points that is shorter than all other possible curves. Such a curve, the shortest of all possible curves, is called a *geodesic*. The assumption that nature operates in the most efficient fashion therefore becomes: "Freely falling bodies move along geodesics in curved space-time." When the concept of a geodesic is expressed mathematically, the *geodesic equations* are obtained. Finally, scientists have a method to do hard-core calculations and can predict where the moon is or calculate the orbits followed by planets. First of all, the field equations are utilized to tell how space-time is curved. The geodesic equations then are used to calculate how objects move in that curved space-time. Aside from a lot of complicated mathematics, the fundamental ideas are really incredibly simple and beautiful.

"What nonsense!" you may say. "Suppose I shoot a bullet so that it just skims only a fraction of an inch above the ground and lands 30 feet in front of me. During its flight the bullet, which is moving at about 1,500 feet per second, is falling freely in the earth's gravitational field. Its path is drawn in Figure 11–3. At the same time, I also throw a baseball into the air with a forward speed of 15 feet per second. My baseball reaches a height of about 15 feet in the air and also falls to the ground 30 feet in front of me. During its flight, the baseball also is falling freely in the earth's gravitational field. Both the bullet and the baseball start off at the same place and end up at the same place. But they obviously travel along very different paths. How can it be that *both* paths are geodesics? Any fool can see that the two curves are very different!"

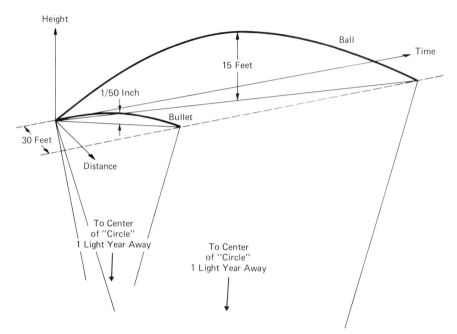

Figure 11–4. The Ball and the Bullet. Although the paths of the ball and the bullet look very different in space, their paths through space-time are very similar. Both paths are arcs of circles each having a diameter of about two light years. (*Adapted with permission from* Gravitation *by Charles W. Misner, Kip S. Thorne and John A. Wheeler. W. H. Freeman and Company. Copyright © 1973*)

Although the paths followed by objects like bullets and baseballs are very different in space, we must always remember that in general relativity we speak in terms of space-time. To examine the bullet-baseball problem properly from the viewpoint of relativity, we must draw a space-time diagram. Figure 11–4 shows the flight of the bullet and the baseball in space-time. Both paths, as viewed in space-time, are virtually identical. Both are, essentially, small pieces of huge circles, each having a diameter of about 2 light years. The curve for the path of the bullet is shorter than the curve for the path of the baseball. The bullet travels faster and arrives at its destination much sooner than the baseball. The baseball spends a much longer time in the air. Nevertheless, both curves have identical shapes. Therefore, although the paths of these freely falling objects as viewed in space look very different, as seen in space-time they look very similar.

During the nineteenth century, the brilliant German mathematician Georg Friedrich Bernard Riemann, made a bold attempt to arrive at a geometrical description of gravitation. But he did his mathematical work in space, not in space-time. It was left to the genius of Albert Einstein, with his profound insight into the nature of physical reality, to extend the mathematics of *Riemannian geometry* to four-dimensional space-time. The final result was a totally new approach to gravitation, the general theory of relativity, and only recently have astrophysicists begun to explore its fullness, richness, and beauty.

11.2 Classical Tests

For centuries, Newtonian mechanics and the universal law of gravitation were unshakable pillars in the very foundation of all modern science. It was the force of gravity that holds things to the ground and keeps the moon in orbit about the earth. It was the force of gravity that prevents the solar system from flying apart. It was the force of gravity that dominates the interactions between stars and galaxies. The approach worked very well. During the seventeenth and eighteenth centuries the successes and triumphs of Newtonian gravitation mounted steadily. After all, before Newton, mankind really had no understanding of why the planets move the way they do. Following the work of Newton, incredibly accurate predictions of the motions of comets, planets, satellites, and asteroids could be made. Even today, when scientists calculate the trajectories of Apollo astronauts to the moon or unmanned spacecrafts to Mars, Venus, and the planets beyond, all that is needed is ordinary, traditional Newtonian mechanics.

In spite of the obvious successes of Newtonian mechanics, by the mid-nineteenth century it had become apparent that something was wrong. Mercury was not precisely following its predicted orbit. According to Newton's law of gravity, after perturbations of all the outer planets have been accounted for, the orbit of Mercury around the sun should be a perfect ellipse with the sun at one focus. That's just Kepler's first law. But careful observations made during the late 1800s showed that even after all the perturbing effects of the outer planets have been subtracted away, Mercury's orbit actually is a *precessing ellipse*. In other words, as Mercury tries to move along an elliptical orbit, the orbit itself slowly moves. Therefore, Mercury traces out a rosette figure as it orbits the sun, as shown in Figure 11–5.

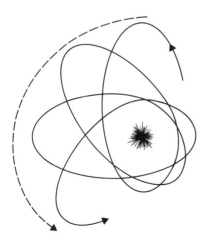

Figure 11–5. The Precession of Mercury's Perihelion. Mercury's orbit precesses by a very small amount each year. Mercury therefore traces out a rosette figure in space. The general theory of relativity fully accounts for this unusual motion of Mercury.

This effect is *very* small. A point on Mercury's orbit such as the perihelion (the point where Mercury is closest to the sun) moves through 43 seconds of arc in one century. This effect, which cannot be explained by Newtonian mechanics, is called, therefore, the *precession of Mercury's perihelion.*

People tried all kinds of gimmicks to explain the anomalous precession of Mercury's perihelion. For example, less than a century earlier, Uranus had not been following its predicted orbit. The deviations of Uranus from its predicted orbit led to the discovery of Neptune (see Section 2.4). Just as Neptune was pulling Uranus away from its expected path in the solar system, it seemed there might be another planet very close to the sun that might be perturbing Mercury's orbit. Astronomers, therefore, started looking for Vulcan, a hypothetical planet whose orbit lies inside that of Mercury. Vulcan, however, does not exist.

In short, nothing worked. By the turn of the century it had become clear that the 43 seconds of arc per century that shift in Mercury's orbit could not be explained in the framework of Newtonian mechanics and gravitation. Newton's ideas, which had been so incredibly successful everywhere else, could not completely account for the observed motions of the innermost planet in the solar system.

Shortly before the beginning of World War I, Albert Einstein proposed a radically new theory of gravitation called the general theory of relativity. According to this new theory, the gravitational field of an object such as the sun manifests itself by warping the very fabric of space and time. Objects such as planets then move along the shortest possible paths, along geodesics, in that curved space-time.

After Einstein succeeded in formulating the general theory of relativity, he naturally wanted to do some hard-core calculations. He focused his attention on the motion of planets about the sun. First, he solved the field equations, which told him how much space-time is curved near the sun. He then solved the geodesic equations, which told him how planets move in that curved space-time. His conclusion was that planets move in precessing ellipses! Indeed, when he plugged in numbers for the case of Mercury, he found that its orbit should be an ellipse whose perihelion precesses at the rate of 43 seconds of arc per century. Einstein's new theory succeeded triumphantly where Newton's theory had failed miserably!

On December 15, 1915, Einstein wrote to a colleague in Poland:

I am sending you some of my papers. You will see that once more I have toppled my house of cards and built another; at least the middle structure is new. The explanation of the shift in Mercury's perihelion, which is empirically confirmed beyond a doubt, causes me great joy.*

So, according to this new theory of gravitation, the orbits of planets around the sun are precessing ellipses. Does this mean that Kepler's laws and Newtonian

---

*Albert Einstein to Polish mathematician and physicist Wladyslaw Natanson.

mechanics are wrong? Not really. Remember that the old-fashioned physics worked just fine in every single case except that of Mercury. And Mercury's orbit deviates from Kepler's first law by only a very tiny amount. Even today, when scientists calculate the orbits of satellites they use Newtonian mechanics, not general relativity.

What all this really means is that from the viewpoint of relativity theory, the gravitational field of the sun is very weak. Since Newton's theory was so successful in virtually every conceivable case, Einstein realized that general relativity must look exactly like Newtonian gravitation in the so-called *weak field limit*. In any problem where the curvature of space-time is slight, where space-time is almost perfectly flat, general relativity must give the same answers as Newtonian gravitation. Therefore, Kepler's laws and Newton's gravitation are "special cases" of general relativity. General relativity reduces to Kepler's laws and Newton's gravitation in the case of very weak gravitational fields where space-time is almost perfectly flat.

When applied to planets moving about the sun, general relativity is important only in the case of Mercury. Although, ideally, all planets should move in precessing ellipses, Venus, Earth, and all the other planets are so far from the sun that relativistic effects are unimportant. Beyond the orbit of Mercury, space-time is so very nearly flat that relativistic precession is hardly noticed at all. For all practical purposes, ordinary Newtonian mechanics can be used.

The fact that only weak gravitational fields are encountered in the solar system has been made apparent in the previous section. In the discussion the paths of the bullet and the baseball, in space-time were essentially pieces of circles having a diameter of 2 light years. For all practical purposes, a 30-foot section of a circle 2 light years in diameter is virtually a straight line. It is, therefore, painfully obvious that general relativity is not going to predict any tremendously dramatic or surprising effects, at least as far as the solar system is concerned.

When Dr. Einstein discovered that his new theory of gravitation could explain the mysterious advance of Mercury's perihelion, he naturally wanted to see if general relativity could explain or predict anything else of interest. He wanted further *tests* of his new theory.

As discussed earlier, gravitational fields of familiar objects, such as those found in the solar system, are very weak by relativistic standards. Newtonian theory had been eminently successful with everything except the fine details of Mercury's orbit. Therefore, to test general relativity further, it was necessary to devise new experiments, or observations, that no one had ever thought of before.

The sun is by far the most massive object in the solar system. The sun, therefore, has the strongest gravitational field of any object in the solar system. Since, according to general relativity, the curvature of space-time is a direct measure of the strength of a gravitational field, the obvious place to look for relativistic effects is near the sun. Anything passing near the surface of the sun moves through the warped space-time around the sun and should exhibit relativistic effects.

Every star in the galaxy sends beams of light near to the sun's surface. Figure

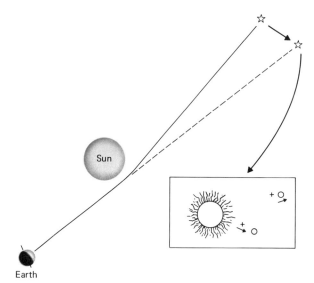

Figure 11-6. The Deflection of Light by the Sun. As a beam of light from a distant star passes through the curved space-time near the sun, the beam is deflected slightly from its usual "straight" path.

11-6 shows one such typical star sending one beam of light past the sun's surface to the earth. As the beam of light grazes the sun, it is moving through the warped space-time of the sun's gravitational field. Thus, the beam should be deflected slightly from a "classical" straight line. It should appear that the beam, as it moves along a *lightlike geodesic*, follows a slightly curved path.

Prior to the work of Dr. Einstein, no one had ever considered the possibility that light rays could be *deflected* by gravity. Photons have no mass, and since Newtonian gravitation applies only to objects with mass, it seemed reasonable that light rays should always travel along straight lines. However, once scientists realized that gravity warps space-time and that light moves along geodesics in that curved space-time, it became evident that light must be deflected by a gravitational field. Figure 11-7 shows the path of two light rays passing near the sun in an embedding diagram that graphically illustrates this so-called *deflection of light*.

In performing detailed calculations on the deflection of light by the sun, Dr. Einstein found that a beam of starlight just grazing the sun's surface should be deflected through an angle of 1.75 seconds of arc. An incredibly tiny angle! For light rays passing the sun at larger distances, this deflection angle is even smaller. How could scientists ever hope to observe such an effect?

The first obvious problem encountered in trying to detect the deflection of light by the sun is that the stars cannot be seen during the daytime. The sun is just too bright. But during a total eclipse of the sun, the moon blocks out the blinding solar disc and the stars in the sky can be seen.

In 1919, a team of English astronomers observed a total eclipse of the sun hoping to detect the deflection of light as predicted by general relativity. They took photographs of the stars seen near the sun's surface during totality and compared these pictures with photographs taken months earlier when the sun was in another part of the sky. The positions of the stars had indeed shifted from their usual locations. The observations agreed entirely with Einstein's theory. Not only had Einstein explained the baffling mystery surrounding Mercury's orbit, he also had predicted and explained a totally new phenomenon, the deflection of light, which no one had ever thought of. General relativity again emerged triumphant!

The precession of Mercury's perihelion and the gravitational deflection of light were the only two experimental tests of general relativity available during Einstein's lifetime. He did, however, predict a third test, the *gravitational redshift*.

Since a gravitational field of an object, such as the sun or the earth, warps *both* space *and* time, it might be expected that clocks located at different positions in a gravitational field should run at different rates. In fact, general relativity states that clocks in a strong gravitational field should run slower than clocks in a weaker gravitational field. Gravity slows down time. (This effect should *not* be confused with the dilation of time in the special theory of relativity.) In other words, a clock in the basement of a building should tick more slowly than a clock in the attic. The basement is closer to the center of the earth and is, therefore, in a slightly stronger gravitational field.

As with the previous two tests of general relativity, this effect is very small. If two clocks were placed in a building 100 feet tall, one clock in the basement and one in the attic, after 10 years the two clocks would differ by only one millionth of a second. One decade after starting the experiment, the clock in the basement would have lost only one millionth of a second compared to the clock in the attic. During Einstein's life, clocks with the necessary accuracy to test this prediction simply did not exist.

Figure 11–7. Light Rays in an Embedding Diagram. Light rays travel along the shortest possible paths (called "geodesics") in curved space-time. Since the space-time around the sun is not flat, light rays are deflected. (*Adapted with permission from* Gravitation *by Charles W. Misner, Kip S. Thorne and John A. Wheeler. W. H. Freeman and Company. Copyright © 1973*)

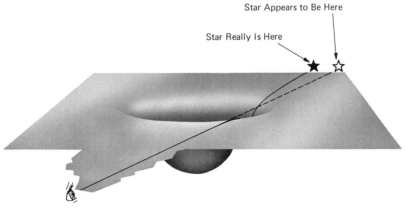

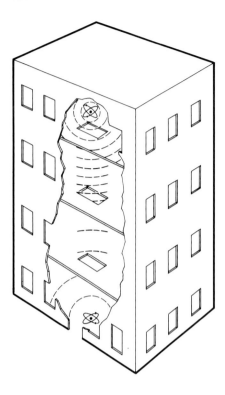

Figure 11-8. *The Gravitational Redshift.* Einstein predicted that gravity slows down time. This effect, the so-called gravitational redshift, was first measured in 1960 by physicists who used radioactive nuclei as accurate clocks. The clock in the basement "ticks" slower than the clock in the attic.

During the 1950s, the German nuclear physicist R. Mössbauer discovered a phenomenon associated with radioactive substances that could be used as extremely precise clocks. The Nobel Prize in physics was awarded to him. His discovery, the *Mössbauer effect*, has numerous useful applications. In particular, a team of scientists at Harvard University headed by Dr. R. V. Pound realized that the Mössbauer effect could be used to measure the slowing of time by the earth's gravitational field. They placed some radioactive cobalt in the basement of a campus building. The wavelengths of the γ rays emitted by the cobalt act like the ticking of a very precise clock. In the penthouse of the building 73 feet above the basement, they placed some iron, which absorbs γ rays from the radioactive cobalt. In essence, the cobalt in the basement acts like a ticking clock; the iron in the penthouse listens to how fast the clock in the basement is ticking. Dr. Pound and his associates found that the rate at which the radioactive cobalt "ticks" was slowed down by precisely the amount predicted by Einstein's theory. Time slows down in a gravitational field!

This slowing down of time is called the *gravitational redshift*. The atoms emitting light (or radioactive nuclei emitting γ rays) can be thought of as tiny clocks. The frequency, or wavelength, of the emitted radiation is a measure of how fast these clocks tick. The slower they tick, the longer is the wavelength; light emitted from inside a gravitational field is *redshifted*.

Chapter 11   The General Theory of Relativity

So, it looks as though general relativity really isn't very interesting at all. Oh sure, it's nice to have a whole new way of looking at gravitation. But in virtually every conceiveable situation encountered in the solar system, the old Newtonian approach is quite adequate. And doing mathematical calculations in general relativity is extremely complicated; it's a lot easier to solve the equations of Newtonian mechanics if scientists want to grind out some answers. So why give a damn about general relativity anyway? For all practical purposes, general relativity looked very uninteresting.

11.3 The Schwarzschild Black Hole

Following the formulation of the general theory of relativity, there was a tremendous amount of excitement in the scientific community. Classical Newtonian mechanics had stood as a major pillar at the very foundations of science for three full centuries. Yet, suddenly a dramatically new and different approach to gravitation appeared that worked even better. But after the initial excitement died down, scientists realized that for all practical purposes they could continue right on using ordinary Newtonian mechanics. By the 1920s, interest in general relativity had all but vanished. For decades following, theoretical research was confined to a few hard-core fanatics who possessed the insight to realize the profound possibilities offered by a *geometrodynamical* approach to gravitation.

With the development of high-speed computers in the early 1960s, astrophysicists finally succeeded in working at many of the details of stellar evolution. In particular, information gradually developed indicating that a dying star must eject all but $2\frac{1}{2}$ solar masses in a nova or supernova explosion *if* it is to become a white dwarf or neutron star. If the dead core of a star that has used up all of its nuclear fuel contains more than $2\frac{1}{2}$ solar mass, *no physical forces can hold up the star*. The star simply gets smaller and smaller and smaller.

Recall from the earlier discussion of gravity (see Section 2.4) that if the earth were squeezed down to a smaller size, someone standing on that compressed earth would weigh considerably more than usual. Similarly, as all the matter of the collapsing star is squeezed to smaller and smaller size, the intensity of the gravitational field increases. As gravity gets stronger, the collapse proceeds more and more rapidly, which further increases the strength of the gravitational field. It is an incredibly rapid and vicious cycle from which there is absolutely no escape. In fact, as seen by someone standing on the surface of a dead star, once this catastrophic collapse begins, *the entire star is crushed down to zero size in less than a thousandth of a second!*

Remember, however, that Newtonian gravitation works only for weak gravitational fields, such as those encountered here in our own solar system. In strong gravitational fields, such as those involved in the collapse of a star, the general

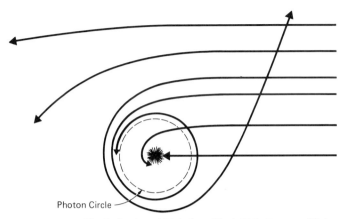

Figure 11-9. *The Deflection of Light by a Black Hole.* Beams of light passing near a black hole are deflected through large angles. In fact, a beam of light approaching a black hole at the appropriate distance will go into a circular orbit about the black hole.

theory of relativity is used for a correct analysis of what happens. According to general relativity, the gravitational field of an object manifests itself by warping space-time. The stronger the field, the greater is the curvature of space-time. As a massive dead star undergoes catastrophic gravitational collapse, the intensity of the gravitational field gets higher and higher. This means that the warping of space-time above the star gets more and more severe. In fact, at one critical stage in the collapse of such a star, the warping of space-time becomes so great that *space and time fold in over themselves and the star disappears from the universe!* What is left is called a *black hole.*

To understand the properties of black holes more fully, imagine one of these black holes floating in space. An experiment can be performed with a flashlight shining beams of light past the black hole at various distances, as shown in Figure 11-9. Very far from the black hole, the strength of the gravitational field is quite weak. For example, far from a black hole—a black hole formed from a dead 10 solar mass star—the strength of the gravitational field still corresponds to 10 solar masses. In such a region, the curvature of space-time is small and light rays are deflected only slightly. But, as the flashlight is aimed more directly at the black hole, the beam of light will move into regions of space-time that are more highly warped. As shown in Figure 11-9, beams of light passing closer to the black hole are deflected through larger angles. In fact, it is possible to aim the beam of light toward the black hole at precisely the right distance so that the light goes into circular orbit about the black hole. Such an orbit is called the *photon sphere.* Every star in the universe contributes light to the photon sphere surrounding the (idealized) black hole. Finally, beams of light aimed more directly at the black hole get sucked in by the tremendous warping of space-time.

Still further insight into the nature of black holes can be gained by following the collapse of a dead star. Imagine a (foolhardy) astronaut sitting on the surface of a dying star that is just about to begin collapsing. He has a flashlight (or radio transmitter) and finds that he can shine a beam of light away from the star just by pointing the beam in any upward direction. But as the collapse proceeds, the intensity of the gravitational field rises dramatically, as does the curvature of space-time. Pretty soon, the astronaut riding on the surface of the collapsing star realizes that only beams of light aimed near the vertical direction manage to escape. Beams of light at angles only a few degrees above the horizontal direction always fall back to the surface of the star, as shown in Figure 11–10. The warping of space-time is beginning to make itself felt. As the star collapses even further, the curvature of space-time becomes greater. The astronaut must aim his flashlight very near to the vertical direction in order for the light to get to the outside universe.

We can think of an imaginary cone, the so-called *exit cone*, centered about the vertical direction, as shown in Figure 11–10, that has this property: all beams of light emitted from the star at angles inside the cone will manage to escape to the outside universe. All beams of light emitted from the surface of the star at angles outside the exit cone will *not* escape; they go up and they come back down.

As the star gets smaller and smaller, as all the matter of the star gets squeezed into a smaller and smaller volume, the curvature of space-time increases dramatically. This means that the ill-fated astronaut must aim his flashlight closer and closer to the vertical direction if he wants the light ever to get out. As the collapse proceeds to its inevitable end, the exit cone gets narrower and narrower. Finally, at a critical stage in the collapse, the exit cone closes up completely! No matter in what direction the astronaut aims his flashlight, *no light gets out*. Beyond this stage in the collapse, absolutely nothing can ever escape from the star. The astronaut turns on his radio transmitter and finds that radio waves cannot escape from the star. He no longer can communicate with Mission Control in Houston. Space-time has folded over him! He has disappeared from the universe! He has passed the *event horizon*.

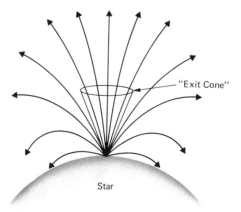

Figure 11–10. Beams of Light from the Surface of a Collapsing Star. Beams of light emitted at angles near the horizon fall back to the star. Only beams of light emitted very near the vertical direction will manage to escape.

When a star has collapsed to the stage where nothing—not even light rays—can get out, the star has fallen inside its event horizon. For a 3 solar mass dead star, the diameter of the event horizon is about 11 miles. When a 3 solar mass star collapses down to a diameter of 11 miles, it literally disappears from the universe. The term event horizon is really quite appropriate. It is a place in space and time beyond which we, in the outside universe, can know no events. No information whatsoever can get out to us. When a star collapses inside its event horizon, everything in the star has passed over a horizon in space-time beyond which we can see no events.

A star that has collapsed inside its event horizon still has a finite size. For example, a 3 solar mass dead star would have a diameter of $10\frac{7}{8}$ miles just after falling inside its event horizon. But there are still no physical forces that can stop the collapse. The star continues to shrink in size until all the matter of the star is crushed out of existence at a single point, the *singularity*. At the singularity at the center of a black hole, there is infinite pressure, infinite density, and infinite curvature of space-time.

If we were to travel into a black hole we would, therefore, encounter the following. First we would pass through the photon sphere, a thin shell of light orbiting above the black hole. Then we would pass through the event horizon and completely disappear from the outside universe. Finally, in about a hundred thousandth of a second, we would be dragged into the singularity where everything is crushed out of existence by infinitely warped space-time. This arrangement of the photon sphere, event horizon, and singularity is shown schematically in Figure 11–11.

The type of black hole discussed in this section is the simplest of all possible black holes. The black hole is spherically symmetric (shaped like a sphere) and is not rotating. It is described by a solution of the Einstein field equations first discovered by K. Schwarzschild in 1916, only a few months after Einstein published his theory. This mathematical solution to the field equations is, therefore, called the *Schwarzschild solution*. All static, spherically symmetric,

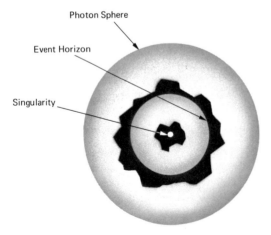

Figure 11–11. *The Structure of a Black Hole.* The structure of a Schwarzschild black hole is shown schematically in this diagram. The singularity is surrounded by an event horizon which is in turn surrounded by the photon sphere.

uncharged, nonrotating solutions to the field equations are essentially equivalent to the Schwarzschild solution. The *Schwarzschild black hole* is, therefore, the archetype of the simplest possible black hole.

Imagine a dying star whose mass is the critical limit of $2\frac{1}{2}$ solar masses (that is, it cannot become a white dwarf or a neutron star) and that is in the process of collapsing to become a Schwarzschild black hole. What might we expect to observe? What would such a star look like as gravitational collapse proceeds? First of all, time slows down in a gravitational field. Since the atoms involved in the emission of light may be thought of as little clocks, when light is emitted from inside a gravitational field, it will have a longer wavelength than usual. The light will be redshifted. Therefore, the light from the collapsing star will be redshifted. As the star collapses to smaller and smaller sizes, it will appear redder and redder. This is the first important observable effect.

In addition to a constant reddening of the star, once the star's surface has collapsed inside the photon sphere, the exit cone begins to close up. This means that part of the light emitted from the star's surface does not escape. Therefore, the star gets dimmer and dimmer. Finally, when the star has collapsed down to the event horizon, the exit cone closes up completely and no light at all escapes. For example, for a 3 solar mass black hole, the photon sphere is about 16 miles in diameter and the event horizon is 11 miles in diameter. As a dying star of 3 solar masses shrinks in size from 16 to 11 miles, the star dims rapidly.

So there are two simultaneous effects in the formation of a black hole. If astronomers were lucky enough to catch a star just as it was beginning gravitational collapse, they would see the star very rapidly get *red* and *dim* as it disappears from the universe. However, with extremely sensitive instruments, astronomers watching the formation of the black hole would notice an additional effect. Since time slows down in a gravitational field, as the star shrinks into its event horizon, the collapse appears to slow down. In fact, at the event horizon, *time stops completely!* Hence, the last few atoms at the star's surface never make it to the event horizon as seen by distant astronomers. As seen by astronomers very far from the collapsing star, it takes an infinite amount of time for the star to be completely swallowed by the gravitational field. Thus, there is always just a little bit of light emitted by those few atoms, which take an infinite number of years to reach the event horizon. This light, however, is so severely redshifted that for all practical purposes the collapsing star becomes "black" in a comparatively short period of time—in less than a thousandth of a second. A "hole" in space-time is left behind. Quite literally, a black hole is formed.

There are several things that might happen as the astronaut plunges into a black hole. As seen by an outside observer (that is, an astronomer back on earth), the astronaut would *never* reach the event horizon because as he gets closer and closer to the event horizon, time slows down more and more. As seen from far outside the black hole, time stops completely at the event horizon and so does the astronaut.

The astronaut sees an entirely different situation. First of all, suppose he is falling *feet first* into the black hole. Long before he gets anywhere near the event horizon, it becomes painfully obvious that his feet are closer to the black hole than his head. His feet start falling faster than his head! He gets stretched out longer and longer. Instead of being 6 feet tall, he soon finds that he is 6 miles tall, as his feet plunge faster toward the black hole than his head. The astronaut finds that he looks like a long thin wire.

To overcome this difficulty, an astronaut who wants to arrive successfully at the event horizon must first of all magically shrink himself to the size of an atom, or even smaller. Thus, the distance between his head and feet will be so small that all the parts of his body will fall at the same speed. The intensity of the gravitational field will be essentially the same over-all dimensions of his tiny body. Let's assume that our astronaut has studied enough relativity to take this precaution.

Although time appears to slow down as seen by an outside observer, the astronaut falling into the black hole notices no such effect. He compares his wrist watch with other clocks in his (tiny) spaceship and notices that everything is just fine. In a very short period of time, he plunges right through the event horizon.

Upon entering the event horizon, two important things occur. First, the astronaut finds that he cannot communicate with the outside universe. Radio signals from his transmitter will never get out. He, nevertheless, still can see the outside universe. Incoming light rays still reach him. The CARE packages we drop in after him still get there. But now the astronaut knows he is in big trouble. Inside the event horizon, the roles of space and time become interchanged! It is almost as if clocks measure length and rulers measure time. In the outside universe, we are powerless to halt the relentless passage of time. Everyone on the earth is dragged forward in time, from age twenty to forty, from forty to eighty. Similarly, inside the event horizon, our ill-fated astronaut is powerless to halt the relentless passage of space. He is dragged into the singularity. There is absolutely no escape. At the singularity he experiences an infinite warping of space-time. No matter how small he is, no matter what he tries to do, he is crushed out of existence at the center of the black hole.

This discussion of black holes only scratches the surface of one of the most fantastic and bizarre topics in all of modern science. For example, it can be shown that the full geometry of the Schwarzschild black hole actually connects *two* separate universes! The development of a black hole exhibits an increasing curvature of space, as shown in Figure 11–12. This is an embedding diagram of space about a collapsing star on its way to becoming a black hole. The shaded area shows where the matter of the star is. However, as the embedding diagram

Figure 11–12. The Embedding Diagram of a Collapsing Star. The curvature of space around a collapsing star is visualized with the aid of an embedding diagram. The shaded area indicates the location of the matter in the star.

Chapter 11 The General Theory of Relativity

Figure 11–13. *The Einstein-Rosen Bridge.* The Schwarzschild black hole actually connects our universe (the upper sheet) with another universe (the lower sheet) as shown in this embedding diagram.

evolves, space opens up into a second universe. An embedding diagram showing this double-universe nature of the Schwarzschild solution is given in Figure 11–13. This is called an *Einstein-Rosen bridge* or a *wormhole* in space-time.

In the case of a Schwarzschild black hole, which is not rotating and does not have any electric charge, it is impossible to go from our universe into the "other universe." The astronaut always must hit the singularity after passing through the event horizon. However, during the 1960s, astrophysicists discovered that if the black hole is rotating, or if the black hole does have an electric charge, then you *can* get to a multitude of other universes! The possibility then exists that matter and energy from some other universe might be gushing up into our own universe through one of these strange objects! This is called a *white hole.*

If the astronaut were to plunge into an idealized rotating or charged black hole, he could emerge in another universe without every encountering any singularities! But he can also use one of these black holes connecting a multitude of universes to return to our own universe! He could return to our universe at some distant star or galaxy. Or he could return to our earth—a billion years in the future or a billion years in the past! This is called a *time machine.*

There are all kinds of arguments against time machines. One astronaut could return to the earth 5 minutes before he left. He could meet himself and tell himself what a nice trip he had. Then *both* of him could climb aboard the rocket ship. They could do it again. And again. Not only would this result in an overpopulation problem of astronauts, it violates causality.*

The solution to Einstein's field equations, which give these incredible results, is highly idealized. Research during the late 1970s has indicated that under realistic conditions, such phenomena could not possibly occur.

11.4 The Kerr Solution

The previous section dealt with one of the most incredible of the predictions that came from Einstein's geometrodynamical approach to the gravitational field: the black hole. Attention was focused on the simplest of all black holes.

---

*The reader who desires further discussion of spaceflight to other universes and time machines in relativity is referred to this writer's *Relativity and Cosmology* (New York: Harper & Row, 1973). An extensive discussion of various types of black holes is found in this writer's *The Cosmic Frontiers of General Relativity* (Little, Brown & Company, 1977).

The Schwarzschild black hole is static, spherically symmetric, nonrotating, and uncharged. Such an object has a number of very peculiar properties. But could a nonrotating black hole actually exist in nature? Is it at all possible that astronomers someday will find a Schwarzschild black hole in space? In all probability: No! The reason should be obvious.

Black holes are one of the three possible final states of stellar evolution. Dead stars with masses greater than about $2\frac{1}{2}$ suns must undergo catastrophic gravitational collapse. No known physical forces can prevent the complete implosion of such stars. But as far as is known, all stars rotate. Our sun rotates once a month; many stars rotate once every few hours. As discussed earlier in connection with white dwarfs and pulsars, when a dying star collapses, its rate of rotation must speed up. Just as an ice skater doing a pirouette on the ice speeds up as she pulls in her arms, the law of the conservation of angular momentum dictates that a *real* black hole is rapidly rotating. Even if the dying star rotated as slowly as once a year, by the time it collapsed to a black hole it would have to be spinning at a furious rate. Real black holes in space must be rotating black holes.

Up until the 1960s, astrophysicists discussed only nonrotating, static black holes such as those described by the Schwarzschild solution. The reason for this is embarrassingly simple. Nobody could solve the Einstein field equations to include rotation. The mathematics was too complicated. It was not until 1963 that the Australian mathematician Dr. Roy P. Kerr succeeded in discovering a complete solution to the field equations that properly takes rotation into account. This important mathematical discovery is called the *Kerr solution*. It fully describes the properties of space-time around a rotating black hole.

The *Kerr black hole* has numerous fascinating and unbelievable properties. The full, idealized solution has *two* event horizons, not one. In its most complete form, the Kerr black hole connects our universe to an infinite number of parallel universes through wormholes bounded by singularities separated by pairs of event horizons. We will, however, turn our attention only to those properties of the Kerr solution that may be astrophysically important.

To appreciate the difference between static (Schwarzschild) and rotating (Kerr) black holes, an experiment can be performed. Imagine a static black hole in front of you. Place a number of flashbulbs, such as those commonly used to take indoor photographs, around the black hole. Set off the flashbulbs and watch where the light goes.

In ordinary, flat space-time, such as here on earth, if you set off a flashbulb, a spherical shell of light expands outward from the light equally in all directions.

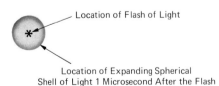

Figure 11-14. A Flashbulb in Flat Space. A flash of light, such as that emitted by a flashbulb, gives rise to an expanding spherical shell of light. The "star" indicates the location of the flash while the circle designates the location of the expanding shell of light one microsecond after the flash.

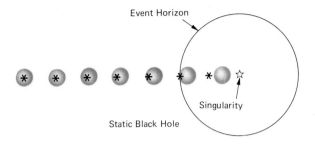

Figure 11-15. *Flashes Near a Schwarzschild Black Hole.* Flashbulbs set off near the black hole give rise to expanding shells of light which are drawn towards the black hole. The static limit is at the event horizon.

As shown in Figure 11-14, in flat space-time a little "star" can be drawn marking the location of the flashbulb at the moment of the flash. Around the star we can draw a small circle indicating the location of the expanding shell of light 1 microsecond after the flash. Since the light goes off equally in all directions, the circle will be centered on the location of the flashbulb.

To explore the nature of a static black hole, set off several flashbulbs at a variety of distances from the black hole, as shown in Figure 11-15. Very far from the black hole, where space-time is essentially flat, the expanding shell of light expands equally in all directions about the location of the flashbulb. But, as you get closer and closer to the black hole, the expanding shell of light is drawn toward the black hole. The light falls toward the black hole. In Figure 11-15 this effect is indicated by the fact that the circles, representing the locations of the shells of light 1 microsecond after each flash, are *not* centered on the locations of the flashbulbs. As you get closer and closer to the black hole, the expanding shells of light are drawn more and more toward the black hole. In fact, a flashbulb set off at the event horizon has its expanding shell of light entirely *inside* the event horizon. This means two things. First of all, it means that light cannot escape from the event horizon. But it also means that at the event horizon, if you wanted to stay at the same place for a long time (for example, not fall inward) you would have to travel outward at exactly the speed of light. In the Schwarzschild black hole, therefore, the event horizon is also the location of the so-called *static limit*. Outside the static limit (that is, above the event horizon) you could hover above the black hole just by having a powerful rocket that would prevent you from falling in. The closer you were to the black hole, the more powerful your rocket would have to be. At the event horizon, however, your rocket would have to be traveling at the speed of light outward to prevent you from falling into the singularity. Inside the event horizon, it is impossible to ever remain in the same location. Since your rocket cannot go faster than the speed of light, you must constantly fall inward. Once inside the static limit, you must always be moving.

With a Kerr black hole, the situation is a little more complicated. Again, perform the experiment setting off flashbulbs around the rotating black hole. As expected, far from the black hole, shells of light expand equally in all directions about the locations of the flashbulbs. Also as expected, nearer to the black hole

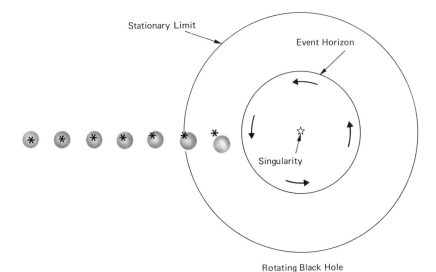

Figure 11-16. *Flashes Near a Kerr Black Hole.* Flashbulbs set off near the black hole give rise to expanding shells of light which are drawn towards *and* around the black hole. The static limit is *above* the event horizon.

the expanding shells are offset from the locations of the flashes. Now, however, two simultaneous effects are noticed. Not only is the light dragged toward the black hole, the light also is dragged slightly around the black hole. Gravity not only causes things to fall inward, but also *around,* in the direction in which the black hole is rotating. This phenomenon is called the *dragging of inertial frames.* Space-time is dragged around the black hole as the black hole rotates. The final result of these simultaneous effects, namely the inward *and* sideways falling of light, means that the static limit occurs *above* the event horizon. Before reaching the (outer) event horizon in a Kerr black hole, you arrive at that region where it is impossible to stand still. The existence of a static limit outside the event horizon has some interesting implications.

The region inbetween the static limit and the event horizon in a rotating black hole is called the *ergosphere.* A cutaway diagram of a rotating black hole showing the ergosphere is given in Figure 11-17. Planets or astronauts can travel into and out of the ergosphere with little difficulty. As long as they do not cross the event horizon, which forms the inside boundary of the ergosphere, they are not forced to collide with singularities or go into other universes. They can always manage to stay in our universe.

In 1969, the brilliant astrophysicist Roger Penrose discovered a process by which energy could be extracted from a rotating black hole through its ergosphere. Imagine an object that falls into the ergosphere of a Kerr black hole. When the object is inside the ergosphere it splits in two, as shown in Figure 11-17. One piece goes into the event horizon and vanishes from our universe.

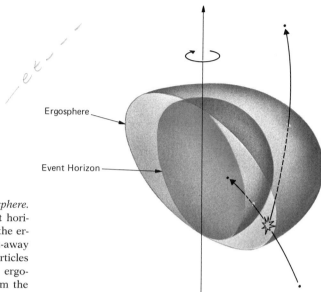

Figure 11-17. *The Ergosphere.* The region between the event horizon and static limit is called the ergosphere, shown here in a cut-away view of a Kerr black hole. Particles which split in two inside the ergosphere can extract energy from the black hole.

The other piece comes flying back out. Even though the piece that comes out of the ergosphere is smaller than the object that fell in, this piece can be moving at an incredibly high speed; it, thus, possesses an enormous amount of energy. The object has extracted energy from the black hole and the black hole rotates a little slower. Enormous amounts of energy can be extracted from rotating black holes in this fashion. In fact, as the Caltech physicists Drs. Press and Teukolsky pointed out in 1972, it is possible to construct a *black hole bomb*. Imagine a rotating black hole surrounded by a spherical mirror. Light rays bounce around inside the mirror, each time passing in and out of the ergosphere as shown in Figure 11-18.

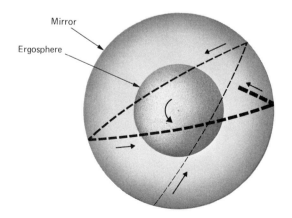

Figure 11-18. *The Black Hole Bomb.* A spherical mirror is constructed around a Kerr black hole. Light reflecting around inside the mirror is "superradiantly scattered" by the black hole. The light eventually gains so much energy that the mirror is blown apart.

The light rays will acquire so much energy from the black hole by this process of *superradiant scattering* that the mirror eventually must be blown apart by the enormous radiation pressures that build up. This effect might be important in astronomy if a black hole were located in a cloud of ionized gases—that is, inside a so-called *plasma*. The plasma could act as a mirror and would reflect light around the black hole until the entire cloud blew up.

The idea that energy can be extracted from a rotating black hole by means of the ergosphere brings to mind a variety of applications. For example, consider the superecological city first proposed by the theoretical physicists Drs. C. W. Misner, K. S. Thorne, and J. A. Wheeler in the early 1970s. Suppose that an advanced civilization were to find a Kerr black hole. Suppose they were to construct a city above the black hole, as shown in Figure 11–19. Everyday garbage trucks carrying refuse from the city would back up to a conveyor belt and dump trash into containers on the belt. The conveyor belt extends down into the ergosphere, as shown in the diagram. Deep inside the ergosphere, the containers dump the garbage into the event horizon. This is analogous to an object splitting in two, part of it falling into the event horizon and part of it flying back out with a lot of energy. As each container dumps its contents into the event horizon, the container receives a strong tug. This causes the conveyor belt to move faster and faster. An electric generator and a powerhouse could be attached to the conveyor belt, thereby providing electricity for the entire city! Each time some garbage is dropped into the black hole, it rotates a little slower. The energy from the rotation of the black hole is converted into electrical energy to power the city.

Returning to the mundane realities of the twentieth century, astronomers ask if they could ever hope to discover a Kerr black hole in the universe. Depending on whom you ask, answers will range from "Maybe" and "Why not?" all the way to "We already have!" As will be seen in the next section, this writer holds the latter opinion.

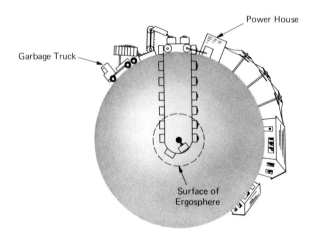

Figure 11–19. *The Super-Ecological City.* A city built around a Kerr black hole could extract energy from the black hole by lowering garbage into the ergosphere on a conveyor belt. As the garbage is dropped into the event horizon, the conveyor belt gets a "tug" which could be used to run electric generators.

Recall that scientists rejected the possibility of ever finding a Schwarzschild black hole because it is not rotating. Scientists have every reason to believe that black holes formed from real stars *must* be rapidly rotating. The Kerr black hole meets this requirement. There is only one additional possible complication: the black hole might have some magnetic or electric charge. During World War I, two physicists discovered the solution to the Einstein field equations for a charged black hole. This mathematical discovery, the *Reissner-Nordstrøm solution*, was named in their honor. Like the Kerr solution, the Reissner-Nordstrøm solution in its most complete, idealized form, connects a multitude of parallel universes with our own universe. However, magnetically or electrically charged black holes are probably not very important in astronomy. If a black hole with a large charge existed in nature, it seems reasonable that electric charges (protons and electrons) floating in space would rapidly "neutralize" it. If a black hole carried a huge positive charge, it would strongly attract lots of (negatively charged) electrons from interstellar gas. Similarly, if a black hole carried a huge negative charge, it would strongly attract lots of (positively charged) protons. In a very short time, the black hole's charge would be reduced substantially. When, and if, astronomers ever succeed in conclusively discovering a black hole in space, we can feel quite confident that it will be a Kerr black hole.

11.5 Discovering Black Holes

At first glance, it might seem that finding a black hole in space is a hopeless task. Since nothing ever escapes from a black hole—not even light—astronomers cannot ever actually see such an object through their telescopes. Indeed, up until the mid-1960s, not one single concerted effort of any kind had been made by astronomers to find black holes in the sky. After all, the "party line" was that all stars, even the most massive, become white dwarfs after ejecting most of their matter in a supernova explosion.

In 1964, two Soviet astrophysicists, Drs. Zel'dovich and Guseynov hit on an interesting idea. Since about half of the stars seen in the sky are binary stars (see Section 7.5), is it possible that black holes exist as "unseen companions" in double star systems? As stated in Chapter 7, one common type of double star is the *single-line spectroscopic binary*. Such a system is too far away to see two separate stars. In addition, one star is too dim to be detected at all. The astronomer knows that he must be looking at a binary because we see the spectral lines of the visible star shift back and forth as the two stars orbit each other. In most cases, a single-line spectroscopic binary simply consists of two ordinary stars fairly close together. Only one star is detected because it just happens to be considerably brighter than its companion.

Binary stars are important in astronomy because it is possible to estimate the masses of the individual stars in the system. Suppose we could find a single-line spectroscopic binary in which the mass of the "unseen companion" exceeded 3 solar masses. Suppose, furthermore, that we could prove that this massive "unseen companion" *should* be bright enough to be observed. After all, from the mass-luminosity relation (see Section 7.4) we know that the more massive a star is, the brighter it is. Then we would have a good case for having discovered a black hole.

Of the hundreds of spectroscopic binaries listed in catalogues, Drs. Zel'dovich and Guseynov were able to narrow the possibilities down to five good candidates. In 1968, Drs. Trimble and Thorne at Caltech revised and extended the work of their Soviet colleagues. The result was eight good candidates. Unfortunately, however, none of these eight binaries presented a truly convincing case. It was always possible to conjure up some excuse why an unseen, massive companion was invisible without resorting to black holes. Thus, by 1970, the search for black holes looked like a dead-end street.

One way of thinking about black holes in space is that they act like giant "cosmic vacuum cleaners!" Far from a black hole, space-time is essentially flat. Thus, a few million miles from a 10 solar mass black hole, the gravitational field looks just like the gravitational field of any ordinary 10 solar mass star. But if anything travels very near to the black hole, it will be scooped up by the highly warped space-time. In the mid-1960s, astrophysicists began calculating what would happen if gases were to fall onto a black hole and be sucked into the hole. As these gases are compressed and crushed by the intense gravitational field, they will be heated to extremely high temperatures of millions of degrees. At such temperatures, the infalling gases should emit lots of X rays. Therefore, if any of the eight binary stars examined by Drs. Thorne and Trimble also are emitting X rays, the case for the existence of a black hole is strengthened enormously. Unfortunately, until the early 1970s, astronomers simply did not have the equipment to study the X-ray properties of stars properly.

In December 1970, the first astronomical X-ray satellite was launched. By using this satellite, *Uhuru*, astronomers finally had the ability to make a detailed examination of the appearance of the X-ray sky (see Section 6.5). Within a year, more than a hundred X-ray sources had been discovered. Six are members of binary systems. The following table lists all six binaries. In all cases, careful analysis of the data from *Uhuru* has permitted accurate determinations of the locations of the X-ray sources. In all cases, astronomers were able to turn their telescopes to these locations and discover the visible star in each of the binary systems. The names and apparent magnitudes of the visible stars are given in Table 11–1.

In this table, the "name" column gives the designation of the X-ray source from the *Third Uhuru Catalogue*. The series of numbers following the designation 3U give the approximate right ascension and declination of the X-ray source. Thus, 3U1700 − 37 is located at $\alpha = 17^h00^m$ and $\delta = -37°$ in Herculus. The so-called common name comes from the pre-*Uhuru* days when X-ray sources were discovered from brief rocket flights. Thus, SMC X-1 refers to the first X-ray source discovered in the Small Magellanic Cloud. The column giving

Chapter 11 The General Theory of Relativity

Table 11-1. Identified X-Ray Binaries (as of 1975)

Name	Common Name (if any)	Orbital Period of Binary (in days)	Name of Visible Star	Apparent Magnitude of Visible Star	Probable Mass of X-Ray Star or Comment
3U0115−73	SMC X-1	3.9	Sanduleak 160	13	2 solar masses
3U0900−40		8.9	HD 77851	7	$1\frac{1}{2}$ solar masses
3U1118−60	Centaurus X-3	2.1	(Unnamed)	13	X-ray pulsar (period = 4.8424 seconds)
3U1653+35	Hercules X-1	1.7	HZ Herculis	14	X-ray pulsar (period = 1.2378 seconds)
3U1700−37		3.4	HD 153919	7	$2\frac{1}{2}$ solar masses
3U1956+35	Cygnus X-1	5.6	HDE 226868	9	8 solar masses

the orbital period of the binaries shows that they are all very short. In each case it takes only a few days for the two stars to revolve about each other.

Two of the six X-ray binaries turned out to be pulsars emitting regular bursts of X rays. Nothing associated with black holes could possibly produce this kind of regular behavior. No off-axis magnetic field could be anchored to a black hole. Centaurus X-3 and Herculus X-1 are probably neutron stars.

In the remaining four cases, painstaking observations of the X-ray sources from *Uhuru*, along with optical observations of the visible stars with telescopes, have permitted astronomers to obtain good estimates of the masses of the X-ray stars. Cygnus X-1 stands out as an obvious candidate for a black hole! Even if one assumes the role of the devil's advocate and presents the worst case analysis of the data, the mass of Cygnus X-1 is still well above the $2\frac{1}{2}$ solar mass upper limit for a neutron star. In the mid-1970s, it was fashionable for astronomers to dream up all kinds of schemes to get the mass of Cygnus X-1 below the $2\frac{1}{2}$ solar mass limit. As evidence mounted, however, these schemes looked more and more artificial, if not ridiculous. In all probability, Cygnus X-1 is a black hole.

Beginning in 1971, teams of astrophysicists around the world started doing calculations to explain the details of how X rays are emitted by Cygnus X-1. Important initial contributions were made by Drs. Shakura and Sunynaev at Moscow and Drs. Pringle and Rees at Cambridge, England. There are several other X-ray sources discovered by *Uhuru* that look suspiciously like Cygnus X-1 most notably Circinus X-1 (3U1516-56), so a good theoretical model for Cygnus X-1 might be useful in predicting what astronomers should look for in future searches for black holes.

Imagine two stars in a binary system, as shown in Figure 11–20. It is possible to draw a figure eight around the stars, called the *Roche limit*, to denote the location in space at which the inward gravitational forces are balanced by the

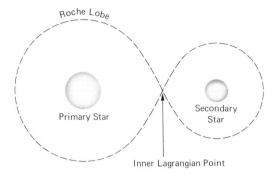

Figure 11–20. *The Roche Lobes.* A special figure-eight can be drawn around the two stars in a binary. Gas floating around inside one of the lobes belongs to the star in that lobe.

outward centrifugal forces. Any gases floating around inside one loop of the figure eight, inside one of the so-called *Roche lobes*, belongs to the star in that lobe. Any gas floating around outside of the figure eight is free to wander off into interstellar space.

The importance of the Roche limit has been recognized for many years. In binary systems containing ordinary stars, one star sometimes expands to such an extent that it *overflows* its Roche lobe. In becoming a red giant, for example, the star's surface could extend beyond the Roche lobe. When this happens, the star loses its outer atmosphere, which just floats away into space. Figuratively speaking, the astronomer is then looking at a star that has lost its skin. Also, when a star overflows its Roche lobe, it can dump great quantities of matter on its companion. Thus, the companion, which started out as the less massive of the two stars, can end up being the more massive star.

In the case of Cygnus X-1, the visible star (HDE 226868) is probably not large enough to fill its Roche lobe completely. But just as the sun is constantly ejecting matter in the form of high-speed particles that produce the *solar wind*, all stars are probably leaking gas into space in what could be called a *stellar wind*. In other words, it is reasonable to assume that HDE 226868 is ejecting a small amount of gas at a fairly constant rate. Under normal circumstances, this stellar wind would be totally unnoticeable. However, when these gases reach the vicinity of the crossing point in the figure eight, the so-called *inner Lagrangian point*, they are free to flow into the Roche lobe of the other star. Thus, it is

Figure 11–21. *Overflowing the Roche Lobe.* A star in a binary can "overflow its Roche lobe" by simply expanding. However, a star which is smaller than its Roche lobe can still eject matter into space (and onto its companion) by a stellar wind.

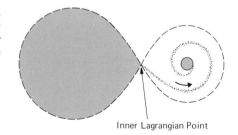

Figure 11–22. HDE 226868. The star indicated with the arrow is the visible star in the binary which contains Cygnus X-1. *(Hale Observatories. Copyright by the National Geographic Sky Survey)*

expected that some of the matter from the visible star is flowing past the inner Lagrangian point onto the invisible X-ray star.

If Cygnus X-1 is a black hole, it must be a rotating black hole described by the Kerr solution. As gases fall toward the rotating black hole, they are captured into circular orbit about the black hole. The final result is a disk about 2 million miles in diameter. As shown in Figure 11–23, some of the infalling gases will be ejected

Figure 11–23. The Cygnus X-1 Binary. The stellar wind from HDE 226868 is captured into an accretion disk about the black hole. As the gases spiral into the black hole they are heated to high temperatures and emit X rays.

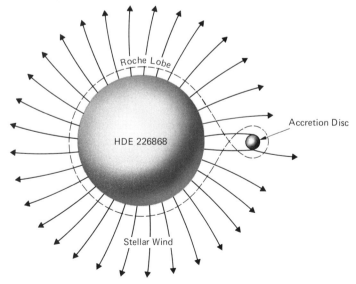

341

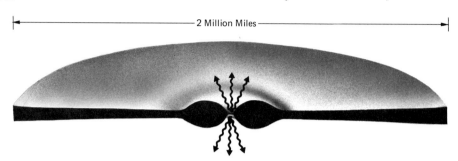

Figure 11-24. *The Accretion Disk.* The structure of the accretion disk around the black hole in Cygnus X-1 is shown in cross-section. The X rays are emitted less than 200 miles from the black hole. (*Adapted from Dr. Thorne*)

into space through the Roche lobe surrounding the X-ray star. Some of the infalling gases might be returned to the visible star. But most of the gas ends up in the *accretion disk* orbiting the black hole. In certain respects, this disk is just like the rings around Saturn, but on a much larger scale.

Just as Mercury goes around the sun faster than Pluto (Kepler's third law), the inner parts of the accretion disk must be rotating faster than its outer parts. This means that various parts of the disk will rub against each other because they are rotating at different speeds. The resulting friction will heat up the gases as they slowly spiral inward toward the black hole.

Detailed calculations by physicist Dr. K. S. Thorne and his associates at Caltech reveal that by the time the gases have spiraled into a distance of only 200 miles above the black hole, friction will have heated the gases to temperatures in excess of 1 million degrees! In the last 75 miles, the temperature of the gas is probably hotter than 10 million degrees! Anything this hot must emit great quantities of energy in the form of X rays. Based on these calculations, it can be concluded that the X rays from Cygnus X-1 must come from very hot gases spiraling into a rotating black hole!

Figure 11-24 shows the structure of the accretion disk based on the calculations of Dr. Thorne and his colleagues. The entire disk is about 2 million miles in diameter. At edges it is about 60,000 miles thick. Rotation of the gases and the gravitational field of the black hole tend to flatten the disk as we move inward. But at a distance of about half a million miles from the black hole, thermal pressures in the disk cause it to thicken substantially. In the innermost few thousand miles, the disk again becomes very thin as the gases spiral toward the black hole. X rays are emitted at distances less than 200 miles from the black hole when the disk is only about 2 miles thick. The hole itself is approximately 30 miles in diameter.

All of the known properties of Cygnus X-1 can be explained completely in terms of an accretion disk of gases spiraling into a Kerr black hole. Of course this does not *prove* that Cygnus X-1 *must* be a black hole. Nevertheless, as the details

Chapter 11 The General Theory of Relativity

of the calculations are worked out and compared to the observations, it becomes clear that this is the most natural and obvious explanation. Finally, therefore, one of the most fantastic predictions arising from Einstein's geometrodynamical approach to the gravitational field has been confirmed!

Questions and Exercises

1. In your own words, describe the central ideas in the general theory of relativity.
2. According to the general theory of relativity, what effect does the gravitational field of an object have on space-time?
3. What can be calculated from the field equations of general relativity?
4. What can be calculated from the geodesic equations?
5. What is meant by a geodesic?
6. What is an embedding diagram?
7. Compare and contrast the orbit of a planet about the sun according to general relativity and to classical Newtonian theory.
8. Why is the precession of Mercury's perihelion considered to be an experimental confirmation of the general theory of relativity?
9. Why must general relativity reduce to classical Newtonian gravitation in the weak field limit?
10. In your own words, present an argument why light rays must be deflected by the sun's gravitational field, even though light does not contain any mass.
11. When and how was the deflection of light by the sun first measured?
12. What is meant by the gravitational redshift?
13. When and how was the gravitational redshift first experimentally observed?
14. In your own words, present an argument suggesting that black holes might be very common objects in space.
15. With the aid of a diagram, sketch the paths of light rays in the vicinity of a black hole.
16. What is meant by the photon sphere?
17. What is meant by the exit cone?
18. What is meant by the event horizon?
19. What is meant by the singularity?
20. What is the diameter of the event horizon of a 3 solar mass black hole?
21. Compare and contrast the Schwarzschild solution and the Kerr solution.
22. Why is it reasonable to suppose that a real black hole, such as that which an astronomer might detect in the sky, must be a Kerr black hole rather than a Schwarzschild black hole?

23. How would the appearance of a collapsing star change as it forms a black hole?
24. What would happen to your body as you fell into a black hole?
25. What is the Einstein-Rosen bridge?
26. What is meant by the static limit?
27. Compare and contrast the locations of the static limit in Schwarzschild and Kerr black holes.
28. What is meant by the dragging of inertial frames?
29. What is the ergosphere?
30. Briefly discuss a way in which energy can be extracted from a rotating black hole.
31. What is meant by superradiant scattering?
32. What kinds of black holes are described by the Reissner-Nordstrøm solution?
33. Discuss in general how astronomers might actually discover black holes in space.
34. What is the Roche limit?
35. Describe the accretion disk thought to be surrounding the black hole in Cygnus X-1.
36. What happens that causes X rays to be emitted from the accretion disk in Cygnus X-1?

The Realm of the Galaxies

12.1 Our Galaxy

Gazing out into the nighttime sky, we are treated to a wide range of astronomical phenomena. Stars, the moon, planets, meteors, and an occasional comet are familiar sights. Gradually, with the development of modern science, an understanding of all these objects has become possible. But of all the things that can be seen with the naked eye, it was the *Milky Way* that remained a mystery for the longest period of time. This hazy band of light that completely encircles the sky was a source of myths and legends in ancient and primitive civilizations.

When Galileo first turned his telescope toward the Milky Way he discovered that it was not really an irregular diffuse band of light 20° wide, but rather consisted of millions upon millions of very dim stars close together. Astronomers were quick to realize that most of the stars are located in or near the *plane* of the Milky Way. It seemed that all the stars in space are distributed in an enormous flattened disk or "grindstone." This huge flattened disk of stars is called the *Milky Way Galaxy*.

A question arose in the minds of eighteenth-century astronomers concerning the location of the sun in our galaxy. If all the stars are located in a huge disk of stars, and if our sun is a typical star, where are we? Near the center? Near the edge?

The first person to tackle this problem was the great astronomer Sir William Herschel. Herschel, who also discovered the planet Uranus, took a very straightforward approach. Imagine standing in a forest surrounded by trees. Near the center of the forest, you will see an equal number of trees in all directions. However, if you happen to be near the edge of the forest, this will not be true. In one direction (toward the center of the forest) you will see many trees, but in the opposite direction (toward the edge of the forest) there will be a fewer number of trees. Toward the edge of the forest, a "thinning" in the number of trees will be observed. With careful observations you might be able to get a good idea of your location in relation to the rest of the forest.

In 1785, Herschel published his observations of stars he had counted in 683 selected regions of the sky. Using this method of *star gauging*, he hoped to determine our location in the Milky Way. By analogy with the trees in the forest,

Figure 12-1. The Milky Way. Through a telescope, the Milky Way is seen to consist of millions of stars. This mosaic of photographs extends from Sagittarius in the south to Cassiopeia in the north. (*Hale Observatories*)

Herschel reasoned that in those directions in which he saw the greatest number of stars, the Milky Way extended the farthest. In those directions in which he saw few stars, the Milky Way thinned out in a comparatively short distance. Based on these observations, Herschel concluded that the sun is located at the center of a flattened disk of stars. These observations were repeated by the Dutch astronomer J. C. Kapteyn, in 1922, with essentially the same result.

Return for a moment to the analogy of counting trees in the forest. This method of determining our location will work *only* if (1) the forest is not too big and (2) observations are done on a clear day. If the forest is huge and extends in all directions for hundreds of miles, the number of trees in any direction will appear roughly the same, no matter where we are. Thus, unless we are very near the edge, we will always conclude that we are at the center. Secondly, suppose it's a foggy day. Due to the fog floating between the trees, we will not be able to see very far through the forest. Again, unless we are very near the edge, we will see roughly the same number of trees in all directions and will conclude again that we are at the center of the forest. Both of these effects play an important role in negating Herschel's and Kapteyn's conclusion. We are *not* at the center of the Milky Way Galaxy.

In 1930, the American astronomer R. J. Trumpler noticed an interesting effect in studying star clusters in the Milky Way. Many of these clusters of stars appeared unusually dim for their observed angular sizes. Ordinarily, if an astronomer sees a very dim cluster of stars, he is led to believe that the cluster is far away. But if Trumpler assumed that the clusters he observed were far away, then they would have to have been incredibly huge, much larger than the typical sizes of nearby clusters. This dilemma can be resolved if it is assumed that interstellar space is *not* a perfect vacuum. If the Galaxy contains dust, then the starlight from clusters in the Milky Way will quite naturally be dimmer than usual. Since starlight is absorbed by this interstellar dust, this phenomenon is

Chapter 12 The Realm of the Galaxies

called *interstellar absorption*. Interstellar absorption acts just like fog floating between the trees in our hypothetical forest. Due to interstellar absorption we simply cannot see very far through the Milky Way Galaxy. This is why the Galaxy appeared to be heliocentric and small from the observations of Herschel and Kapteyn.

A method far better than star gauging for determining our location in the Galaxy was developed by the famous American astronomer Harlow Shapley during World War I. Shapley spent a lot of time observing globular clusters. Recall from Chapter 7 that globular clusters are spherical clusters containing between 100,000 and 1,000,000 stars. About 120 such globular clusters are known to be scattered around our Galaxy. Most of them are not in the Milky Way but, rather, lie far above or below the *galactic plane* where interstellar absorption is unimportant. Shapley obtained the distances to these clusters by several techniques, such as main sequence fitting (Section 7.3) and from the apparent magnitudes of certain well-known types of stars such as RR Lyrae variables (Section 7.6). In addition, most globular clusters are nearly the same size, so that their apparent angular diameters are a measure of their distance. The smaller they look, the farther away they must be.

When Shapley plotted the locations of globular clusters, he found that they are scattered around the Galaxy inside a huge sphere. This sphere is about 100,000 light years in diameter and is centered on a point 30,000 light years away, in the direction of Sagittarius. A cross-section view of our Galaxy, including globular clusters in the *galactic halo*, is shown in Figure 12-2.

Shapley's observations gave astronomers their first truly correct picture of the Galaxy. Shapley's work can be thought of as being analogous to trying to determine our location in the solar system if—for some peculiar reason—the sun could not be seen. By examining the locations of the planets and asteroids, we would conclude that they are centered about a point some 93 million miles away. Just as the earth goes around the sun once each year, it takes the sun 200 million years to go once around the Galaxy. This huge period of time is called the *galactic year*. Astronomers estimate that our Galaxy must contain at least 10 billion stars like our sun.

One of the obvious problems astronomers face in studying our Galaxy is that, due to interstellar absorption, our view of the Galaxy is severely obscured. In fact, virtually all of the stars that can be seen in the sky are at most only a couple of thousand light years away. With ordinary optical astronomy, we see only a small fraction of this gigantic stellar system in which we live. The next major breakthrough in *galactic astronomy* had to wait for the development of radio telescopes.

On a cloudy, rainy, foggy day, you can turn on your radio or TV set and receive transmissions from a broadcasting station with no difficulty at all. Even though you might not be able to see 50 feet in front of you, radiowaves have no difficulty penetrating the fog. By analogy, it is, therefore, apparent that radio astronomy holds great promise in determining the detailed structure of our Galaxy.

In 1944, the Dutch astronomer H. C. van de Hulst did some calculations that

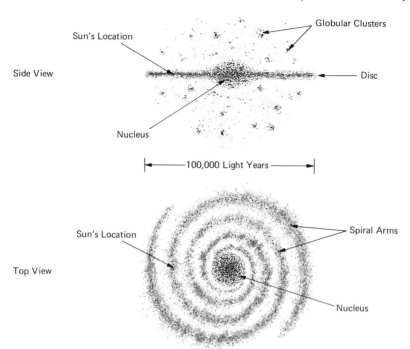

Figure 12–2. *Our Galaxy.* The position of the sun in the Galaxy was deduced from observing globular clusters. The spiral structure of the Galaxy was discovered from radio observations.

showed that astronomers should be able to detect radio waves from hydrogen gas floating in space. This radio light is not produced in the usual fashion, as was described in Chapter 5, but rather comes from a so-called *hyperfine transition* in the structure of hydrogen atoms. It turns out that atomic particles such as protons and electrons possess *spin.* Crudely speaking, they may be thought of as little rotating balls of matter and electric charge. In the hydrogen atom, which consists of one proton orbited by one electron, there are two possibilities. Either the proton and the electron are spinning in the same direction (in which case their *spins are parallel*) or the proton and the electron are spinning in opposite directions (in which case the *spins are antiparallel*). These two cases are shown schematically in Figure 12–3. When the electron in a hydrogen atom "flips" from parallel to antiparallel spin, the atom emits a small amount of radiowaves. This radio light has a wave length of about $8\frac{1}{4}$ inches, which equals 21 centimeters. This radiation is, therefore, called the *21-centimeter line of hydrogen.*

By the early 1950s, radio astronomy finally had progressed to the extent that a 21-centimeter emission from the cool hydrogen gas in our Galaxy could be detected. Since the clouds of hydrogen gas in the Galaxy are moving at different speeds—some toward earth and some away—the 21-centimeter radiowaves are Doppler shifted. A cloud coming toward earth has its radiation shifted toward

Figure 12-3. The 21-cm Line. The electron in a hydrogen atom can have its spin either parallel or antiparallel to the spin of the proton. When the electron flips over, it emits radio radiation having a wavelength of 21 centimeters.

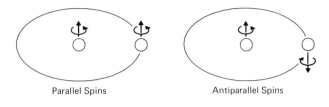

Parallel Spins   Antiparallel Spins

shorter wave lengths, whereas a cloud moving away has its radiation shifted toward longer wave lengths. Just as Mercury goes around the sun faster than Pluto, it can be assumed that gas near the *galactic nucleus* revolves about the center of the Galaxy faster than gas farther away. Therefore, from observing the Doppler shifts in the 21-centimeter line of hydrogen, astronomers found that they actually could map the structure of the Galaxy. The resulting maps showed that our Galaxy has a *spiral structure,* as shown schematically in Figure 12–2. Astronomers, therefore, say that we live in a *spiral galaxy.*

The *spiral arms* of our Galaxy encircle the galactic nucleus in such a way that they wind up as the Galaxy rotates. The sun is located on the inner edge of the *Carina-Cygnus arm* near the *Orion spur.* This arm includes such features as the Orion nebula and the North America nebula, as well as dark regions of dense interstellar obscuration such as the famous Coalsack (near the Southern Cross) and the Cygnus rift. During summer nights, when the constellation of Sagittarius is high in the sky, the view is toward the center of our Galaxy. The portion of the Milky Way that can be seen passing through this constellation is part of the *Sagittarius arm.* Conversely, in the winter sky, when looking away from the galactic nucleus, the portion of the Milky Way we see is part of the *Perseus arm.* The Sagittarius and Perseus arms are each located about 10,000 light years inside and outside the sun's position, respectively. A map of those sections of the Galaxy that we see as the Milky Way is shown in Figure 12–4.

One of the most fascinating regions of our Galaxy is unquestionably its very center, or nucleus. Surrounding the galactic nucleus is the *central bulge,* which contains most of the mass of the Galaxy. This central bulge has a perpendicular thickness of about 3,000 light years. For comparison, at the location of the sun,

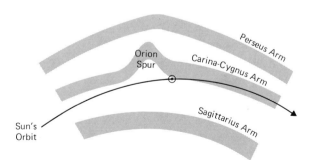

Figure 12–4. The Solar Neighborhood. The sun is located on the so-called Carina-Cygnus arm of the Galaxy. The portion of the Milky Way seen in the winter sky is the Perseus arm. The portion of the Milky Way seen in the summer is the Sagittarius arm.

Figure 12–5. *The Galactic Center.* This view of the Milky Way in Sagittarius includes the nucleus of the Galaxy. The small rectangle is centered about Sagittarius A. (*Hale Observatories*)

the Galaxy is only 1,500 light years thick. The great quantities of dust and gas in the central bulge and intervening spiral arms completely obscure our view of the galactic nucleus. Only with the aid of radio and infrared telescopes can astronomers examine what is going on at the center of the Galaxy.

The center of our Galaxy is emitting great quantities of radiation. One of the strongest radio sources in the sky, Sagittarius A, is located at the galactic nucleus. Coincident with Sagittarius A is a powerful source of infrared radiation. From the small angular diameter of this infrared source, astronomers conclude that the galactic nucleus is only a few light years in size. What could be going on at the center of our Galaxy to produce this intense radio and infrared radiation is inadequately understood. In addition, the mystery of the galactic nucleus deepened when it was discovered that great quantities of gas are being ejected from the region of the galactic center. It seems quite probable that violent explosions occur at the galactic nucleus.

In order to explain these violent events, astronomers such as Dr. Donald Lynden-Bell of Cambridge University have proposed the possibility of supermassive black holes at the galactic center. Such black holes might have masses equal to 100 million suns and would be about half a billion miles in diameter! The British astronomers Drs. D. Lynden-Bell and M. Rees have calculated that such black holes would suck gas in from the regions surrounding the galactic nucleus to form a gigantic accretion disk analogous to that around Cygnus X-1. The resulting accretion disk would be a strong source of radio and infrared radiation.

Finally, one of the most controversial topics in astronomy during the 1970s

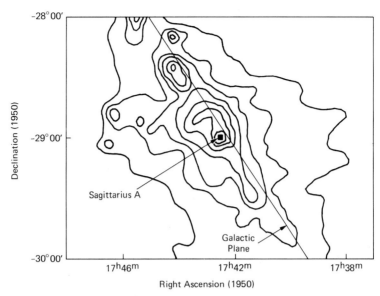

Figure 12–6. *Radio Light from the Galactic Center.* This radio map covers the area in the small rectangle shown in Figure 12–5. The map was made by Drs. Whiteoak and Gardner using the 210-foot radio telescope at Parkes, Australia.

Figure 12–7. *Sagittarius A.* This radio map of Sagittarius A covers the area in the small square shown in Figure 12–6. The map was made by Dr. Bruce Balik at the National Radio Astronomy Observatory. The little stars indicate the locations of five intense infrared sources discovered by Drs. Low and Rieke.

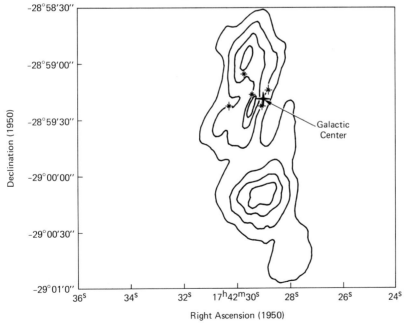

centers about the observations of Dr. Joseph Weber at the University of Maryland. Dr. Weber claims to have detected *gravitational waves* from the galactic center. According to general relativity, whenever matter moves around, the geometry of space-time must readjust itself. Thus, a bouncing ball or a man waving his arms causes small ripples in the over-all geometry of space-time. These ripples, which move at the speed of light, are called gravitational waves.

Compared to electromagnetic waves (for example, radiowaves, light, X rays, and so on), gravitational waves are very, very weak. In order to detect gravitational waves, Dr. Weber makes use of huge aluminum cylinders on which special crystals have been glued. These sensitive crystals respond to any changes in the shape of the aluminum cylinder by producing a small electric current. If a gravitational wave passes through the aluminum cylinder, it will flex and oscillate very slightly. Dr. Weber claims that his "telescopes" do indeed oscillate in a way that can be explained only by gravitational waves.

Although Dr. Weber's observations have been called into question, if he is correct in his assertion that gravitational waves have been detected, then incredible events must be occurring at the galactic nucleus. Dr. Weber claims to detect half a dozen gravitational waves each day. Since gravitational waves are so weak, and since Dr. Weber's telescope is comparatively crude, a truly enor-

Figure 12–8. A Gravitational Wave Antenna. Dr. Weber is shown here with one of the aluminum cylinders used in his experiments to detect gravitational waves. It is possible that the Galactic nucleus is an intense source of gravitational radiation. (*Courtesy of Dr. Weber*)

mous amount of energy must have gone into producing these waves at the galactic nucleus.

Astrophysicists are at a complete loss to explain how such gigantic quantities of energy could be produced. If a supermassive black hole at the center of our Galaxy is vibrating, it will emit gravitational radiation. Or perhaps, if there are, instead, numerous smaller black holes at the galactic center that often pass near each other, they will produce *focused* gravitational waves, as suggested by Drs. Thorne and Kovacs at Caltech. Or maybe there is a *naked singularity* at the galactic center. A naked singularity is just the "center" of a black hole, not surrounded, or clothed, by an event horizon. Such an object also would focus gravitational waves into the plane of the Galaxy, thereby making them more easily detectable. This final (and perhaps very weird) proposal, however, violates the law of cosmic censorship proposed by the British physicist Dr. Roger Penrose that states "Thou shalt not have naked singularities!" In any case, it is quite clear that astronomers really do not understand what is going on at the center of our Galaxy. It is certainly one of the most fascinating places in the universe.

12.2 The Mystery of the Nebulae

Comet hunting was a really hot topic back in the late eighteenth century. Anybody who owned a telescope spent his evenings searching the skies for new comets. Substantial cash prizes and medals were awarded to successful comet hunters. Looking for new comets on their way toward the sun was the fashionable thing to do.

When a new comet is first sighted by an astronomer, it looks like nothing more than a dim, fuzzy blur through his telescope. As discussed in Section 4.6, only after the comet is very near the sun do we see the characteristic long flowing tail. One problem apparent to comet hunters in the late 1700s was that there are numerous fuzzy objects in the sky that can be mistaken for distant comets. Such objects, called *nebulae,* had to be very far from the solar system. A comet, no matter how far away it is, will appear to move slowly against the background of distant stars from night to night. It is from observing such motion that the comet's orbit can be calculated. But the nebulae are fixed with respect to the stars in the sky, and they kept getting in the way of serious comet hunters who would mistake them for new comets.

To deal effectively with this problem, the French astronomer Charles Messier prepared a catalogue of all known nebulae that kept interfering with the work of comet hunters. This catalogue, first published in 1781, contained slightly more than a hundred objects. An up-dated version of the *Messier Catalogue* is given in Appendix 8.

The *Messier Catalogue* contains many well-known and familiar objects. M1, the

Figure 12–9. The Andromeda Nebula. This object, also called M31 or NGC 224, looks like a fuzzy patch of light through a telescope. The nature of "nebulae" such as M31 was hotly debated for many years. (*Lick Observatory*)

first object on Messier's list, is the Crab nebula in Taurus. M42 is the Orion nebula, which can be seen just barely with the naked eye as the middle "star" in Orion's sword. The catalogue contains planetary nebulae such as M27, the Dumbbell nebula in Vulpecula, and M57, the Ring nebula in Lyra. Many open (galactic) clusters are included, such as M44, the Praesepe, and M45, the Pleiades, both of which can be seen on a clear night with the naked eye. Twenty-seven globular clusters are listed in Messier's catalogue, the most famous of which is probably M13 in Hercules. There are also a few mistakes. M40, M91, and M102 do not correspond to anything of interest; maybe they really were comets.

About one third of the objects catalogued by Messier are fundamentally different from all the others. These thirty-two nebulae are not supernova remnants or HII regions. They are not planetary nebulae, nor are they galactic or globular clusters. These objects, which include M31—the famous Andromeda nebula, which can be seen with the naked eye—remained a complete mystery for more than one and a half centuries.

The discovery and cataloguing of all kinds of nebulae were in full swing during the nineteenth century. William Herschel and his son, John Herschel, together made a detailed survey of both the Northern and Southern Hemispheres. The final result of their efforts was published in 1888 as the *New General Catalogue* (NGC), compiled and prepared by J. L. E. Dreyer. This catalogue supplanted several earlier catalogues published by the Herschels. Objects in the *New General Catalogue* are listed according to their *NGC numbers*. Thus, the Crab nebula, M1, is the same as NGC 1952, and the Andromeda nebula, M31, is the same as NGC 224.

In 1895 and again in 1908, Dreyer found it necessary to publish supplementary catalogues listing objects that had been missed in the NGC. These new lists are in the *Index Catalogue* (IC); objects appear in them according to their *IC numbers*. Thus, the open cluster M25 in Sagittarius is also called IC 4725. With the publication of the second *Index Catalogue* in 1908, nearly fifteen thousand nebulae had been discovered and described.

As early as 1755, the great philosopher Immanuel Kant hypothesized that many of the fuzzy patches of light astronomers find in the sky could be huge aggregations of millions or billions of stars just like our own Milky Way. He theorized that these gigantic collections of stars were like "island universes" floating in space. He wrote that "the system of stars in which we find ourselves . . . is in perfect agreement with the idea that these elliptical objects are just universes—in other words, Milky Ways."* In spite of Kant's prophetic (and correct) interpretation, the true nature of many of the nebulae in the NGC and IC was destined to remain shrouded in mystery until the mid-1920s.

By the time of World War I, both the 60-inch and 100-inch telescopes were in operation on Mount Wilson. With the simultaneous development of high-quality astronomical photography, astronomers were finally able to take long time-exposure photographs of the nebulae. It was immediately apparent that many of the objects listed in the NGC and IC were fairly nearby, definitely within the confines of our own Milky Way. These objects included gaseous nebulae (planetaries, HII regions, supernovae remnants, and so on) as well as galactic and globular star clusters. There was, however, a second class of objects. These objects were either elliptical or showed a pinwheel or spiral structure. They were the island universes of which Kant had spoken 150 years earlier.

By 1920, a violent controversy was taking place among professional astronomers. The two major protagonists in the sometimes-noisy debate were H. Shapley of Mount Wilson Observatory and H. D. Curtis of Lick Observatory. Shapley believed that these elliptical and spiral nebulae were really quite near our Milky Way. Perhaps they formed part of the galactic halo along with the globular clusters. Curtis, on the other hand, staunchly defended the idea of island universes. According to Curtis, the elliptical and spiral nebulae were separate *galaxies* like our own Milky Way.

Figure 12–10 shows these two opposing views schematically. Actually, Shapley had a good idea of the correct size of the Milky Way Galaxy, but then he placed

---

*From *Universal Natural History and Theory of the Heavens*, 1755.

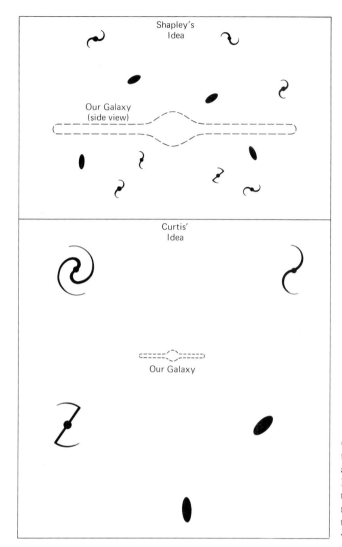

Figure 12–10. *Shapley vs. Curtis.* In the early 1920s, there were two major theories about the "spiral nebulae." Dr. Shapley believed that these nebulae were small and nearby. Dr. Curtis believed that they were very large and very far away.

the elliptical and spiral nebulae far too close to us. Curtis severely underestimated the size of our Galaxy but had a more correct picture of where the other nebulae are. So the debate could only be settled if astronomers could somehow directly measure the actual distances to individual nebulae. Only then could they really know precisely where the nebulae are.

If someone takes a 50-watt light bulb and stands somewhere away from you on a dark night, when the light bulb is turned on, you can easily figure out how far away it is. If the light bulb appears bright, it must be nearby. If the light bulb appears dim, it must be far away. In other words, as emphasized in Section 7.1, there is a unique relationship between (1) the apparent brightness of a source of light, (2) the absolute or real brightness of a source of light, and (3) the distance to the source of light. The issue of the distances to the elliptical and spiral nebulae, therefore, could be resolved if astronomers could discover some stars in these nebulae for which they know the absolute magnitudes. If these mysterious and controversial nebulae contain well-known, recognizable stars whose absolute magnitudes are known, by comparing the absolute magnitudes with the apparent magnitudes, the distances can be calculated easily.

Although galaxies have been mentioned occasionally in this text, it is remarkable to realize that only half a century ago astronomers had no clear understanding of where these objects are. As will be detailed in Chapter 13, an equally important and noisy controversy is occurring today with regard to mysterious objects called *quasars*. Some astronomers are convinced that the quasars are fairly nearby, whereas others present evidence suggesting that they are incredibly remote. Only with the resolution of these controversies can we obtain a true understanding of man's place in the universe.

12.3 Hubble and Galactic Distances

At southern latitudes, and especially from places like Australia and South America, it is possible to see two hazy patches of light in the night sky only 20° from the South Celestial Pole. These two hazy objects, which look like detached pieces of the Milky Way, are actually two nearby companion galaxies to our Galaxy. They are called the Large Magellanic Cloud (LMC) and the Small Magellanic Cloud (SMC).

In the years just prior to the beginning of World War I, some of the first detailed astronomical observations were made of the Magellanic Clouds. The principle investigator in this pioneering work, which twelve years later would profoundly affect our concept of the universe, was Henrietta Leavitt from the Harvard College Observatory.

Through even the smallest of telescopes, the Magellanic Clouds are seen to consist of millions upon millions of stars, just like the Milky Way. Many of the brightest stars in the Clouds can be easily identified and have been carefully catalogued. Since the Magellanic Clouds are far away, and since their over-all sizes are small compared to their distance, the apparent magnitudes of stars in the Clouds are a direct measure of their absolute magnitudes. Stars that appear to be dim really are dim; stars that appear to be bright really are bright. Years later, after the distances to the Magellanic Clouds had been measured by techniques such as main sequence fitting, the actual numerical values of the absolute magnitudes of individual stars could be calculated easily.

One type of bright, easily recognizable star in the Magellanic Clouds is the Cepheid variable. Recall from Section 7.6 that Cepheid variables are stars that change their luminosity in a very specific fashion. They brighten rapidly and dim slowly with a period that ranges from about 1 day to 3 months, depending on the particular star. In 1912, Leavitt succeeded in identifying twenty-five such Cepheid variables in the Small Magellanic Cloud. In studying these stars she noticed that the period of pulsation (or light variation) is directly related to the average absolute magnitude of the Cepheid variable. Dim cepheid variables pulsate rapidly, whereas bright Cepheid variables pulsate much more slowly.

Harlow Shapley immediately recognized the importance of this discovery and promptly set about determining the distances to Cepheids. Since 1912, several dozens of Cepheid variables have been discovered in both Magellanic Clouds whose distances are known. Once the actual distance to a Cepheid is known, the numerical value of its average absolute magnitude can be correlated with its pulsational period. Years later, John B. Irwin developed a method for finding the average absolute magnitudes of Cepheid variables in open star clusters. All the resulting data from the work of astronomers such as Shapley and Irwin can be plotted on a diagram, as shown in Figure 12–11. The average absolute magnitude of the variable stars is plotted vertically, but their periods are plotted horizontally. The graph shows that there are *two* types of Cepheid variables.

Stars that are very ancient must have formed early in the history of the universe. Since hydrogen and helium were the only two abundant elements during the early universe, these stars will show unusually weak (if any) metallic lines in their spectra. Even though nucleosynthesis has been making lots of heavy elements at their cores, these elements have not migrated to the star's surfaces. Stellar atmospheres of such stars are almost pure hydrogen and helium; they are *metal-poor*. These stars are called *Population II* stars and are usually found in the galactic halo, either individually or in globular clusters. By way of contrast, stars that have been formed only recently show strong metallic lines in their spectra. The gases that produced these stars have been enriched with heavy elements that were created inside earlier generations of stars. These younger stars are *metal-rich*. They are called *Population I* stars and are usually found in the disk of the Galaxy.

The chemical composition of a star affects its over-all luminosity. In particular, metal-rich Population I Cepheid variables have a *higher* average absolute magnitude than the metal-poor Population II Cepheid variables. The metal-rich Cepheids are called *Type I Cepheids* or *Classical Cepheids;* they are the Cepheids that Leavitt studied. On the other hand, metal-poor Cepheids are called *Type II Cepheids;* they are dimmer by almost $1\frac{1}{2}$ magnitudes from the Type I Cepheids. An astronomer can figure out which type of Cepheid he is observing by simply examining its spectra to see the strength of metallic lines.

When the average absolute magnitudes and periods of Cepheids are plotted on a graph, as shown in Figure 12–11, all the data fall into two well-defined regions. As was anticipated from Leavitt's work, there is a pronounced correlation between the magnitudes and periods of Cepheids. The brighter the star, the longer is its period. This is called the *period-luminosity relation.*

Incidentally, the periods and magnitudes of the RR Lyrae variables also are

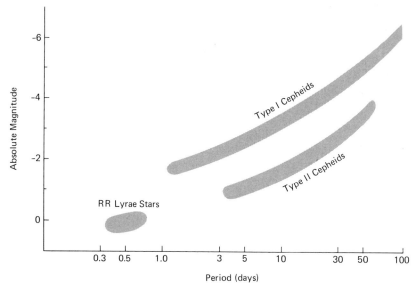

Figure 12-11. *Period-Luminosity Relation.* The period of a Cepheid variable is directly correlated with its absolute magnitude. Population I Cepheids are about $1\frac{1}{2}$ magnitudes brighter than Population II Cepheids.

shown in Figure 12-11. Such stars often are found in globular clusters and are, therefore, sometimes called *cluster-type variables.*

It turns out that the period-luminosity relation is an incredibly powerful tool for the astronomer. Suppose an astronomer finds a new Cepheid variable in the sky. He recognizes the star as a Cepheid from its characteristic light curve; the star brightens rapidly and fades slowly. He then takes a spectrum of the Cepheid to see if it is Type I (metal-rich) or Type II (metal-poor). With a clock he measures the period of pulsation. Turning to the period-luminosity relation in Figure 12-11, he immediately knows the star's average absolute magnitude. By measuring the star's average apparent magnitude and comparing the apparent and absolute magnitudes, he immediately obtains the distance to the star. The period-luminosity law can be used to find the distances to the stars!

In the 1920s, a very heated controversy was taking place among professional astronomers. At issue was the location of certain types of elliptical and spiral nebulae. Were they fairly nearby and, perhaps, members of the galactic halo as Shapley believed? Or were they incredibly remote stellar systems, each like our own Milky Way Galaxy as Curtis argued? The debate could only be settled by somehow actually measuring the distances to individual nebulae.

A final resolution of the controversy came about through the work of a young astronomer, Edwin Hubble, at the Mount Wilson Observatory. During 1923 and 1924, Hubble discovered a number of variable stars in some of the nebulae. In particular, from examining the light curves of variables in M31, M33, and NGC

Figure 12–12. Variables in M31. Dr. Hubble identified Cepheid variables in M31. These Cepheids appear so dim (two variable stars are indicated in this view of one "arm" of the galaxy) that Hubble correctly concluded M31 must be very far away. Figure 12-9 shows the entire galaxy. (*Hale Observatories*)

6822, he found that these nebulae contained Cepheids. Cepheid variables are intrinsically very bright stars. A typical Cepheid with an absolute magnitude of −5 emits ten thousand times more light than our sun, whose absolute magnitude is +5. Yet, the Cepheid variables that Hubble discovered appeared very faint—near magnitude 18. Figure 12–12 shows a photograph of a region of M31. The Cepheid variables are so dim in this photograph that they can hardly be seen. Obviously the nebulae containing these stars must be incredibly remote! Measuring the periods of the Cepheids in M31 and utilizing the period-luminosity relation (Figure 12–11) enabled an immediate determination of the absolute magnitudes of these stars. By comparing their absolute magnitudes with their apparent magnitudes, astronomers learned that M31 is 2 million light years away! This is far beyond the confines of our own Milky Way Galaxy. The elliptical and spiral nebulae that spawned such a heated debate are not "nebulae" at all. They are individual *galaxies*, just as Kant had hypothesized almost two centuries earlier. With the announcement of Hubble's discovery of Cepheid variables in M31, M33, and NGC 6822 in December 1924, the issue was decided once and for all.

Figure 12–13. The Galaxy M33. Dr. Hubble also discovered Cepheid variables in M33. Since Cepheids are intrinsically very bright, and since they appear so very dim in M33, Hubble concluded that M33 must be very far away. (*Lick Observatory*)

There are an incredible number of galaxies out in space. No matter in what direction we look, as long as our view is not obscured by interstellar dust, we always observe countless galaxies. There must be literally billions upon billions of galaxies in the universe, most of which are beyond the limits of detection of even the world's most powerful telescopes. Figure 12-14 is a photograph of a region of the sky in the constellation of Corona Borealis that has a typical rich concentration of galaxies.

Naturally, the astronomer is interested in knowing the distances to the galaxies he observes. In this regard, Cepheid variables are still the astronomer's most important tool. Unfortunately, there are only about thirty galaxies that are near enough so that their Cepheid variables can be seen at all. Most galaxies are so far away that individual stars cannot be resolved even through the 200-inch telescope at Palomar. Astronomers must, therefore, turn to other methods.

The brightest Cepheid variables have an average absolute magnitude of -6. Yet, there are stars with absolute magnitudes of -9 and even -10. In cases in which the Cepheid variables are just below the limits of detection, the astronomer can make use of these *supergiants*. If any individual stars can be seen at all, they must be supergiants that emit about twenty times more light than the brightest Cepheids. Even though supergiants allow astronomers to determine the distances to a few more galaxies, in most cases, no individual stars can be seen at all.

It is reasonable to suppose that most galaxies contain globular clusters, just as globular clusters are an important constituent of our own galactic halo. Typically, globular clusters contain about 100,000 stars. To a high degree of approximation, astronomers can assume that the over-all *integrated magnitudes* of globular clusters are roughly the same. For galaxies in which individual stars cannot be resolved, it is often possible to see globular clusters surrounding the galaxy. Figure 12-15 shows a photograph of M87, a spherical galaxy in the

Figure 12-14. A Cluster of Galaxies. There are countless billions of galaxies scattered throughout the universe. This photograph of a region of the sky in Corona Borealis shows many galaxies. (*Hale Observatories*)

Figure 12-15. Globular Clusters around M87. Careful examination of this photograph of this galaxy reveals many faint, fuzzy patches surrounding the galaxy. Each little fuzzy patch is a globular cluster. (*Hale Observatories*)

constellation of Virgo. Careful examination of the photograph reveals numerous faint, fuzzy spots around the galaxy. These are globular clusters. From studying nearby globular clusters in our own Galaxy, astronomers have a good idea of their integrated absolute magnitudes. Assuming that all globular clusters are the same, astronomers can compare the observed apparent magnitudes with the assumed absolute magnitudes, thereby obtaining the distances to galaxies. However, this only works for galaxies in which globular clusters can be identified positively.

A similar method of distance determination involves the use of HII regions. HII regions are huge clouds of hot hydrogen gas floating in space. The famous Orion nebula is a good example of an HII region. Other familiar nearby HII regions include the North America nebula in Cygnus and the Lagoon nebula in Sagittarius. If the astronomer assumes that *all* HII regions are nearly the same size and emit nearly the same amount of light, then they can be used as distance indicators. This assumption is like saying "All cars are the same size." Oh, surely, there are Volkswagens and there are Cadillacs, but typical cars on the road are roughly the same size. When an astronomer finds HII regions in a distant galaxy, he can compare their apparent magnitudes with the known absolute magnitudes (obtained from studying nearby HII regions) and, thereby, deduce a distance to the galaxy. Alternatively, he can measure the apparent angular size of the HII region in a distant galaxy and compare this size with the known dimensions of nearby HII regions and again estimate the distance to the galaxy.

From time to time, a star in a distant galaxy blows up and becomes a nova or supernova. Astronomers have reason to believe that all nova reach roughly the same absolute magnitude at maximum light, about -8. Thus, if an astronomer is lucky enough to see a nova in a galaxy, which might become slightly brighter than typical bright stars, the nova can be used as a distance indicator. Unfortunately, novae never get extremely bright.

Even though novae are not very bright, supernovae can become extremely luminous. Depending on the type of supernova, the maximum absolute magnitudes of these violently exploding stars can reach as high as -20. This means that the supernova can become as bright as all the other stars in a galaxy combined. If the astronomer assumes that all supernovae reach the same maximum absolute magnitude, they can be used as distance indicators. This is like saying "All the headlights on typical cars have nearly the same brightness." If an astronomer is lucky enough to see a supernova in a remote galaxy, he can compare its apparent magnitude with the assumed absolute magnitude and, thereby, arrive at a distance to the galaxy.

Unfortunately, supernovae explosions are rare events. To make some sort of headway in determining distances to the most remote galaxies, the astronomer

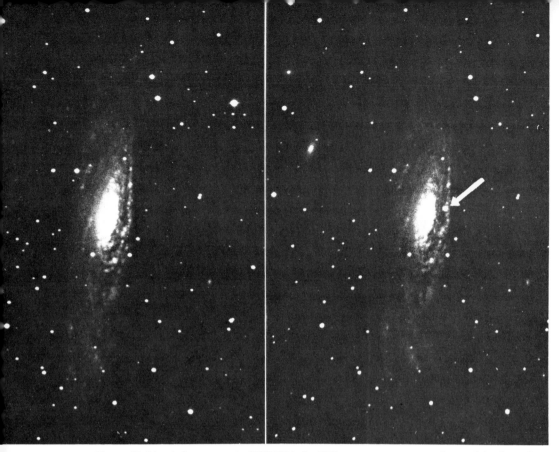

Figure 12–16. A Supernova in NGC 7331. In 1959, a supernova was observed in the galaxy NGC 7331. By observing a supernova in a remote galaxy, astronomers can estimate the distance to the galaxy. (*Lick Observatory*)

can assume that *all* galaxies of a given shape have very nearly the same absolute magnitude. Thus, if he sees a distant spiral-shaped galaxy through his telescope, he can assume that this galaxy has roughly the same total integrated absolute magnitude as nearby spiral-shaped galaxies. Since the distances to nearby spiral-shaped galaxies such as M31 and M33 have been determined by other, more reliable methods (for example, Cepheid variables and the period-luminosity relation), he knows how bright such galaxies really are. By comparing this known absolute magnitude with the observed apparent magnitude of the galaxy in question, he can readily calculate the distance to this galaxy.

Bear in mind that determining the distances to remote galaxies involves many assumptions: (1) *suppose* all bright globular clusters have the same luminosity; (2) *suppose* all HII regions have the same size; (3) *suppose* all supernovae reach the same maximum absolute magnitude, and on and on. There are an awful lot of "supposes." These assumptions are like saying "All trees are roughly the same size" or "All street lights have nearly the same brightness." Obviously, there is room for substantial error and astronomers involved in *observational cosmology* are constantly checking and rechecking their answers. A pioneer in this field has been Dr. Alan Sandage of the Hale Observatories. In the mid-1970s he published a series of papers summarizing years of tedious and laborious observations of galactic distances. With the work of Sandage and his colleagues, the methods of distance determination have been refined to such an extent that most astronomers feel fairly confident that they have a good idea where the galaxies are.

12.4 Hubble Classification

It is remarkable that up until only half a century ago, scientists did not know where the galaxies are. Galaxies are the largest objects in the universe. And, yet, in spite of a preoccupation with the heavens that stretches back thousands of years, astronomers did not truly comprehend what they were observing until Edwin Hubble discovered a few Cepheid variables in some nearby "nebulae." Only then did astronomers realize that galaxies are incredibly remote systems that each contain billions upon billions of stars.

The discovery that galaxies are huge, remote stellar systems had a profound effect on the course of astronomy. Beginning in 1925, astronomers realized that galaxies were really very different from anything they had ever seen before. It was as if a zoologist discovered a new island or continent inhabited by strange and exotic beasts, unlike any of the animals with which he was familiar. Naturally, one of the first things such a zoologist would attempt to do is to examine and classify these new creatures.

In addition to being a gifted astronomer, Hubble had a head start on everyone else. He was the first astronomer to propose a classification scheme for galaxies—it is still widely used. The Hubble classification scheme is based on the *appearance* of galaxies, on what they look like. After examining numerous photographs of galaxies, Hubble discovered that he could classify them into one of four categories:

1. ellipticals
2. spirals
3. barred spirals
4. irregulars

Elliptical galaxies, as their name suggests, have an amorphous elliptical appearance. They look like blobs. No features or structure such as spiral arms are seen. Some of the most luminous galaxies known are ellipticals.

The category of elliptical galaxies can be subdivided according to how flattened they look. Elliptical galaxies that are perfectly circular in appearance are called E0 galaxies. The flattest appearing ellipticals are called E7 galaxies. The number following the letter *E* is used to designate the degree of flattening. For example, an E5 galaxy is slightly flatter than an E4 galaxy. These numbers run from 0 (perfectly round ellipticals) to 7 (the flattest ellipticals).

Some of the most beautiful galaxies seen in the sky are *spiral galaxies.* Again, as their name suggests, these galaxies have a characteristic spiral or pinwheel structure. Depending on how tightly wound the spiral arms are, this class of

ELLIPTICAL GALAXY
(M 87 also known as NGC 4486)

SPIRAL GALAXY
(M 74 also known as NGC 628)

BARRED SPIRAL GALAXY
(NGC 1300)

PECULIAR or IRREGULAR
GALAXY (NGC 2623)

Figure 12–17. *The Four Hubble Types.* Dr. Hubble classified all galaxies into four basic categories: ellipticals, spirals, barred spirals, and irregulars. (*Hale Observatories*)

galaxies can be subdivided further into at least three subclasses: Sa, Sb, and Sc galaxies. Sa galaxies have very tightly wound spiral arms around a large, bright nucleus. At the other extreme, Sc galaxies have very loosely wound spiral arms emanating from a very small nucleus. In between the Sa and Sc subclasses are Sb galaxies, which have moderately tightly wound spiral arms originating from a medium-sized nucleus. Our own Galaxy is an Sb spiral.

About one quarter of all galaxies with spiral arms also show a *bar* running through their nuclei. These are called *barred spirals;* their spiral arms originate from the ends of the bar rather than from their nuclei. Just as with spirals, the barred spirals can be divided further into at least three subclasses: SBa, SBb, and SBc galaxies. The SBa galaxies have very tightly wound spiral arms; SBb galaxies have moderately tightly wound spiral arms; and SBc galaxies have loosely wound spiral arms.

Finally, galaxies that do not fall readily into one of the three previous categories usually look very weird. Such galaxies are called *irregulars*. Some irregular galaxies are obviously in the process of exploding, whereas others have been distorted severely and bent out of shape by a collision or near encounter with a neighboring galaxy. Both the Small and Large Magellanic Clouds are irregular galaxies.

These different types of galaxies have several characteristics that led some astronomers to believe that the Hubble classification might be related to the evolution of galaxies. Ellipticals consist mostly of Population-II stars with some old Population-I stars. In addition, ellipticals do not seem to contain any dust or gas out of which new stars can form. Recalling that Population-II stars are the most ancient stars, it seems that star formation has shut off in elliptical galaxies.

By contrast, spirals and barred spirals contain a mixture of Population-I and Population-II stars. The young Population-I stars are found in the spiral arms, and the older Population-II stars are scattered throughout such galaxies. Both spirals and barred spirals contain gas and dust out of which new stars can be born.

Finally, irregulars usually contain lots of gas and dust and consist primarily of Population-I stars. Some irregular galaxies seem to show some suggestions of spiral structure. Careful examination of the Large Magellanic Cloud, for example, suggests that a bar exists across the nucleus of the galaxy. Other irregulars

Figure 12-18. *The Hubble Classification Scheme.* Elliptical galaxies are classified according to how flattened they look. Spirals and barred spirals are classified according to how tightly wound the spiral arms are.

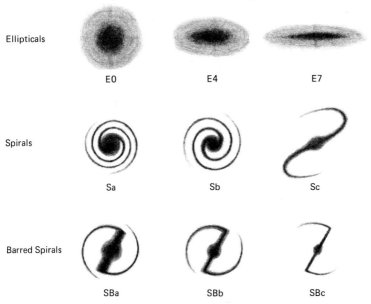

such as M82 in the constellation of Ursa Major appear to be ejecting enormous quantities of matter at speeds of up to 1,000 miles per second.

Could it be that irregulars are very young galaxies? Could it be that they eject matter into space that condenses into stars to form spiral arms? Could it be that spiral galaxies (both normal and barred) are middle-aged, and that as they rotate their spiral arms get wound tighter and tighter around their nuclei? And is it possible that after their spiral arms have completely wound up they turn into ellipticals? After all, irregulars contain lots of young stars, whereas ellipticals are composed mostly of ancient stars. And ellipticals do not seem to contain any dust or gas out of which new stars could be born. Perhaps this is because all the gas and dust have simply been used up in star formation.

These are some of the hotly debated questions in astronomy today. Unfortunately, too little is known about *galactic evolution* to make any definitive statements. In Chapter 13, some startling new discoveries will be discussed concerning galaxies that seem to support evolutionary picture proposed here. Such discoveries, if supported by further observations, will have incredibly profound implications. In this connection it has been suggested that the birth of galaxies is perhaps the creation of matter through a *white hole,* the time reversal of a black hole, where matter and energy are gushing into our universe from some other universe. Such matter would be fundamentally different from the "ancient" matter with which scientists are familiar. Only after this newly created matter "ages" does it begin to behave like the stuff out of which people, the solar system, and our Galaxy are made. Either these revolutionary ideas are total insanity (the majority opinion as of 1975) or science is at the brink of some new and fundamental breakthroughs in understanding physical reality and the nature of the universe.

12.5 Clusters of Galaxies

By the time of Word War II, astronomers had discovered that galaxies were not uniformly distributed through space. Instead, some galaxies seemed to be grouped together in *clusters.* During the 1940s, several dozen clusters of galaxies were discovered by accident on astronomical photographs. How common are clusters of galaxies? Could it be that most, if not all, galaxies are members of clusters? To explore such questions, a young graduate student at Caltech, George O. Abell, undertook a massive survey of the sky in the 1950s, looking for clusters of galaxies. Dr. Abell published the results of his observations in the now-famous *Abell Catalogue,* which lists thousands of clusters. His work shows that most galaxies are found in clusters.

In examining numerous photographs taken with the Schmidt telescope at Mt. Palomar, Abell found that there are basically two types of clusters of galaxies: *regulars* and *irregulars.* Regular clusters show a high degree of spheri-

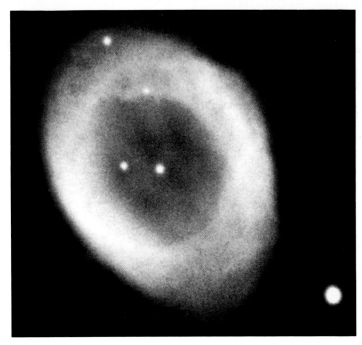

Plate 9. The Ring Nebula. This planetary nebula (also called M57 or NGC 6720) is in the constellation of Lyra. About 1000 planetary nebulae are known to astronomers. (*Hale Observatories*)

Plate 10. The Dumbbell Nebula. This planetary nebula (also called M27 or NGC 6853) is in the constellation Vulpecula. Planetary nebulae last for only 30,000 to 40,000 years. (*Hale Observatories*)

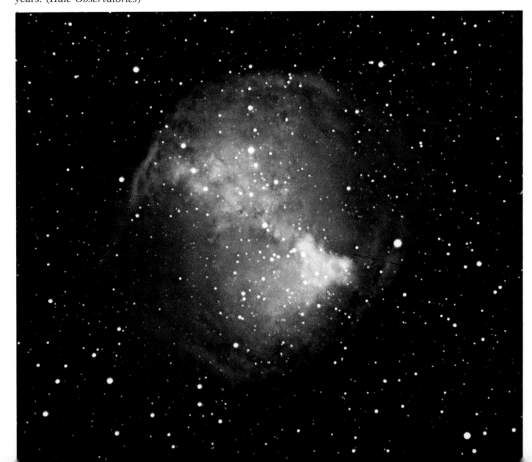

Plate 11. *The Crab Nebula.* This supernova remnant (also called M1 or NGC 1952) is in the constellation of Taurus. The supernova which produced this nebula was observed by Chinese astronomers in 1054 A.D. (*Hale Observatories*)

Plate 12. *The Veil Nebula.* This section of a supernova remnant (also called NGC 6992) is in the constellation of Cygnus. The complete supernova remnant is shown in its entirety in Figure 7-35. (*Hale Observatories*)

Plate 13. A Spiral Galaxy in Andromeda. (NGC 224) (*Hale Observatories*)

Plate 14. A Spiral Galaxy in Sculptor. (NGC 253) (*Hale Observatories*)

Plate 15. A Spiral Galaxy in Pegasus. (NGC 7331) (*Hale Observatories*)

Plate 16. An Irregular Galaxy in Ursa Major. (NGC 3034) (*Hale Observatories*)

Chapter 12   The Realm of the Galaxies

Figure 12-19.   *The Coma Cluster.* This huge cluster of galaxies in the constellation of Coma Berenices contains over 1000 members. It is a rich, regular cluster dominated by two giant cD-ellipticals at its center. (*Hale Observatories*)

cal symmetry along with a strong central concentration. Some astronomers refer to them as *globular clusters of galaxies.* They tend to be very "rich" in that they often contain at least one thousand extremely bright galaxies, most of which are ellipticals. There are probably thousands of additional dimmer galaxies in typical regular clusters.

The nearest rich regular cluster is found in the constellation of Coma Berenices. This is the so-called *Coma cluster.* It is about 300 million light years away and about 10 million light years in size. In spite of its incredible distance of almost a third of a billion light years, more than one thousand galaxies have been discovered in the Coma cluster. Near the center of the Coma cluster are two huge elliptical galaxies. These are called *cD-galaxies* and have an absolute magnitude of about -24. This makes them many times brighter than a typical galaxy, such as our own, which has an absolute magnitude of -21. Detailed examination of the Coma cluster reveals that as astronomers look to fainter and fainter magnitudes, they see progressively more and more galaxies. This means that regular clusters, such as the Coma cluster, might actually contain tens of

Figure 12–20. A Dwarf Elliptical. This dwarf elliptical galaxy, NGC 147 in the constellation of Cassiopeia, probably contains fewer than one billion stars. This galaxy happens to be a member of the Local Group. (*Hale Observatories*)

thousands of galaxies. Such galaxies probably are *dwarf ellipticals*, which typically have absolute magnitudes of around -10 and contain slightly fewer than a billion stars. At the distance of the Coma cluster, dwarf elliptical galaxies are too faint to be seen even with the most powerful telescopes.

Just as regular clusters are sometimes called globulars, irregular clusters are referred to as *open clusters of galaxies*. Irregulars show no spherical symmetry or central concentration. They consist of an amorphous grouping of galaxies scattered helter-skelter over a region of space. While the overwhelming majority of bright galaxies in regular clusters are ellipticals, irregular clusters contain all kinds of galaxies. Irregular clusters are far more common than regular clusters. Irregular clusters can range from being very rich, in which case they contain more than a thousand members, to very poor, with only a couple of dozen galaxies or less.

A good example of a nearby, rich, irregular cluster is found in the constellation of Virgo. This *Virgo cluster* appears as a great cloud of more than one thousand galaxies covering an area of the sky measuring 10 by 12°. The distance to the center of the cluster is about 40 million light years. This distance is obtained from the apparent faintness of supergiant stars, as well as from the apparent magnitudes of the globular clusters surrounding M87, a giant cD galaxy near the center of the Virgo cluster. Many faint dwarf ellipticals can be seen in this nearby irregular.

Our own Galaxy is a member of a small, irregular cluster affectionately known

Chapter 12 The Realm of the Galaxies

as the *Local Group*. The Local Group contains about twenty galaxies, many of which are dwarf ellipticals.

It is interesting to note that the Local Group contains only four very bright galaxies having absolute magnitudes brighter than -18. This means that far from the Local Group, astronomers on some distant planet in the Coma cluster would see only four galaxies: M31, Milky Way Galaxy, M33, and LMC. They might not ever detect all the dwarf ellipticals. This, in turn, very strongly suggests that astronomers here on Earth probably do not see many of the fainter galaxies in the remote clusters they observe. Surely thousands of dwarf ellipticals in the Virgo and Coma clusters have escaped detection.

The discovery of clusters of galaxies is a fairly recent development in the field of *observational cosmology*. The fact that clusters are very commonplace and the notion that astronomers probably miss a lot of the fainter, dwarf galaxies make it reasonable to suppose that *all* galaxies are members of clusters.

Even more recent developments suggest that clusters of galaxies group together in *superclusters*. These *clusters of clusters* may be hundreds of millions of light years in diameter. For example, our Local Group along with the Virgo cluster and a few smaller irregular clusters make up the *local supercluster*. Our local supercluster is roughly 150 million light years and probably has a total mass of nearly a million billion suns!

Finally, from observations of galaxies, it is possible to make some very important general statements about the distribution of matter throughout the universe. On the very small scale, matter is concentrated in stars, which are located in

Figure 12–21. An Irregular Cluster. This cluster of galaxies in Hercules is a good example of an irregular or open cluster. (*Hale Observatories*)

Figure 12-22. *A Very Distant Cluster.* The galaxies in this cluster (marked by little white lines) are more than a billion light years away. (*Hale Observatories*)

Figure 12-23. *The Distribution of Rich Clusters.* Each dot on this map of the sky indicates the location of a rich cluster. The areas of the sky which cannot be seen from Palomar Observatory and which are obscured by the Milky Way are indicated. (*Adapted from Dr. Abell*)

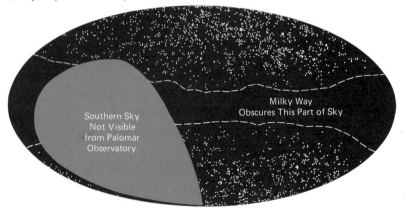

galaxies, which are grouped into clusters. Therefore, on the small scale, the universe looks very lumpy; the distribution of matter is inhomogeneous. But when astronomers look at the universe on the large scale, they find a very different picture. Figure 12–23 shows the distribution of rich clusters of galaxies over portions of the sky that can be seen from Mt. Palomar. One glance at this map of the sky shows that rich clusters are scattered rather uniformly through space. Thus, on the large scale, the universe looks very smooth. On the large scale, the universe is, in fact, homogeneous. this is an important observational result for theoretical astrophysicists, who construct mathematical models of the universe. In *theoretical cosmology*, it is entirely appropriate to assume that on the largest scales the matter in the universe is evenly distributed throughout space.

12.6 The Redshift

Whenever an astronomer sees something of interest in the sky—a planet, a star, or a galaxy—his immediate inclination is to attach a spectograph to his telescope and obtain a spectrum. As discussed in Chapter 5, a spectrum contains an incredible wealth of information. The prism or diffraction grating inside the astronomer's spectrograph breaks up the incoming light into the colors of the rainbow. In and among the colors, thin dark *spectral lines* often are observed. These spectral lines are formed by the atoms in the star or galaxy from which the light is being emitted. From detailed examination of the spectral lines, astronomers are able to deduce many of the properties of the source of light.

In the years between 1912 and 1925, Dr. V. M. Slipher at the Lowell Observatory obtained spectra of about forty spiral and elliptical "nebulae." He was, however, unaware of the true nature of the objects he was observing; Edwin Hubble had still not proved that these nebulae are in fact individual galaxies located at incredibly remote distances. In virtually *all* cases, Slipher discovered that the spectral lines in the spectra of these forty galaxies were *not* where they should be. Instead, the spectral lines were shifted toward the red end of the spectra.

From the discussion of spectra in Section 5.4, recall that the locations of spectral lines are affected by the motion of the source of light. According to the *Doppler effect*, if a source of light is moving toward you, its spectral lines will be shifted to shorter wave lengths—toward the blue end of the spectrum. On the other hand, if a source of light is moving away, its spectral lines will be shifted to longer wavelengths—toward the red end of the spectrum. the higher the speed, the greater is the shift. By measuring precisely how much a given spectral line is shifted from its usual position, the astronomer can calculate exactly how fast the source of light is approaching him or receding from him. From Silpher's observations, it was concluded that many of the galaxies he observed were actually moving away from us at speeds up to 1,125 miles per second!

By the mid-1920s, astronomers had begun to suspect that there might be a relationship between the distances to galaxies and the sizes of their redshifts. Galaxies that appear big and bright in the sky—and, therefore, presumably are nearby—show spectral lines shifted by only a small amount toward the red end of the spectrum. By contrast, galaxies that appear faint and dim—and, therefore, presumably are far away—show much larger redshifts in the locations of their spectral lines.

Figure 12–24 shows photographs of five elliptical galaxies along with their spectra. Each spectrum consists of the spectrum of the light from the galaxy surrounded on top and bottom by a comparison spectrum. These comparison spectra were artificially placed on the spectrographic plate by the astronomer to serve as reference marks. In all cases, two spectral lines due to calcium atoms in the galaxies (the so-called H and K lines) are easily visible. This, of course, does not mean that galaxies are made out of calcium. Rather, conditions found in galaxies are ideal for the formation of the H and K lines. Under these ideal conditions, the H and K lines appear very strong in the spectra of galaxies.

In the elliptical galaxy in the Virgo cluster, the H and K lines are almost exactly where they should be. These lines are shifted by only a very small amount toward the red end (to the righthand side in Figure 12–24) of the spectrum. Thus, the speed with which this galaxy is receding from us is fairly low, only 750 miles per second. But as we move to small and fainter galaxies in the Ursa Major, Corona Borealis, and Bootes clusters, the redshifts get bigger and bigger. In fact, in the case of the tiny, dim elliptical galaxy in Hydra, the redshift is so huge that the H and K spectral lines, which normally appear among the blue colors of the spectrum, are shifted all the way across the spectrum into the red colors. This galaxy has a recessional velocity of 38,000 miles per second. Finally, just from looking at the photographs of the five galaxies in Figure 12–24, one immediately suspects that the galaxy in Virgo is the nearest, the galaxy in Ursa Major is somewhat farther away, the galaxy in Corona Borealis is still more remote, and so on.

By 1929, Edwin Hubble had succeeded in determining the distances to a number of galaxies by methods outlined earlier in this chapter. In addition, spectra of these galaxies also were available from which their recessional velocities were measured. Hubble discovered that there is a simple and direct relationship between the distances and speeds of galaxies.

During the early 1930s, Hubble collaborated with M. L. Humason at the Mount Wilson Observatory to obtain reliable observations of very distant galaxies. The resulting measurements of distances and redshifts conclusively proved that nearby galaxies are moving away from us slowly, while more distant galaxies are moving away from us more rapidly. In fact, the distances and recessional speeds of galaxies are in direct proportion to each other. Double the distance to a galaxy and the speed doubles. Triple the distance and the speed triples. A galaxy at a distance of 600 million light years is moving away from us at a speed twice as fast as a galaxy that is only 300 million light years away. This relationship between distance and speed is called the *Hubble law* or the *law of redshifts*.

Chapter 12 The Realm of the Galaxies

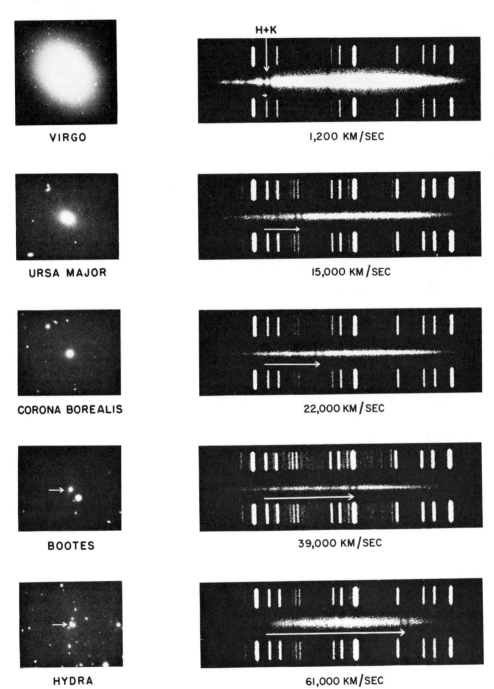

Figure 12-24. Five Galaxies and Their Spectra. Photographs of five elliptical galaxies and their spectra are shown here side-by-side. The H and K lines of calcium are seen in each spectrum (*Hale Observatories*)

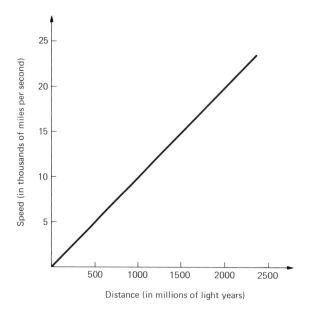

Figure 12–25. *The Hubble Law.* The distances and redshifts (i.e., recessional speeds) of galaxies are directly related, as shown on this graph. Nearby galaxies are moving away slowly while distant galaxies are moving away more rapidly.

The best way to display the Hubble law is with the aid of a graph. As shown in Figure 12–25, the distances to galaxies are measured on the horizontal axis, whereas the recessional speeds of the galaxies are measured on the vertical axis. All the data from Hubble, Humason, and later investigators fall along a straight line. The farther away the galaxies are, the more rapidly they are moving.

One of the most important projects in modern observational cosmology has been the determination of the *slope*, or orientation, of the line in Figure 12–25, which displays the Hubble law. Since the 1930s, astronomers have refined the methods of measuring galactic distances. Hubble and Humason had the line in their graphs of the Hubble law tilted at a somewhat different angle. The line plotted in Figure 12–25 is based on the most recent and reliable data obtained by Dr. Alan Sandage at the Hale Observatories.

The slope, or orientation, of the line in Figure 12–25 is given by the *Hubble constant*. In the early 1970s, Dr. Sandage announced the results of many years of careful and patient observations. His data tell us that the Hubble constant is equal to *10 miles per second per million light years*. This means that—ideally—a galaxy 1 million light years away is receding from us with a speed of 10 miles per second. Since the distances and speeds are in direct proportion, a galaxy 100 million light years away has a recessional speed of 1,000 miles per second; a galaxy 2 billion light years away is receding from us at a speed of 20,000 miles per second; and so forth.

A very important application of the Hubble law is that it can be used as a *distance indicator* for remote galaxies, especially in those cases where Cepheid variables and globular clusters are too faint to be observed. For example, suppose an astronomer discovers a dim galaxy in the sky. He can take a spectrum of the galaxy, measure its redshift and, thereby, discover exactly how fast the galaxy is moving away from us. Let's suppose he comes out with 15,000 miles per second as the recessional speed. He then simply turns to the Hubble law in Figure 12–25 and from the graph "reads off" what the distance to the galaxy must be. Figure 12–25 shows that a speed of 15,000 miles per second corresponds to a distance of $1\frac{1}{2}$ billion light years.

The idea that the Hubble law can be used as a distance indicator stands as one of the major pillars in modern observational cosmology. The traditional astronomer of the 1970s is convinced absolutely that the direct relationship between the distances and speeds of the galaxies is a fundamental property of the universe. But there are a few maverick astronomers. As will be discussed in the next chapter, there are a small number of very competent astronomers who seem to have evidence to suggest that the Hubble might not apply to certain types of exploding galaxies and quasars. Thus, the astronomical community is again embroiled in a fierce debate, very similar to the Shapley-Curtis debate of the 1920s.

Shortly after Albert Einstein proposed his general theory of relativity, he applied his new ideas to the problem of calculating theoretical models of the universe. Einstein was surprised to discover that his resulting cosmological models were *not* stable. The gravitational attraction between all the galaxies in the universe made them want to collapse in on each other. Of course, back around 1920, the Hubble law was still unknown. It was not known that all the galaxies in the universe are rushing away from us with speeds that are proportional to their distances. Einstein, therefore, proceeded to fiddle with his equations. He added a number to his field equations—the *cosmological constant*—that resulted in stable theoretical models of the universe. What a tragedy! If Einstein only had trusted his equations and believed in what he was doing, he could have predicted the true inner meaning of the Hubble law. As will be seen in the final chapter, he would have predicted that the universe is expanding!

Questions and Exercises

1. Name the astronomer who first tried to deduce the location of the sun in the Milky Way Galaxy.
2. Briefly describe the method of star gauging.
3. What is interstellar absorption and how did it affect the star-gauging observations of Herschel and Kapteyn?
4. How did Harlow Shapley use globular clusters to deduce the location of the sun in the Milky Way Galaxy?
5. What, approximately, is the diameter of our Galaxy?

6. Approximately how far is the sun from the galactic center?
7. Briefly describe the mechanism that produces 21-centimeter radiation from hydrogen.
8. Why are radio observations more fruitful than optical observations in determining the structure of our Galaxy?
9. Briefly describe, with the aid of a drawing, the over-all structure of our Galaxy.
10. How long does it take the sun to go once around the Galaxy?
11. What is a gravitational wave?
12. Who was Charles Messier?
13. What is the *NGC*?
14. In your own words, briefly describe what the Shapley-Curtis debate was all about.
15. Who was Henrietta Leavitt?
16. What is the period-luminosity relation?
17. What is the difference between Population-I and Population-II stars?
18. In your own words, briefly describe how the period-luminosity relation can be used to determine distances.
19. How did Hubble measure the distance to M31?
20. How can globular clusters be used to determine the distances to remote galaxies?
21. How can supernovae be used to determine the distances to remote galaxies?
22. Briefly describe the Hubble classification scheme for galaxies.
23. Contrast and compare regular and irregular clusters of galaxies.
24. What is a cD-galaxy?
25. What is a dwarf elliptical?
26. What is the Local Group?
27. What is meant by superclustering?
28. How is the redshift of a galaxy related to its recessional speed?
29. What is the Hubble law?
30. How can the Hubble law be used as a distance indicator?
31. What is the Hubble constant?

Quasars and Exotic Galaxies

13.1 The Discovery of Quasars

The field of radio astronomy got off to a feeble start just before World War II when Karl Jansky, an engineer at Bell Telephone Laboratories, discovered weak radio emission coming from the Milky Way. Nobody took Jansky seriously, and his results were not followed up. However, in 1944, the physicist Grote Reber discovered a discrete radio source in the constellation of Cygnus. By the end of World War II, astronomers had found that many of the recent developments in electrical engineering could be applied to making good radio telescopes. In 1946, the English astronomers J. S. Hey, S. J. Parsons, and J. W. Phillips confirmed Reber's discovery of the extragalactic source known as Cygnus A. The announcement of their observations marks the birth of extragalactic radio astronomy.

In 1948, one of the first radio astronomers J. G. Bolton in Australia fully confirmed the existence of Cygnus A and announced the discovery of six additional discrete radio sources. These included Taurus A (the Crab nebula) and Centaurus A (associated with the peculiar galaxy NGC 5128). These successes inspired other astronomers to begin making general surveys of the sky. In 1950, one of the first catalogues of the radio sky was published. This catalogue, which was prepared by M. Ryle, F. G. Smith, and B. Elsmore in England, is known as the *First Cambridge Catalogue*. It contained a listing of more than fifty radio sources known as the 1C objects. Unfortunately, there were large probable errors in the right ascensions and declinations of the sources. Radio astronomers still had a long way to go in perfecting the art of determining the precise locations of the sources they observed.

Five years later, in 1955, the *Second Cambridge Catalogue* was published. This so-called *2C Catalogue* contained 1,936 sources and constituted the first truly extensive survey of the radio sky. Unfortunately, the *2C Catalogue* also had its problems. There are so many dim radio sources in the sky that many of the 2C objects were not properly resolved. As a result, the *2C Catalogue* contains a lot of radio sources that do not really exist. So it was back to the proverbial drawing board and work began on the *Third Cambridge Catalogue*.

The *3C Catalogue* in use today was published in its final form in 1962 by the radio astronomer S. A. Bennett. There are very few errors in the *3C Catalogue*, which lists 328 sources in a region of the sky from the north celestial pole down to a declination of $-5°$. Only recently have the 4C and 5C surveys begun to take over. Nevertheless, the *3C Catalogue* has been in use for so long that astronomers still refer to objects according to their 3C numbers.

While radio astronomers were studying the skies with their new telescopes, nuclear physicists were hard at work building new and more powerful accelerators (popularly called atom smashers) in their laboratories. In particular, at the General Electric Company in Schenectady, New York, nuclear physicists had succeeded in building a new type of accelerator called a *synchrotron*. In a synchrotron, electrons travel in circular orbits in an intense magnetic field. The machine is designed so that the magnetic field "pumps" energy into the electrons, which can achieve speeds in excess of 99 per cent of the speed of light. When the synchrotron at General Electric was put into operation, physicists looking through a small glass window in the side of the machine noticed that an intense light was being emitted. High-speed electrons moving in a magnetic field emit electromagnetic radiation. This form of light is called *synchrotron radiation*.

This discovery of synchrotron radiation had important implications for radio astronomers. Objects in the sky emitting radiowaves possess magnetic fields and have lots of electrons floating around. Of course, the magnetic fields in nebulae and galaxies are much weaker than the intense fields produced at General Electric. But the natural magnetic fields in galaxies are spread out over billions of cubic light years. As electrons in these galaxies spiral around the magnetic fields, synchrotron radiation is emitted, but at very long wave lengths. The synchrotron radiation comes out as radiowaves. This is the primary explanation of the radio noise astronomers detect from nebulae and galaxies.

Since the magnetic fields in galaxies are generally quite weak, and since the numbers of electrons per cubic foot floating in interstellar space are very low,

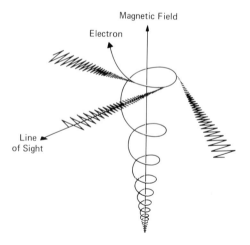

Figure 13–1. Synchrotron Radiation. Electrons moving at high speeds in a magnetic field emit synchrotron radiation, as shown schematically in this diagram.

Figure 13-2. The First Quasar. Accurate determination of the radio source 3C 48 led to the identification of the first quasi-stellar radio source. Hundreds of quasars are known today. (*Hale Observatories*)

astronomers were led to expect that the radio sources they observe must be huge. They thought these radio sources must be spread out over a large volume so that the combined effect of many electrons spiraling around weak magnetic fields can produce a "signal" that can be detected here on earth. In virtually all of the 3C objects, this suspicion was born out. In almost every case, the 3C objects cover a relatively large region of the sky. They are *not* "point sources" like stars but rather are spread out over areas comparable to the sizes of nebulae and galaxies that optical astronomers observe.

In around 1960, astronomers realized that they were on to something really peculiar. At the Jodrell Bank Observatory in England radio astronomers had undertaken the task of measuring the angular size of the brighter objects in the 3C catalogue. Four objects were especially interesting because they seemed to have extremely small angular diameters. These unusual objects were 3C 48, 3C 286, 3C 196, and 3C 147.

By the fall of 1960, the position of the radio source 3C 48 (that is, the forty-eighth object in the *Third Cambridge Catalogue*) had been sufficiently established so that optical astronomers at the 200-inch telescope could begin searching for unusual optical objects near the location of the radio source. Photographs of the region surrounding 3C 48 studied by the astronomer T. A. Matthews revealed something that looked like a star. In October, Alan Sandage took a spectrum of

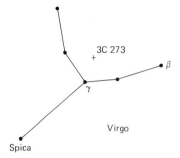

Figure 13-3. *The Location of 3C 273.* In the early 1960s, radio astronomers discovered that 3C 273 has an unusually overall angular diameter. Since the moon passes through Virgo, a lunar occultation was used to pin-point the location of this unusual radio source.

this "star." No one could identify the spectra lines! The spectrum of 3C 48 consisted of four broad, bright emission lines that no one had ever seen before in this part of the spectrum.

The mystery of 3C 48 deepened when photoelectric observations revealed that this object was varying in brightness during periods of a day. If something varies in brightness in 24 hours, then its size must be *less* than a "light day"—the distance light can travel in 24 hours, or about 17 billion miles. Anything larger than 17 billion miles in size cannot change its brightness in less than 24 hours. Although 17 billion miles might sound like a huge distance, 1 light day is very tiny compared to the size of a galaxy. After all, our own Galaxy is 100,000 light *years* in diameter. In view of its starlike appearance, as well as its small size, some astronomers suggested that 3C 48 just might be a new breed of star in our own Galaxy.

The next major breakthrough came in 1962. By this time, it had become apparent that another 3C object, 3C 273, also had a very small angular diameter. This radio source happens to be located in the constellation of Virgo, a constellation of the zodiac, as shown in Figure 13-3. Of course, the sun, moon, and

Figure 13-4. *The Lunar Occultation of 3C 273.* When the edge of the moon arrived at the location of 3C 273, the radio noise abruptly shut off. This abrupt shut-off (and later turn-on) along with the diffraction patterns clearly indicated to astronomers that 3C 273 was a point source of radio radiation.

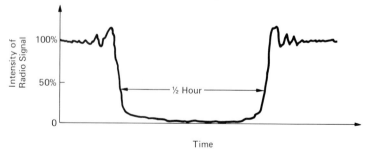

Chapter 13 Quasars and Exotic Galaxies

Figure 13–5. The Quasar 3C 273. As in the case of 3C 48, the spectrum of 3C 273 shows spectral lines which initially no one could identify. With a visual magnitude of 13, 3C 273 is the brightest-appearing quasar in the sky. (*Hale Observatories*)

planets move through the zodiac and, fortunately, the moon was scheduled to pass directly in front of 3C 273 three times in 1962 as seen from Australia. Such a situation is called a *lunar occultation*. A team of radio astronomers, Drs. Hazard, Mackey, and Shimmins, observed a lunar occultation of 3C 273 with the Parkes radio telescope in Australia on August 5, 1962. Everyone expected that the radio noise from 3C 273 would gradually decrease as this source was gradually covered by the moon. Since intelligent astronomers knew that 3C 273 must have some angular diameter, it couldn't possibly be a point source. Then, as the moon moved along in its orbit, everyone expected the radio noise to gradually pick up as the moon uncovered more and more of 3C 273. It didn't happen! The moment the limb of the moon arrived at the expected location of 3C 273, the radio noise shut off abruptly! Total silence reigned for half an hour while 3C 273 was behind the moon. Then, at the moment 3C 273 was scheduled to appear from behind the moon, the radio noise promptly shot back up to its original intensity. Figure 13–4 shows a record of the intensity of the radio emission observed at Parkes. The wiggles in the intensity just before and after the occultation are the clincher. These so-called *diffraction patterns* (see Section 5.1) conclusively prove that 3C 273 is a point source!

Astronomers know exactly where the moon is at any given moment. From the location of the moon, with respect to the background stars during the occultation, it was easy to calculate the exact location of 3C 273. It coincided with a bluish thirteenth magnitude "star."

This identification of 3C 273 with a visible object prompted M. Schmidt at Palomar to obtain a spectrum. As with 3C 48, the spectrum consisted of broad emission lines that no one could identify. This is an incredibly troublesome state of affairs for astronomers. All the spectral lines of all the chemical elements are well known. There really should not be any surprises. The astronomer can go to massive books in the library and look up the wavelengths of spectral lines of virtually any chemical. Every spectral line in every star and galaxy ever observed can be explained in this fashion. Yet, here were two objects, both of which are bright, highly compact radio sources, whose optical spectra were a total mystery!

To make matters more confusing, 3C 273 is sufficiently bright in visible light that searches through old photographs at the Harvard and Pulkova Observatories revealed that 3C 273 had been photographed for the past 70 years. As in the case of 3C 48, variations in the visible light were again apparent. Whatever these objects are, they must be fairly small.

Since these strange objects look like stars, they soon became known as *quasi-stellar objects* (QSOs), since the prefix *quasi* means "kind of like." This somewhat cumbersome term has since been contracted to *quasars*. Early in 1963, only two quasars had been positively identified. Both were compact, bright radio sources coincident with bluish stars whose spectral lines were unidentifiable. Astronomy was ready for an incredibly historic moment. The mystery of the spectra would be resolved by one swift blow of sheer genius. The era of the quasars was at hand!

13.2 The Redshifts of Quasars

The historic moment came early in 1963. The observational results of the lunar occultation of 3C 273 at Parkes, Australia, had just been published and Dr. M. Schmidt of Caltech had succeeded in obtaining a spectrum of the visible "star" associated with this quasar. The big stumbling block was that most astronomers suspected that the strange stars associated with 3C 48 and 3C 273 were *inside* our own Galaxy. Now, in our Galaxy, most stars move around the sun in circular orbits at fairly low speeds. These stars are called *low-velocity stars*. Our sun is one of these low-velocity stars that typically have speeds (measured from earth) of roughly 25 miles per second. On the other hand, there are some stars that are in highly elliptical orbits and move at comparatively high speeds of roughly 50 miles per second. These are called *high-velocity stars*. The point is that neither of these two types of stars are really moving very fast; their redshifts are quite small. Thus, no one ever considered the possibility of large redshifts for the quasars.

In examining his spectrum of 3C 273, Dr. Schmidt found four broad emission lines. Although lines such as these had never been seen before in this part of the spectrum, they did look slightly familiar to Schmidt. In particular, their wave lengths were in precisely the same ratios as one would expect for the ordinary, familiar lines of the hydrogen atom (electrons jumping from the third, fourth, fifth, and sixth orbits down to the second orbit). Schmidt, therefore, very boldly suggested that perhaps these mysterious lines *are* familiar hydrogen lines that have been enormously redshifted. His efforts met with immediate success. The redshift of 3C 273 corresponds to 16 per cent of the speed of light, or 30,000 miles per second!

Figure 13–6. *Four Quasars.* Quasars have enormous redshifts. For example, the quasar 3C 147 has a redshift corresponding to a velocity greater than half of the speed of light! (*Hale Observatories*)

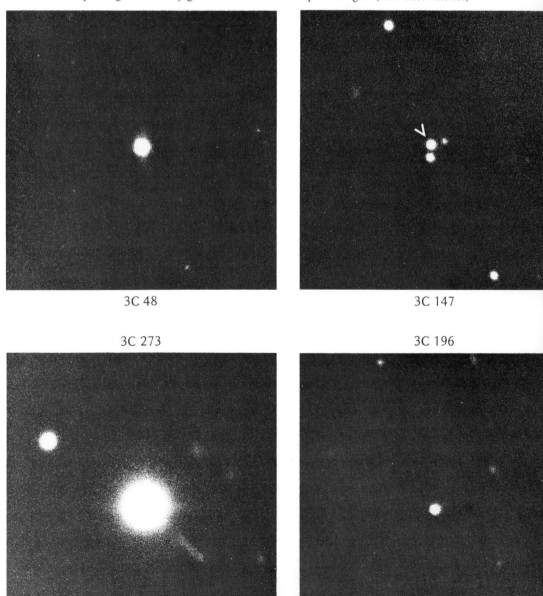

Inspired by Schmidt's discovery, Caltech astronomers J. L. Greenstein and T. A. Matthews tackled the spectrum of 3C 48. All of its spectral lines can be explained easily by an enormous redshift corresponding to 37 per cent of the speed of light. That's almost 70,000 miles per second! Certainly 1964 was a year of consolidation. Eight more quasars were discovered and Schmidt and Matthews went about the business of obtaining spectra. They found that all the spectral lines in the quasar 3C 147 could be explained by a redshift 55 per cent of the speed of light. Prior to this discovery, the largest known redshift of anything belonged to the galaxy 3C 295, a galaxy that is also a strong radio source receding from us with a velocity equal to 46 per cent of the speed of light. It was becoming clear that quasars have the highest redshifts of anything in the universe.

Another significant development during this year dealt with the colors of quasars. Martin Ryle in England and Alan Sandage in the United States noticed that quasars are always bluish and are very bright in the near ultraviolet. They all seem to have a large *ultraviolet excess,* compared to usual objects found in the sky. This discovery permitted rapid optical identification of quasars. All the astronomer has to do is take two photographs of the region of the sky surrounding the radio location of a suspected quasar. One photograph is taken with an ultraviolet (U) filter and the other with a blue (B) filter. (Recall the discussion of the standard U, B, V system in Section 6.3.) The quasar then stands out as unusually bright on the U photograph compared to the B photograph. Using this method, Ryle and Sandage identified the quasars 3C 9, 3C 216, and 3C 245.

In 1965, astronomers continued to identify spectral lines of quasars with intermediate and large redshifts. Believe it or not, the redshifts that had been measured up to this time are actually quite *low* as far as quasars are concerned. The major difficulty that faced Schmidt at this point was the fact that the spectra

Figure 13–7. *The Quasar 3C 9.* The redshift of this quasar is so great that spectral lines from the invisible ultraviolet are shifted all the way into the visible part of the spectrum. The redshift of 3C 9 corresponds to 80% of the speed of light. (*Hale Observatories*)

Figure 13-8. A Radio-Quiet Quasar. In the mid-1960s, it was discovered that some quasars are not radio sources. This quasar, BSO-1 (BSO stands for *blue stellar object*), is a good example. Its redshift corresponds to 67% the speed of light. (*Hale Observatories*)

he was observing were so highly redshifted that he was seeing spectral lines normally observed only in the ultraviolet. These quasars had to be moving so fast that spectral lines from the invisible ultraviolet were shifted all the way into visible wavelengths of the spectrum. Astronomers were unfamiliar with spectral lines in the ultraviolet. Schmidt turned to the work of Donald E. Osterbrock at Lick Observatory, who had been working on the ultraviolet spectra of nebulae. With the aid of Osterbrock's work, Schmidt was able to identify spectral lines of singly ionized magnesium, doubly ionized carbon, and triply ionized carbon in the spectrum of 3C 287. From these identifications, Schmidt showed that this quasar has a redshift corresponding to 60 per cent of the speed of light, or 110,000 miles per second!

Using the fact that lines of triply ionized carbon appear in the spectra of quasars, Schmidt was able to measure even higher redshifts. The particular spectral line of carbon in question is usually located at 1,550 Å in laboratory experiments. That's in the *far* ultraviolet. In the spectrum of 3C 9, this spectral line appears at a wavelength of about 4,670 Å, among the blue-green colors of the visible spectrum. This huge redshift corresponds to the fantastic velocity of 80 per cent of the speed of light, or 150,000 miles per second! Also easily seen in the spectrum of 3C 9 is a very short wavelength line produced by hydrogen atoms. This spectral line, called *Lyman* α, arises when electrons jump from the second orbit down to the first orbit, or ground state. In the laboratory, where the hydrogen atoms are at rest, this spectral line has a wavelength of 1,216 Å. In 3C 9, Lyman α appears at 3,666 Å, which corresponds to the same redshift (80 per cent of the speed of light) as in the case of the carbon line. Today, the proper identification of spectral lines in the spectra of high redshift quasars often depends on observing the line of triply ionized carbon (rest wavelength = 1,550 Å) and Lyman α of hydrogen (rest wavelength = 1,216 Å), both of which appear in the visible part of the spectrum.

In 1965, another major advance occurred when Alan Sandage at the Hale Observatories discovered that there are quasars that do not emit radio noise. They are "radio-quiet" quasars. Using the Ryle-Sandage method of comparing U and B photographs, both Sandage at Palomar and C. R. Lynds at Kitt Peak found that there were many "stars" with ultraviolet excesses nowhere near any 3C radio sources. Some of these objects really were stars. But in two of the first six cases Sandage and Lynds examined, they were quasars. One quasar, called Ton 256, has a redshift of 13 per cent of the speed of light. The other, called BSO 1, has a redshift corresponding to 67 per cent of the speed of light. In other words, there are quasars that are *not* radio sources. Many such radio-quiet quasars are known today. However, it should be noted that their radio quietness is only relative. As radio astronomers examine these objects with their most sensitive instruments, some weak radio noise often is heard. Indeed many of the quasars that were thought to be radio quiet in the mid-1960s actually show up as very weak sources in the 4C catalogue.

The main properties of quasars follow:

1. They look like stars.
2. They are bluish in color.
3. Some are strong radio sources.
4. They all have *large* redshifts.
5. Many exhibit rapid optical and radio variability.

In the late 1960s, literally hundreds of quasars were discovered by looking for objects that have these five properties. Redshifts corresponding to velocities of half the speed of light turned out to be quite common. In all cases, the quasars were bluish, starlike objects with substantial redshifts. It also became apparent that there were not too many quasars with superhigh redshifts.

As discussed earlier, 3C 9 was discovered to have a redshift corresponding to 80 per cent of the speed of light. Although hundreds of quasars were found with lower redshifts, 3C 9 held a record that seemed hard to beat. By 1971, however, two quasars with higher redshifts had been discovered. First place went to 4C 05.34, whose redshift corresponds to 88 per cent of the speed of light. Second place was held by 5C 2.56, which has a redshift of 84 per cent of the speed of light. In 1972, another high redshift quasar was discovered, and second place was taken over by PHL 957, which seems to be moving away from us at 86 per cent of the speed of light. By this time many astronomers felt that 90 per cent of the speed of light was an upper limit. There probably were no quasars moving away from us faster than 4C 05.34. Wrong again!

Beginning in the late 1960s, radio astronomers at Ohio State University began publishing massive catalogues summarizing an extensive survey of the radio sky. Thousands upon thousands of new radio sources had been discovered. By the end of 1974, more than 19,000 radio sources had been observed of which almost

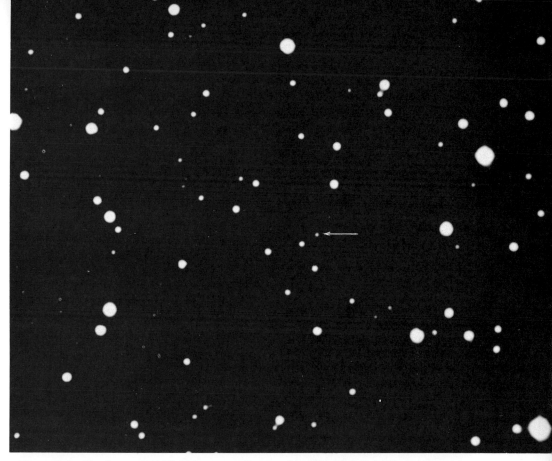

Figure 13-9. The Quasar OH 471. This quasar, discovered in 1973, has a redshift slightly greater than 90% the speed of light. (*Hale Observatories. Copyright by the National Geographic Sky Survey*)

13,000 had never previously been catalogued. Naturally, optical astronomers started working very hard to discover visible objects associated with these new radio sources.

In 1973, P. A. Strittmatter announced that the starlike object associated with OH 471 (*O* stands for Ohio survey; everything else designates the location in the sky) has a redshift corresponding to $90\frac{1}{2}$ per cent of the speed of light. A few months later, E. J. Wampler discovered that the starlike object associated with OQ 172 has an even higher redshift. OQ 172 has a redshift corresponding to 91 per cent of the speed of light, or nearly 170,000 miles per second! Both of these superhigh redshift objects are shown in Figures 13-9 and 13-10.

One year later, in 1974, J. B. Oke at Hale Observatories published her photoelectric observations of OQ 172 and OH 471. This work explained why it is easy for astronomers to miss quasars with extremely high redshifts. Lyman α in the spectra of these objects appears in emission; it is a bright line. Since a large fraction of all the light emitted from these quasars is contained in this emission line, the precise location of this line in the visible spectrum will affect the color of the object. In both these quasars, Lyman α is shifted by more than 4,000 Å into the yellow and orange colors. These quasars are *not* as bluish as more familiar

Figure 13-10. The Quasar OQ 172. This quasar has the highest known redshift. Its redshift corresponds to 91% the speed of light. (*Hale Observatories. Copyright by the National Geographic Sky Survey*)

quasars of lower redshifts. In fact, quasars with redshifts of 95 per cent of the speed of light will have Lyman α appearing in the red colors of the visible spectrum. Such objects would probably look very red! Up until this time, no one had ever looked for red quasars. Could it be that there are many quasars in the sky with redshifts higher than 95 per cent of the speed of light that have been totally ignored just because they are not bluish? The late 1970s could easily witness some dramatic surprises!

Schmidt's identification of the spectral lines in 3C 273 back in 1963 was certainly a major breakthrough in astronomy. Although scientists now understand the spectral lines in quasars in terms of huge redshifts, this understanding gives rise to more questions than answers. In fact, the observed redshifts of quasars have spawned a vicious controversy that is being debated more hotly than any issue in astronomy since the days of Galileo, when people were burned at the stake for daring to think that the Earth goes around the sun!

The central issue is this: *What causes the redshifts of the quasars?* There are three and only three "traditional" answers. Initially, back in 1963, it was thought that perhaps astronomers could be seeing gravitational redshifts as predicted by Einstein's general theory of relativity (see Chapter 11). Indeed, in many respects, it was Schmidt's announcement of the redshift of 3C 273 that resulted in an upsurge in interest in relativity. However, after only a few years of hard work, it was realized that it is virtually impossible to have stable objects with truly huge redshifts. As discussed in Section 11.3, after an object passes a certain point, it collapses rapidly to form a black hole. Astrophysicists have been totally unsuccessful in trying to think of ways in which stable objects could exist with large gravitational redshifts without promptly undergoing gravitational collapse. Of course, this does not mean that such objects can't exist. Perhaps astrophysicists have just not been keen enough. In this connection, it should be mentioned that, in 1975, Dr. J. V. Narlikar at Caltech seemed to be making some progress in using gravitational redshifts to explain quasars. However, his work is far from complete.

The second traditional explanation deals with the Doppler effect. Perhaps many quasars have such high redshifts because they are nearby objects that are rapidly moving away from us. Along these lines, J. Terrell at Los Alamos has suggested that long ago an explosion took place at the center of our Galaxy. This explosion ejected many starlike objects into space at relativistic speeds. In the late 1960s, there was a lot of talk about quasars in this vein, primarily due to the work of Dr. H. C. Arp at the Hale Observatories. Perhaps quasars are violently ejected from the nuclei of galaxies and their redshifts are simply due to their high speeds. For example, in the region of the sky between Cetus and Pisces is a galaxy called NGC 520. Extending toward the southwest of this galaxy are four quasars that are almost in a perfectly straight line that points to the center of NGC 520, as shown in Figure 13–11. A photograph of NGC 520 reveals a very distorted appearance (Figure 13–12), as though the galaxy were recoiling from an explosion. Could it be that these four quasars were ejected long ago from the nucleus of NGC 520? Arp has pointed out numerous similar cases in which quasars seem to be very near certain galaxies.

The primary objection to this "ejection hypothesis" is straightforward and simple. The earth is not located at a special place in the universe; such absurd notions were disposed of centuries ago along with Ptolemy. Therefore, if quasars are ejected from galaxies, we should see roughly half of the quasars coming toward us and half going away. Half of the quasars should have redshifts, but half should have *blue*shifts. No one has ever discovered a quasar with a blueshift. All quasars only have redshifts.

Finally, the last of the traditional explanations involves the Hubble law. Astronomers see lots of galaxies in the sky. All of these galaxies (except for the

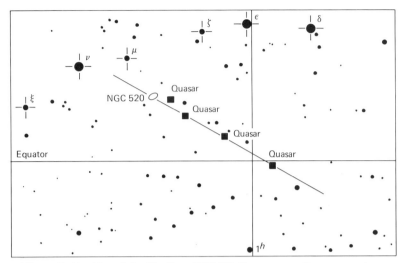

Figure 13–11. Map of the Sky Near NGC 520. Four quasars lie very nearly on a straight line to the southwest of the galaxy NGC 520. Could this mean that the galaxy and the quasars are somehow related? Or is this alignment just due to chance?

Figure 13–12. The Exploding Galaxy NGC 520. This photograph of NGC 520 reveals a very distorted appearance, perhaps suggesting that the galaxy has exploded. Is it possible that quasars are ejected from exploding galaxies? (*Hale Observatories*)

Chapter 13 Quasars and Exotic Galaxies

nearest galaxies such as some of those in the Local Group) have redshifts. Nearby galaxies have low redshifts and, therefore, are moving away from us slowly. More distant galaxies have higher redshifts and are, therefore, moving away from us more rapidly. Perhaps the quasars are out there among the galaxies! The final chapter will explain that although all galaxies (except some nearby ones) have redshifts, this does *not* mean that we are in some special place in the universe. Rather, it means that the universe is expanding. The redshifts of galaxies are, therefore, called *cosmological* redshifts. Perhaps the redshifts of the quasars are also cosmological.

Even though the cosmological interpretation of quasar redshifts is the most widely accepted explanation today, there are some pretty big problems. First of all, as discussed near the end of Section 12.6, the Hubble law can be used as a distance indicator. Traditional astronomers firmly believe that the distance to a galaxy can be determined just by measuring its redshift. From the redshift, an astronomer can calculate the speed with which the galaxy is receding from us. He then goes to the Hubble law, such as the graph in Figure 12–25, and finds the distance that corresponds to his observed velocity. Well, that's all just fine and dandy, except that quasars have *much* higher redshifts than the most remote galaxies ever discovered. For example, the distance from OH 471 and OQ 172 comes out to between 15 and 20 *billion* light years! That's an incredible distance. That's typically ten times more distant than the most distant galaxies astronomers usually observe. Not one single galaxy has ever been seen beyond about $8\frac{1}{2}$ billion light years. Very few galaxies have been observed at distances more than a few billion light years; yet, if we are to believe the cosmological interpretation, there must be lots of quasars out there.

Things rapidly get a lot worse when scientists think carefully about what they are saying. To begin with, up until the 1960s, galaxies were the brightest known objects in the universe. Astronomers knew of nothing brighter than the giant cD ellipticals, which have absolute magnitudes of -23. The point is that, if a bright cD galaxy were located at a distance of 10 or 15 billion light years, astronomers here on earth would *never* see it. At such an incredible distance, it would appear too dim to be detected even with the 200-inch telescope at Mount Palomar. Therefore, *if* quasars are at the distances indicated by their redshifts, they *must* be hundreds—even thousands—of times more luminous than the brightest known galaxies. A quasar, therefore, must emit more light than billions of stars combined.

But that's not all. Recall that many quasars exhibit variability. If physicists understand anything about physical reality, this variability places strict limits on the sizes of quasars. A quasar that changes its brightness in a period of a year cannot possibly be larger than 1 light year across. Yet, galaxies like our own are typically 100,000 light years in diameter. Therefore, quasars *must* be very small. In other words, if the cosmological interpretation is correct, quasars are hundreds of times brighter than the brightest galaxies, yet are only a tiny fraction of the size of typical galaxies. Astrophysicists are at a complete loss to explain how something so small can shine so brilliantly.

Things are a mess. Chaos reigns. Nobody has any real understanding of what quasars are. Half a century ago, astronomers could not explain why the sun shines. It took the discovery of thermonuclear energy to enable scientists to understand how our sun could produce so much light for billions of years. Are astrophysicists on the brink of discovering some new, more powerful source of energy? Or perhaps the cosmological interpretation of the redshifts of quasars is wrong. Maybe they really are not all that far away. This would solve the so-called energy problem, but then very strange and *non*traditional explanations for the high redshifts would have to be invoked. Such explanations would involve fundamentally new laws of physics. No matter what happens, one thing is now clear. When someone finally figures out what quasars are, mankind will have made a major breakthrough in discovering new laws of nature and will have arrived at a much deeper level of understanding physical reality.

13.3 Peculiar Galaxies

In the decades following the pioneering work of Edwin Hubble, astronomers felt that they had a fairly good understanding of what galaxies are. Galaxies are simply large conglomerations of billions of stars all revolving about a common point, not unlike the way in which the planets revolve about the center of the solar system. Furthermore, the redshifts of galaxies are directly related to their distances. And that was pretty much the whole story. However, in the last 10 years, astronomers have begun to have some serious doubts. Things don't look quite so simple. In fact, pessimistic astronomers might even say that the more they learn about galaxies, the more confused they become!

In the 1960s, astronomers began taking a good hard look at galaxies. When familiar objects, such as the galaxies in the *Messier Catalogue*, were scrutinized with great care, some strange discoveries emerged. For example, in the constellation of Ursa Major there is a fairly nearby galaxy called M 82. Photographs of M 82 in the light of the hydrogen atom (that is, with filters that transmit only the light produced by hydrogen atoms in the galaxy) reveal huge filaments of gas shooting out from the center of the galaxy. These filaments of gas extend to distances of more than 10,000 light years above and below the plane of the galaxy, as shown in Figure 13–13. The most distant material in these filaments seems to be moving away from the galaxy at speeds in excess of 1,000 miles per second. You don't have to be a professional astronomer to look at a photograph of M 82 and conclude that the entire galaxy seems to be blowing up! What could possibly be going on inside the M 82 to cause it to explode? No one knows.

M 82 is located in a small, nearby cluster of galaxies that also includes the beautiful spiral galaxy M 81. Close examination of photographs of M 81 (see Figure 13–15) reveals some straight, dark lines. What could possibly be the cause of these straight ripples that are roughly perpendicular to the direction to M 82?

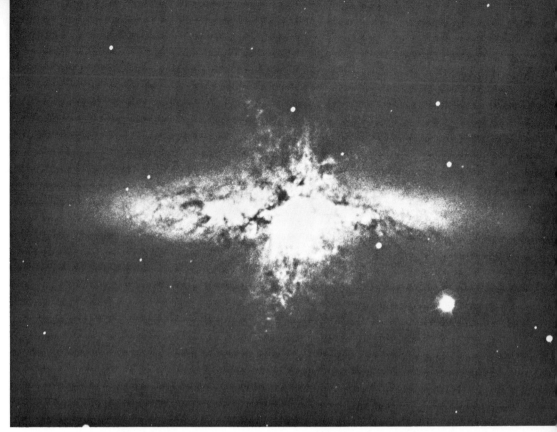

Figure 13–13. The Exploding Galaxy M 82. This galaxy is being blown apart by a huge explosion. Gas is being ejected at speeds over 1000 miles per second out to distances of tens of thousands of light years. (*Hale Observatories*)

Figure 13–14. M 81 and M 82. The exploding galaxy M 82 is in a small cluster of galaxies which includes the spiral, M 81. At first glance, M 81 probably looks very much like our own Galaxy. (*Hale Observatories*)

Figure 13–15. The Spiral Galaxy M 81. Close inspection of M 81 reveals some dark, straight lines on one side of the galaxy. Is it possible that these ripples in M 81 were caused by the explosion in M 82? (*Hale Observatories*)

Could they be *shock waves* from the explosion in M 82 now passing through M 81? Is it possible that the explosion in M 82 was so violent that it produced highly energetic shock waves capable of propagating through the sparce intergalactic medium to M 81? Was the explosion in M 82 so incredibly violent that its echos are now shaking up the guts of M 81? No one knows.

By no means is M 82 an exceptional case. In around 1970, the American astronomer Roger Lynds at Kitt Peak obtained a very fine photograph of NGC 1275 in the light of hydrogen, which is shown in Figure 13–16. This galaxy in the constellation of Perseus looks like the Crab nebula! Huge jets of gas are being ejected thousands of light years out into space at speeds of more than 1,000 miles per second.

NGC 1275 is the brightest galaxy in the Perseus cluster. In the early days of radio astronomy, this exploding galaxy was identified as the strong radio source called Perseus A. In the early 1970s, it was discovered that NGC 1275 is located at the center of an X-ray source, Per X-1, also called 3U0316 + 41 in the *Third Uhuru Catalogue*. In other words, not only is NGC 1275 in the process of blowing up, it also produces lots of radiowaves and X rays.

The fact that these exploding galaxies are often radio and X-ray sources is perhaps not too surprising. After all, if there is enough energy to blow the galaxy

Chapter 13 Quasars and Exotic Galaxies

apart, there should be enough energy to produce the full range of electromagnetic radiation. For example, M 82 produces radiowaves and is listed in the *Third Cambridge Catalogue* as 3C 231.

In certain cases, galaxies that look "normal" at first glance are discovered to be radio and X ray sources. Closer inspection then reveals that these galaxies are really very peculiar. For example, the giant elliptical galaxy M 87 (also called NGC 4486) in the Virgo cluster looks rather ordinary. Figure 12–15 shows a photograph of this galaxy. In the early 1950s, M 87 was discovered to be a strong radio source named Virgo A. Much more recently, it was discovered that M 87 is at the center of a huge X-ray source called 3U1228 + 12, and is more than half a

Figure 13–16. *The Exploding Galaxy NGC 1275.* As seen in this remarkable photograph by Dr. Lynds, NGC 1275 looks more like the Crab Nebula (a supernova remnant!) than a galaxy. This galaxy is a strong source of radio waves and X rays. (*Kitt Peak Observatory*)

Figure 13-17. The Core of M 87. This short time exposure of the cD elliptical galaxy M 87 reveals a huge jet of material surging up out of the galaxy's nucleus. (*Lick Observatory*)

million light years in diameter. A very short time exposure photograph of M 87 reveals a bright, starlike nucleus at the center of the galaxy out of which a huge jet of material appears to be surging (see Figure 13-17). For what it's worth, it should be noted that high-quality photographs of the quasar 3C 273 show a similar-looking jet of gas.

Astrophysicists are at a complete loss to explain the kinds of explosive events observed in galaxies such as M 82, NGC 1275, and M 87 in terms of traditional ideas. So people have started looking at nontraditional stuff. Back in around 1970, Drs. Lynden-Bell and Ryle presented the rather general arguments that whatever is going on at the centers of these peculiar galaxies must involve something that is (1) very massive and (2) very small. Furthermore, they showed that if this "something" is not a supermassive black hole (mass between a million and a billion suns), it certainly becomes one in a period of time that is short compared to the age of the universe. Even though there is not a single shred of observational evidence to prove that supermassive black holes exist at the centers of galaxies, the idea proposed by Lynden-Bell and Ryle is certainly a fascinating one.

As was learned in the discussion of black holes (see Section 11.4), a supermassive black hole would be expected to be rotating. It would be described by the Kerr solution and would be surrounded by an ergosphere. Suppose a globular cluster of stars in a highly elliptical orbit about the center of a galaxy should happen to pass through the ergosphere of one of these supermassive black holes. Some of the stars might get swallowed by the black hole while the remaining ones would get ejected back into space at a tremendous speed. This is a straightforward application of Penrose's suggestions of how energy could be extracted from rotating black holes. Shown schematically in Figure 13–18, this process has been called percolation through the ergosphere by Dr. E. C. Krupp. Could this be the mechanism responsible for the filaments and jets of gas observed surging out of the nuclei of such galaxies as M 82, NGC 1275, and M 87? No one knows.

In the late 1950s, the field of radio astronomy had progressed to such a degree that many of the brightest radio sources could be resolved into "components." In other words, it was discovered that certain big, fuzzy sources actually consisted of several smaller, more compact sources very close together. The first object to be resolved in this fashion was Cygnus A, one of the strongest radio sources in the entire sky. In 1956, Cygnus A was resolved into two bright components separated by 85 seconds of arc. A detailed radio map of Cygnus A obtained in the late 1960s by S. Mitton and M. Ryle is shown in Figure 13–19. The optical object emits no radio noise and, conversely, the radio objects do not show up on visible photographs.

Cygnus A is an excellent example of similar objects discovered in the 1960s and 1970s called *double radio sources*. In all cases, strong radio noise is observed coming from two "blobs" located on either side of a peculiar-looking galaxy. For example, late in 1974, the British radio astronomers G. C. Pooley and S. N.

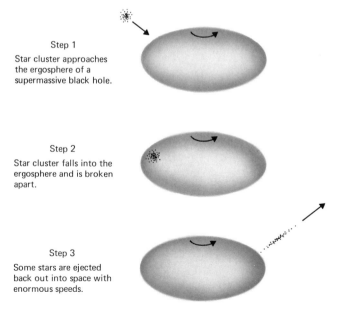

Figure 13–18. "Percolation through the Ergosphere." If the nucleus of a galaxy contains supermassive, rotating black holes, objects passing through the ergosphere of this black hole might be ejected back out into space with enormous speeds.

Step 1
Star cluster approaches the ergosphere of a supermassive black hole.

Step 2
Star cluster falls into the ergosphere and is broken apart.

Step 3
Some stars are ejected back out into space with enormous speeds.

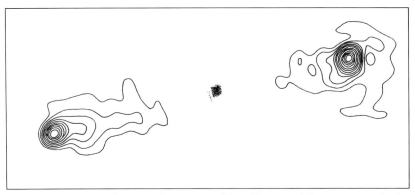

Figure 13–19. A Radio Map of Cygnus A. This radio map of Cygnus A reveals the characteristic "dumbbell" structure of many radio sources associated with peculiar galaxies. (*Adapted from Drs. Mitton and Ryle*)

Henbest at Cambridge published their results of observations of four dozen extragalactic radio sources made at the Mullard Radio Astronomy Observatory. In virtually all cases, their radio maps (see Figure 13–19 for a typical example) revealed this now-familiar appearance of a double radio source approximately centered about an unusual galaxy.

In several interesting cases, radio astronomers have found that the structure of these unusual radio sources is extremely complex, consisting of more than just the simple classical double sources. For example, Centaurus A is one of the brightest radio objects in the southern sky. It consists of two radio sources located on either side of the very strange-looking galaxy NGC 5128. Close examination with high resolution, however, reveals a second pair of radio sources inside the optical image of NGC 5128. Figures 13–21 and 13–22 shows a radio map and optical photograph of Centaurus A.

Back in the 1960s, it was popular to assume that objects like Cygnus A and Centaurus A were colliding galaxies. After all, the optical image of Cygnus A does indeed look like two galaxies, and perhaps Centaurus A is a spiral galaxy seen, edge on, passing through an elliptical galaxy. Such speculations are wrong! First of all, galaxies are so few and far between that collisions would be extremely rare. Secondly, simple collisions between galaxies would not produce the double or multiple radio sources that are located so very far from the optical object.

In the mid-1970s, two very different explanations were put forward to account for double radio sources. In 1974, the British astrophysicists R. D. Blandford and M. J. Rees at Cambridge proposed a "twin-exhaust" model whereby violent

Figure 13–20. The Peculiar Galaxy in Cygnus A. This unusual galaxy is located exactly between the two components of the radio source Cygnus A. Perhaps an explosion occurred long ago inside this object, ejecting two "blobs" in opposite directions which we now see as the radio sources. (*Hale Observatories*)

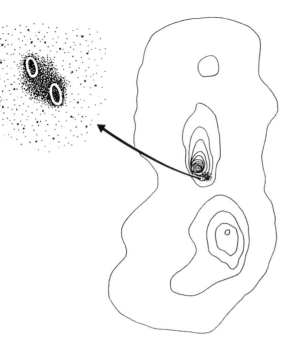

Figure 13–21. *A Radio Map of the Centaurus A.* This strong radio source consists of two sources on either side of a peculiar galaxy. Upon close examination of Centaurus A, radio astronomers discovered another pair of sources inside the visible galaxy. (*Adapted from Dr. Matthews*)

Figure 13–22. *The Exploding Galaxy NGC 5128.* This is the exploding galaxy responsible for the radio source Centaurus A. The distorted appearance of this galaxy clearly suggests violent events occurring at its center. (*Hale Observatories*)

processes in the nuclei of galaxies eject huge amounts of gas at very high speeds in opposite directions. This supersonic flow of hot gas (velocities near half the speed of light and temperatures of a 100 million degrees) produces two blobs on either side of the active galaxy that show up as two radio sources.

A second, alternative explanation was proposed in 1975 by M. J. Rees and W. C. Saslaw. In this "slingshot" model, supermassive black holes are violently ejected in opposite directions out of the nucleus of an active or exploding galaxy. These black holes are expected to have masses between a million and 100 million suns, and would be moving at speeds of roughly 10,000 miles per second. Similar to the case of Cygnus X-1 (see Section 11.5), these black holes would be surrounded by accretion disks that emit the radio noise astronomers observe.

It should be noted that both the twin-exhaust and slingshot models are so new that no one has performed any observations that might tell which model is correct. Maybe both models are wrong. They are mentioned here only as examples of the kinds of ideas that are floating around in the company of astrophysicists today.

The discovery of violent processes in the nuclei of galaxies tends to give credence to those astronomers who believe that quasars are at the incredible distances suggested by their redshifts. Recall from the previous section that if quasars really are located at distances of up to 10 and 15 billion light years, then astronomers are faced with a severe "energy problem." These quasars must then be emitting hundreds of times more energy than the brightest known galaxies. But with the discovery of exploding galaxies (M 82, NGC 1275, and the like), as well as of double and multiple radio sources, it is now accepted that extremely violent processes do, in fact, occur at the centers of galaxies. These processes must involve huge sources of energy that are not understood at the present time. Perhaps similar sources of energy are responsible for the mechanisms by which quasars shine so brightly.

This line of thinking is further supported by the discovery of *Seyfert* and *N-type galaxies*. Seyfert galaxies (named after their discoverer, Carl Seyfert) and N-type galaxies (*N* stands for "nucleus") have extremely bright nuclei. M 77 (also called NGC 1068), shown in Figure 13–23, is an excellent example of a Seyfert galaxy. The center of this galaxy is so bright that on a short-exposure photograph M 77 would look almost like a star. Similarly, if M 77 were extremely far away, it would look like a star—all that could be seen is the bright nucleus of the galaxy. The spiral arms of M 77 would be too faint to be detected.

The list of objects called Seyfert galaxies seems to be getting bigger every day. A Seyfert galaxy is any galaxy with strong, broad emission lines in its spectrum and a small, bright nucleus. All N-type galaxies may be Seyferts. *Compact galaxies* are probably Seyferts. M 82 and NGC 1275 are often called Seyfert galaxies. Even the quasar 3C 120 has been classified as a Seyfert.

Another interesting feature of Seyfert galaxies is that their bright nuclei exhibit rapid light variations. Seyfert galaxies such as NGC 4151 change brightness over a few months or less. This means that the bright, starlike nucleus in Seyfert and N-type galaxies is smaller than a light year in diameter.

Figure 13–23. The Seyfert Galaxy M 77. Seyfert galaxies, such as the one shown here, have very bright nuclei. If M 77 were very far away, it would give a star-like appearance because its center is so very much brighter than its spiral arms. (*Lick Observatory*)

Could it be that Seyfert galaxies are the missing link between quasars and ordinary galaxies? After all, remote Seyfert galaxies would probably look very much like quasars. they would have large redshifts (due to their extreme distances) and would show up on photographs as bright stars exhibiting rapid light fluctuations.

Always remember that looking into space is the same as looking backward in time. If the cosmological interpretation of the redshifts of galaxies and quasars is correct, then the bigger the redshift, the more ancient is the object. Quasars, which have the largest redshifts, must then be the most ancient objects we see in the sky. The light that we are now receiving from the quasars started on its journey billions of years ago. When we look at a quasar we are seeing the way they appeared billions of years ago. Could it be that quasars are just *very* young galaxies? Perhaps when galaxies are "born" they have extremely bright nuclei. Later, as they grow older, they develop spiral arms. As the spiral arms begin to develop, they look like Seyfert and N-type galaxies. Finally, when the violent processes in their nuclei have quieted down and the spiral arms are well developed, they take on the appearance of normal galaxies such as our own. In support of this line of thinking, Seyfert galaxies often have moderately high redshifts, frequently between the very high redshifts of quasars and the lower redshifts of the nearer galaxies. This means that the Seyfert galaxies have ages roughly between the ages of the quasars and the nearby galaxies. In addition, some of the quasars seem to have a fuzzy appearance, as though they are in the process of becoming Seyfert galaxies.

As we look out into space, we are really looking back into the past. Looking to higher and higher redshifts (and thus presumably to greater and greater distances), astronomers might actually be seeing the life cycles of galaxies. This is the general direction in which astronomy seems to be heading in the mid-1970s. But many questions still remain. Although a coherent picture is forming whereby quasars evolve into Seyfert galaxies, which turn into normal galaxies, the energy problem still remains. Even if quasars are really newborn galaxies, astrophysicists are still at a loss to explain why they are so very bright.

Finally, it should be emphasized that everything hinges on the Hubble law. The Hubble law works well for ordinary galaxies. Nearby galaxies have low redshifts and are moving away from us slowly. Distant galaxies have higher redshifts and are moving away more rapidly. But should we believe that the enormous redshifts of quasars mean that they are incredibly remote? Traditional astronomers have a lot of faith in the Hubble law. Their answer is an unqualified "Yes!" But certain observations made in the early 1970s suggest that the Hubble law does not tell the whole story. If this is so, then the high redshifts of the quasars are not necessarily due to their distances, but rather to "something else." If this is so, then a lot of astronomy is in big trouble.

13.4 Intrinsic Redshifts and New Ideas

There are about seven thousand galaxies in the NGC. There are about sixty quasars in the 3C catalogue. Suppose you took a salt shaker filled with 7,000 grains of salt and a pepper shaker containing 60 grains of pepper and sprinkled all this salt and pepper all over a huge map of the sky. You could then ask this question: "What is the probability that a grain of salt lies within 1 millimeter of a grain of pepper?" A mathematician familiar with statistics probably would reply something like "One chance in a thousand."

The seven thousand NGC galaxies and the sixty 3C quasars are sprinkled randomly around the sky like these grains of salt and pepper. In the early 1970s, astronomers such as G. R. Burbidge began asking questions like "What is the probability that a NGC galaxy lies within 7 minutes of arc from a 3C quasar?"

There are five such pairs of galaxies and quasars. Five NGC galaxies are extremely close to five 3C quasars. They are listed in Table 13-1.

Table 13-1.

Galaxy-Quasar Pairs (NGC and 3C Names)	Separation (in minutes of arc)
3C 455 / NGC 7413	0.4
3C 232 / NGC 3067	1.9
3C 268.4 / NGC 4138	2.9
3C 275.1 / NGC 4651	3.5
3C 309.1 / NGC 5832	6.2

The probability of five such "close associations" is very small, perhaps as small as one chance in ten thousand!

In the five galaxy-quasar pairs listed in Table 13-1, the redshifts of the galaxies are all very low, a few per cent of the speed of light or less. Yet, the redshifts of the quasars are all very high; all have redshifts corresponding to velocities *greater* than half the speed of light.

What could be the meaning of this? Is it just sheer luck that these five galaxies and quasars appear so close together in the sky? Or is nature trying to tell us something?

The overwhelming majority of astronomers feel that not much attention should be paid to these "associations." These pairs of galaxies and quasars look close together simply because they lie very nearly along the same line of sight. After all, everyone knows from the Hubble law that the low redshift galaxies must be nearby, and the high redshift quasars must be extremely remote. As far as the "one chance in ten thousand" is concerned . . . well, almost anything can be proved with statistics.

But there are a few astronomers who are not so sure that the majority opinion is right. Astronomers such as G. R. Burbidge, F. Hoyle, and H. Arp seriously consider the possibility that these five pairs of quasars and galaxies look close together because they really *are* close together. To the traditional astronomer, such an opinion is heresy. This would mean that the quasars are *not* at the distances implied by their redshifts. This would mean that the quasars *violate* the Hubble law. This would mean that quasars have *intrinsic redshifts.*

The idea that quasars are not at incredibly remote distances (inferred from their redshifts) means that the energy problem is far less severe (of course, there still is a problem with double radio sources and exploding galaxies). But the

Chapter 13 Quasars and Exotic Galaxies

NGC 3067, 3C 232

NGC 4138, 3C 268.4

NGC 4651, 3C 275.1

NGC 5832, 3C 309.1

Figure 13–24. Four "Close Associations." Four "pairs" of 3C quasars and NGC galaxies are shown here. Some astronomers believe that such "close associations" of galaxies and quasars are highly improbable. (*Courtesy of Dr. Burbidge*)

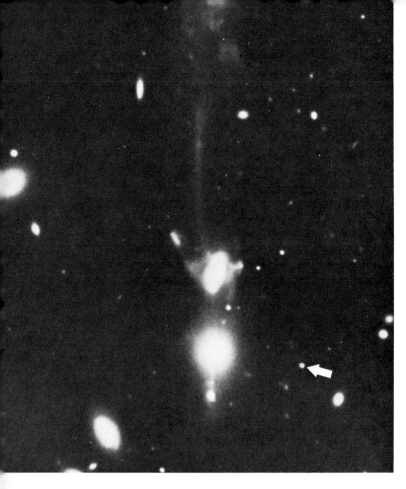

Figure 13-25. *NGC 3561 and a Quasar.* A quasar with a very high redshift is located near the exploding galaxy NGC 3561. Is this a "chance alignment" between a nearby peculiar galaxy and a remote quasar? (*Courtesy of Dr. Arp, Hale Observatories*)

price astronomers must pay if the quasars are nearby is very high. They are then faced with an even more puzzling redshift problem. If the quasars are comparatively nearby, their redshifts cannot be due to their speeds. Quasars must then possess *nonvelocity redshifts*. As was learned in Section 13.2, the only known traditional cause of nonvelocity, or intrinsic, redshifts is the gravitational redshift in general relativity. Yet, after more than a decade of strenuous efforts, astrophysicists have been unsuccessful in all attempts to construct theoretical models of objects with large gravitational redshifts. As soon as an astrophysicist mathematically constructs an object with a high gravitational redshift, he finds that it is unstable against gravitational collapse and very rapidly turns into a black hole. Of course, maybe someday someone will figure out how to make the gravitational redshift work; it's the only remaining hope for a traditional explanation for intrinsic redshifts of quasars. But as of 1975, all such hopes and expectations look grim. Astronomers who believe in the nonvelocity redshifts of quasars, therefore, are led to believe that they must really be searching for new laws of physics. They suggest that quasars will be understood only after astrophysicists have gained a much deeper insight into the nature of physical reality.

Most astronomers think that all this speculation is just a waste of time. They say they know the laws of physics and quasars are where they belong: way out there at the distances given by their redshifts. Therefore, a violent debate over the nature of quasar redshifts is now going on among professional astronomers.

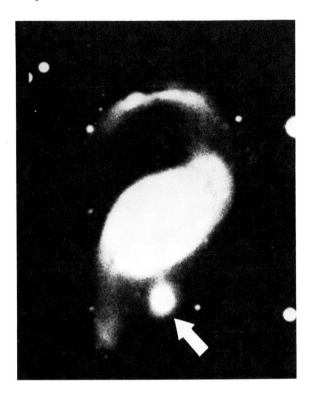

Figure 13–26. NGC 4319 and Markarian 205. This remarkable photograph taken by Dr. Arp with the 200-inch telescope seems to show a bridge of gas connecting the low redshift galaxy and the high redshift quasar. (*Courtesy of Dr. Arp, Hale Observatories*)

At the risk of severe criticism by his colleagues, this writer will outline a series of observations and theories that are alternately referred to by others as garbage, insanity, _____* astrology, the forces of darkness, and heresy.

The central problem is "where are the quasars?" Are their redshifts cosmological? Or are their redshifts intrinsic? A courageous pioneer and crusader of the minority opinion is Dr. Halton Arp at the Hale Observatories. Dr. Arp takes the position that the close association of the five pairs of quasars and galaxies mentioned is *not* due to chance. In support of this position, in 1970, he discovered a galaxy and a quasar that are so close together they seem to be connected! They are the galaxy NGC 4319 and the quasar Markarian 205 (number 205 in a list of starlike objects with strong ultraviolet excess compiled by the Soviet astronomer B. E. Markarian). The galaxy NGC 4319 has a redshift corresponding to a speed of 1,100 miles per second, whereas the quasar Markarian 205 has a redshift corresponding to 13,000 miles per second. The redshift of the quasar is, therefore, more than ten times in excess of the galaxy's redshift.

An excellent photograph of these two objects is shown in Figure 13–26. His observations show a "bridge" connecting the galaxy and the quasar, so Dr. Arp concludes that Markarian 205 was actually ejected from NGC 4319.

Dr. Arp's position seems to be further strengthened by the discovery of

*Expletive deleted.

Figure 13–27. Seyfert's Sextet. Five of the galaxies in this cluster have low redshifts. The sixth (marked with an arrow) has a high redshift. (*Hale Observatories*)

discrepant redshifts. This term usually refers to the situation in which an astronomer finds a small cluster of galaxies and then discovers that one of them has a redshift that is *very* different from the redshifts of all the other galaxies in the cluster. A good example of this is Seyfert's sextet, a cluster of six galaxies in the constellation of Serpens. A photograph of Seyfert's sextet is shown in Figure 13–27. Five of these galaxies have redshifts corresponding to speeds of about 2,800 miles per second. The sixth galaxy (marked with an arrow) has a redshift corresponding to 12,000 miles per second. The majority opinion is that Seyfert's sextet is actually a cluster of *five* galaxies superimposed on a sixth, more distant galaxy. This traditional viewpoint argues that since the sixth galaxy has a redshift four times larger than the other five galaxies, it must be four times more distant. Arp, Hoyle, and Burbidge and Company do not agree.

Another good example of discrepant redshifts is the "chain" of galaxies called VV 172 (number 172 in a list of objects in a catalogue prepared by the Soviet astronomer Vorontsov-Velyaminov) shown in Figure 13–28. Four of these galaxies have redshifts corresponding to 9,800 miles per second. The fifth galaxy (marked with an arrow) has a much higher redshift of 22,000 miles per second. Looking at the photograph of VV 172, it seems quite reasonable to suppose that all five galaxies are located very near each other. Traditional astronomers argue that the high redshift galaxy is twice as far away as the other four galaxies. From the alignment of these five galaxies, Arp, Hoyle, and others argue that the high redshift galaxy possesses an intrinsic redshift.

Obviously, the arguments concerning systems such as Seyfert's sextet and VV 172 would be settled once and for all if astronomers could somehow independently measure the distances to each and every galaxy in these strange clusters. Unfortunately, the clusters in question are too far away for the usual reliable methods (Cepheid variables, globular clusters, and so on).

In the early 1970s, Dr. Arp began making detailed observations of a cluster of five galaxies called Stephan's quintet. The galaxies in Stephan's quintet, shown in Figure 13–29, have the NGC numbers 7317, 7318A, 7318B, 7319, and 7320. The first four galaxies have redshifts of roughly 4,000 miles per second. NGC 7320 has a much lower redshift, corresponding to 500 miles per second. Thus, NGC 7320 has a redshift one eighth as far away as the other four galaxies. The party line is that Stephan's quintet consists of a nearby galaxy superimposed against a background cluster of four distant galaxies.

In the early 1970s, Arp discovered HII regions in two of the five galaxies in Stephan's quintet. From Section 12.3, recall that HII regions are hot clouds of glowing hydrogen gas like the Orion nebula. Assuming that all HII regions are

Figure 13–28. VV 172. Four of the galaxies in this "chain" have low redshifts. But the fifth galaxy (one from the left) has a very high redshift. (*Hale Observatories*)

Figure 13–29. Stephan's Quintet. Four of the galaxies in this cluster have high redshifts. The fifth galaxy, NGC 7320 to the lower left of the photograph, has a low redshift. Dr. Arp discovered that the HII regions in both the low and high redshift galaxies have very nearly the same apparent sizes. (*Lick Observatory*)

roughly the same size, they can be used as distance indicators. This, again, is like saying "All cars are the same size." Of course, there are Volkswagens and there are Cadillacs. However, it is fairly safe to say that "All cars are about 12 feet long."

Arp has found HII regions in NGC 7320 (the low redshift galaxy) and in NGC 7318B (one of the high redshift galaxies). If traditional astronomers are right, then NGC 7318B must be eight times farther away from us than NGC 7320. In addition, if all HII regions are roughly the same physical size, then, on Arp's photographs, the HII regions in NGC 7320 should *look* eight times *bigger* than the HII regions in NGC 7318B. Dr. Arp carefully measured the sizes of all the HII regions on his photographic plates. In 1973, he announced his remarkable discovery: *all* the HII regions he could find in *both* galaxies have the same apparent sizes on his photographs. Since he assumes that all HII regions have the same physical sizes (for example, they have the same diameters measured in miles), it logically follows that NGC 7320 and NGC 7318B are at the *same* distance from the earth! Traditional astronomers have a very difficult time arguing away Arp's observations of HII regions in Stephan's quintet. Indeed, Arp's observa-

Chapter 13 Quasars and Exotic Galaxies

tions constitute powerful evidence for discrepant redshifts. Arp feels that NGC 7320 is probably at the distance indicated by its redshift (about 50 million light years), and that the other four galaxies in Stephan's quintet are also at this distance. These other four galaxies must have high intrinsic, or nonvelocity, redshifts. In this regard, it is interesting to note (see Figure 13–29) that the four high redshift galaxies seem to have bright nuclei and exhibit almost a starlike appearance.

Another troublesome thorn in the side of the traditional astronomer deals with the radio structure of certain quasars. During the early 1970s, radio astronomers succeeded in resolving some quasars into several *components*. In particular, quasars such as 3C 279, 3C 273, and 3C 120 were found to consist of several radio sources very close together. Patient and careful observations extending over many months revealed that these components are moving apart. The angular separation between radio components in these quasars keeps getting bigger and bigger. If the distance to one of these quasars is known, then the rate at which the components are moving apart can be translated into a speed. If the quasars

Figure 13–30. *The Quasars 1548+115a and 1548+115b.* These two quasars have very different redshifts but appear very close together in the sky. The separation between the quasars is only 5 seconds of arc. (*Courtesy of Dr. Gunn*)

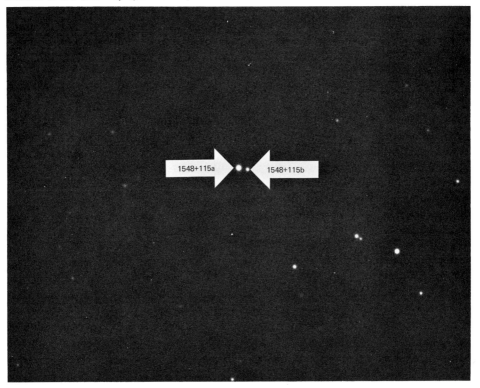

are nearby, the speed at which the blobs are separating must be low. If the quasars are far away, the speed must be high. This is perfectly analogous to the situation in which a nearby airplane appears to be moving very rapidly past the line of sight, while a very distant airplane high in the sky appears to be moving very slowly. If we assume that the quasars 3C 279, 3C 273, and 3C 120 are, in fact, at the very great distances indicated by their redshifts, then the speeds at which the radio components are separating have to be erroneous. The radio blobs are moving apart at velocities as high as 10 light years per year! That's *ten times* the speed of light. That's impossible!

Finally, every once in a while someone proposes some observations that are supposed to "test" whether or not quasars are at the distances indicated by their redshifts. An example is the ingenious observations proposed by J. R. Gott and J. E. Gunn at Caltech involving the two quasars, 1548 + 115a and 1548 + 115b. These two quasars appear *extremely* close together in the sky. They are separated by only 5 seconds of arc (for comparison, the angular diameter of Uranus as seen from the earth during opposition is about 4 seconds of arc). Although they appear close together, they have very different redshifts. The quasar 1548 + 115a has an apparent magnitude of 17 and a redshift corresponding to 35 per cent of the speed of light. The quasar 1548 + 115b has an apparent magnitude of 19 and a redshift corresponding to 79 per cent of the speed of light. Since 1548 + 115b is dimmer and has a higher redshift than 1548 + 115a, Gott and Gunn argued that 1548 + 115b must be farther away than 1548 + 115a. However, Gott and Gunn realized that the small angular separation between these two quasars should allow them to test their traditional interpretation of the redshifts. Suppose these two quasars are, in fact, at the distances indicated by their redshifts. Also, suppose—for lack of anything better—that they are new born galaxies. This means that astronomers should be able to make a good guess about the masses of these quasars. Gott and Gunn make the reasonable assumption that the masses of these quasars are the same as the masses of typical giant elliptical galaxies: about 7 billion solar masses. In view of the assumed masses of these quasars, as the light from the distant quasar passes near the close quasar, the light should be deflected slightly, according to Einstein's general theory of relativity. This is perfectly analogous to beams of starlight being deflected as they pass near the surface of the sun, as discussed in Section 11.2. *However*, due to the almost perfect alignment of the two quasars, the light from the distant quasar (1548 + 115b) should be bent around *both* sides of the nearby quasar (1548 + 115a), as shown in Figure 13–31. This is called the *gravitational lens effect*. The nearby quasar should act as a gravitation lens for the light from the distant quasar. This means that we should see *two* images of the distant quasar, one on either side of the nearby quasar, as shown in Figure 13–31. The "primary image" of the distant quasar is the observed image 5 seconds of arc away from 1548 + 115a. It has an apparent magnitude of 19. Gott and Gunn predicted the existence of a "secondary image" on the other side of the nearby quasar. This secondary image should be about 2 or 3 seconds of arc on the other side of the

Chapter 13   Quasars and Exotic Galaxies   415

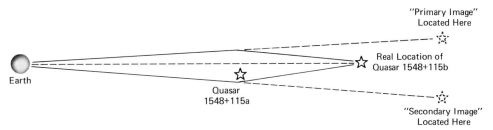

Figure 13-31. The Gravitation Lens Effect. The gravitational field of a nearby massive object focuses the light from a distant source. The geometry of this so-called "gravitational lens" in the case of the quasars 1548+115a and 1548+115b is shown here in detail. (*Adapted by Drs. Gunn and Gott*)

nearby quasar and should have an apparent magnitude of between 20 and 21. Gott and Gunn concluded that "identification of such a secondary image would verify the lens effect and indicate that the two quasars are seen in projection rather than being physically connected."* By late December 1974, in spite of careful searches, it was clear that *no* secondary image exists.

Things are up in the air. Astronomers battle back and forth in the journals. In many respects, the debate is reminiscent of Shapley and Curtis back in the 1920s. The issues will probably not be settled until a genius like Hubble comes along and is able to prove once and for all precisely where the quasars are.

While the astronomers are having a free-for-all, some theoretical astrophysicists have begun to consider the possibility that the Arp-Burbidge-Hoyle ideas of intrinsic redshifts might emerge victorious. If this proves true, a lot of astrophysicists are going to get caught with their pants down. To avoid such an embarrassing situation, some provocative theories have been proposed concerning the causes of intrinsic redshifts.

Imagine sitting on an old-fashioned piano stool, the kind that spins around. By lifting your feet from the floor and pushing on some nearby object you start spinning. You know that you are spinning because you see things passing over and over before your line of sight. As you move your arms in and out, your body experiences forces (called Coriolis and centrifugal forces). Indeed, if you continue this experiment for very long, you may even become quite nauseated. However, suppose that while you are spinning around on the piano stool, you blink your eyes and in that fraction of a second the *entire* universe disappears. Earth, the sun, the stars, and the galaxies completely vanish. Would you still be spinning? Would you still experience forces on your body as you pull your arms in and out? No! Because now there is nothing in reference to which you can describe your motion. The universe is empty and it no longer makes any sense to talk about how you are moving. This thought experiment is closely related to a

*"The Astrophysical Journal" vol. 190 (Univ. of Chicago Press) 1974, p. L105.

somewhat nebulous collection of ideas called *Mach's principle*, named after the nineteenth-century physicist Ernst Mach. In essence, Mach's principle states that the properties of matter somehow must be associated with the distribution of matter in the rest of the universe. In particular, a recent formulation of Mach's principle takes the position that the masses of atoms depend critically on the distribution of matter in the rest of the universe. If the matter in the rest of the universe were to disappear, the masses of the atoms in your body (if you were to stay behind in the empty universe) would go to zero. This is why you would no longer feel any Coriolis and centrifugal forces on your body if you were to pull your arms in and out on a spinning piano stool. The atoms in your body would not have any mass.

These revolutionary ideas were first proposed by the astrophysicists F. Hoyle and J. V. Narlikar in 1971. They take the viewpoint that the source of the masses of atoms lies in the distribution of matter in the rest of the universe. Through long-range interactions, the matter in the rest of the universe communicates its existence to each and every atom, telling it how much mass to have. As the distribution of matter in the universe changes, presumably the masses of protons, neutrons, and electrons also change.

Consider the possibility of newly created matter. Suppose, perhaps through white holes, new matter suddenly appears somewhere in the universe. The atomic particles that compose this matter initially would not know about the existence of the rest of the universe. The rest of the universe would not have had time to communicate its existence to this newly created material. The masses of the protons, neutrons, and electrons in this newly created matter would be very small. Only after many billions of years, after this newly created matter had had a chance to age, would the masses of atomic particles begin to behave like the ancient matter found here on earth. In other words, as newly created particles grow old, their masses get bigger and bigger.

The mass of the electron plays an important role in the structure of atoms. If atoms exist that have low-mass electrons, these atoms must be very large. Their allowed orbits would be spread out. As these low-mass electrons jump back and forth among allowed orbits, the wavelengths of light they emit and absorb would be much longer than usual. The light would be redshifted! Young, recently created matter would be made out of particles with low masses. The light from such atoms would possess an intrinsic redshift. As the matter ages, the masses of atomic particles increase, and this intrinsic redshift will gradually become less and less pronounced. Could this be the cause of the intrinsic redshifts that Arp claims to observe? When scientists look at quasars are they seeing the actual creation of matter? Do the masses of atomic particles really change with time, as predicted by the Hoyle-Narlikar theory? Although we will have to wait many years for definitive answers to these questions, one thing is clear. Either the Hoyle-Narlikar theory is one of the most brilliant works of genius in modern science, or it is a lot of nonsense reflecting the desperation and frustration of astronomers trying to deal with the mysteries of objects at the edge of the universe.

Questions and Exercises

1. Who were Karl Jansky and Grote Reber?
2. What is the *3C Catalogue?*
3. What is synchrotron radiation?
4. In 1960, astronomers obtained a photograph of 3C 48. What was so puzzling about this object?
5. What did astronomers learn from the lunar occultation of 3C 273?
6. Briefly describe what Dr. Schmidt did to resolve the mystery of the spectra of such objects as 3C 48 and 3C 273.
7. Name three general properties of quasars.
8. Are there quasars with redshifts greater than 90 per cent of the speed of light?
9. How and why do the colors of very high redshift quasars differ from the colors of most other quasars?
10. Why does it seem reasonable to rule out the gravitational redshift (from general relativity) as an explanation for the redshifts of quasars?
11. What is meant by the energy problem with quasars?
12. What is so unusual about the galaxy M 82?
13. Briefly describe the galaxy NGC 1275 in Perseus.
14. Present an argument in favor of the idea that there are supermassive black holes at the centers of galaxies.
15. How might a rotating, supermassive black hole at the center of a galaxy be responsible for "jets" of gas ejected from the nuclei of galaxies?
16. Briefly describe Cygnus A.
17. What is a Seyfert galaxy?
18. What is meant by intrinsic redshifts?
19. What are the traditional and nontraditional explanations of the apparent "close associations" of 3C quasars and NGC galaxies?
20. Briefly describe Dr. Arp's observations of HII regions in Stephan's quintet. What was he trying to discover by measuring the sizes of HII regions?
21. What is meant by the gravitational lens effect?
22. What is meant by Mach's principle?
23. Contrast and compare the Shapley-Curtis debate of the 1920s with the modern debate on the distances of the quasars.
24. Contrast and compare the dilemma of scientists 75 years ago facing the question "Why does the sun shine?" with scientists today facing the question "How do quasars shine?"

14

Cosmology and the Universe

14.1 The Expanding Universe

Since the dawn of recorded history, man has wondered about his place in the universe. Every civilization and every religion to come forth on our planet has had a cosmology at the core of its teachings. These cosmologies often were worded in mythical terms and frequently spoke of gods and demons, of men and heros, of battles between cosmic forces. Perhaps these mythical cosmologies reveal far more about the psychology of the people who devised the legends than about the true nature of physical reality.

With the birth of modern astronomy some 400 years ago, the edges of the known universe illusively receded from us with every new discovery. At the time of Newton, the universe was thought to consist of the solar system and stars. By 1785, Herschel proved that our sun was just one star in a huge disk-shaped collection of stars called the Milky Way Galaxy. The next major advance came 140 years later when Hubble succeeded in measuring the distances to the "nebulae," demonstrating that our Galaxy was just one of millions of galaxies scattered through space and time. With every new discovery, the limits of the universe were found to be farther away than anyone had ever before imagined.

The history of *modern* cosmology is incredibly recent. Modern ideas concerning the nature of the universe date back only to the early 1930s when E. Hubble and M. L. Humason at Mount Wilson Observatory announced the *Hubble law*. Based on observations of the distances and redshifts of galaxies, Hubble and Humason found that the nearby galaxies are moving away from us slowly, while the more remote galaxies are moving away at much higher speeds. The resulting linear relationship between the recessional speeds and distances of the galaxies is called the Hubble law. The relationship is said to be linear because on a graph of speed versus distance, all the data fall along a straight line, as shown in Figure 14–1. As was first discussed in Section 12.6, the Hubble law is a powerful tool in modern astronomy. The Hubble law can be used as a *distance indicator* whereby the redshifts of galaxies are assumed to correspond to specific distances in accordance with the graph in Figure 14–1.

But what is the *real* meaning of all this? *Why* do the nearby galaxies move away from us so slowly? And *why* do the distant galaxies seem to be rushing away so much more rapidly? Could it be that nature is trying to say something of profound importance? To answer these questions, imagine an old-fashioned kitchen where Grandma is baking a raisin cake.

To make her raisin cake (according to a famous recipe by Dr. George O. Abell at UCLA) Grandma takes some fresh dough, some yeast, and some raisins and mixes them all together. After thoroughly mixing the dough, yeast, and raisins, Grandma places the raisin cake on her kitchen table. Initially the raisin cake is 1 foot in diameter. Grandma goes back to her knitting while she waits for the dough to rise. Unknown to Grandma, however, there is a very clever bug sitting on "raisin A" inside the raisin cake. This bug measures the distances between his raisin and several other raisins in the cake. The bug then falls asleep. After a 1-hour nap, the bug wakes up and repeats his measurements. He finds, as shown in Figure 14–2, that all the raisins have moved away from him. "Raisin B," which was initially only 2 inches away, has moved to a distance of 4 inches; its speed is 2 inches per hour. "Raisin C," initially 4 inches away, is now 8 inches away; its speed is 4 inches per hour. Similarly, "raisin D" has gone from 6 inches to 12 inches and, therefore, has a speed of 6 inches per hour. And the most distant raisin, "raisin E," has moved from a distance of 8 inches to 16 inches; its speed is 8 inches per hour. None of the raisins came toward the bug; all had moved away. Furthermore, the nearby raisins had moved away from him slowly, whereas the more distant raisins moved away from him much more rapidly. In addition, this clever bug discovers that there is a linear relationship between the distances and the speeds of the raisins. He, therefore, correctly concludes that the raisin cake must be expanding. And he is right! Grandma returns to the kitchen after an

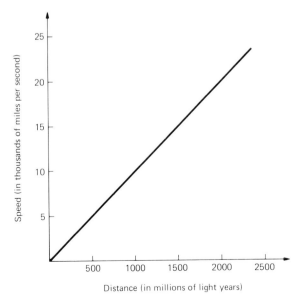

Figure 14–1. The Hubble Law. This graph illustrates the linear relationship between distance and recessional velocity for galaxies. Nearby galaxies are moving away from us slowly; more distant galaxies are moving away from us rapidly.

hour of knitting to find that the yeast has "done its thing." The raisin cake has doubled in size and is now 2 feet in diameter.

Aside from the fact that astronomers should obviously be kept out of the kitchen (What self-respecting Grandmother would make a 2-foot diameter raisin cake with bugs!?), this little story illustrates the true meaning of the Hubble law. Here we are on a galaxy inside the universe. Measuring the distances and speeds of other galaxies shows that the nearby galaxies are moving away slowly, while the more distant galaxies are moving away rapidly. The Hubble law is telling us that *the universe is expanding*. Beginning in the 1930s, mankind began to realize that we live in an expanding universe.

One of the most important questions in astronomy is "How fast is the universe expanding?" Although Hubble and Humason had the right idea, as methods of distance determination improved (that is, methods involving Cepheid variables, sizes of HII regions, bright stars, globular clusters, and so on), astronomers realized that Hubble had used some wrong numbers. All the galaxies Hubble had observed were actually farther away than he had thought. This means that the orientation, or *slope*, of the line in graphs illustrating the Hubble law (such as in Figure 14–1) gradually changed as methods of distance determination improved. As learned in Section 12.6, the slope of this line is given by the *Hubble constant*. The major thrust of observational cosmology in the last four decades has, therefore, been an accurate determination of this Hubble constant.

Beginning in 1974, Drs. A. Sandage and G. A. Tammann of Hale Observatories published a voluminous series of papers in six parts, all of which are entitled, "Steps Toward the Hubble Constant." These papers summarize years of careful and patient observations of hundreds of galaxies. In Part 1 of this classic series of papers, Sandage and Tammann pay special attention to eleven nearby galaxies. Five of these galaxies are in the Local Group and six are in a nearby cluster

Figure 14–2. The *"Raisin Cake."* From the analogy of measuring distances between raisins in a raisin cake (see text for discussion), it is seen that the true meaning of the Hubble law is that the universe is expanding.

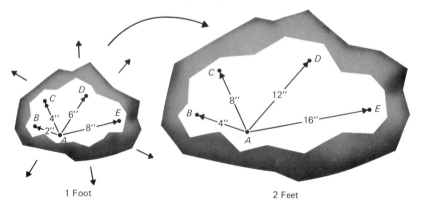

JUNE 9, 1950   FEB. 7, 1951

Figure 14-3. *A Supernova in M 101.* A supernova was observed in the spiral galaxy M 101. Supernova in distant galaxies are used as "distance indicators." (*Hale Observatories*)

that contains M 81 in Ursa Major and NGC 2403 in Camelopardus. The distances to all eleven galaxies are well known from the Cepheid variables and the period-luminosity relation (see Section 12.3). Sandage and Tammann carefully measured the sizes of HII regions in all eleven galaxies. In Part 2, Sandage and Tammann again focused their attention on these same eleven galaxies, but this time they carefully measured the magnitudes of the brightest stars they could find. Since the distances to all eleven galaxies are known, they were able to calculate the real physical sizes of the HII regions and the real, absolute magnitudes of the brightest stars. Using this information, in Part 3 Sandage and Tammann moved on to a more distant cluster of galaxies in Ursa Major that contains M 101. By measuring the apparent magnitudes of the brightest stars and apparent sizes of HII regions in six galaxies in the M 101 group, a distance of 23 million light years was determined. They checked and rechecked their results against other, independent distance indicators. For example, the supernova that appeared in M 101 in mid-1950 (see Figure 14-3) was an important check of their distance measurements. In Part 4, Sandage and Tammann went on to measure the distance to thirty-nine spiral galaxies from the sizes of the HII regions they contained. The whole point of all this work was that from these thirty-nine galaxies and from the galaxies in the Local Group and the M 81–NGC 2403 cluster, Sandage and Tammann were able to determine the average absolute magnitudes of various types of galaxies, in general.

In any cluster of galaxies, each galaxy is moving in a different direction as they all revolve around their common center. If the cluster is very near, this random motion of individual galaxies is so large that the cosmological redshifts due to the expansion of the universe are virtually unnoticeable. Each and every galaxy moving in all different directions in a nearby cluster completely masks the effects of the expansion of the universe. The cosmological redshift is only apparent in *very* distant clusters of galaxies. In clusters at distances greater than several hundred million light years, the cosmological redshifts are very large compared to the Doppler shifts of individual galaxies moving around inside these clusters. These are the clusters for which Sandage and Tammann were aiming. Only at these remote distances does the cosmological redshift stand out. Only by comparing the redshifts and distances for the remote clusters can astronomers hope to obtain the true value of the Hubble constant. Unfortunately, these galaxies are so far away that there is no hope of seeing any HII regions, bright stars, or Cepheids. The best astronomers can do is identify the kind of galaxy at which they are looking. They can tell the difference between an elliptical galaxy and a spiral galaxy from tiny, faint images on a photographic plate. But they cannot see any individual stars or HII regions.

In their first four papers, Sandage and Tammann succeeded in accurately determining the absolute magnitudes of various types of galaxies. In the final two papers (Parts 5 and 6), they applied these known absolute magnitudes to remote clusters of galaxies. By comparing the apparent brightness of certain types of distant spiral galaxies with the known absolute brightness, they were able to obtain accurate distances. By comparing the distances of these remote galaxies with their redshifts, Sandage and Tammann arrived at the most reliable value of the Hubble constant ever determined: 10 miles per second per million light years.

The importance of an accurate determination of the Hubble constant cannot be overemphasized. It determines the age, the size, and the scale of the universe. It shows precisely how fast the universe is expanding and when this expansion started. From Sandage's value of the Hubble constant, astrophysicists conclude that the universe was "created" about 20 billion years ago, if it was created at all. If the universe has a size, it must have dimensions of about 20 billion light years.

Finally, it should be emphasized that the work of astronomers like Sandage epitomizes the traditional approach. The redshifts of distant galaxies are cosmological and directly reflect the expansion of the universe. Period! All this talk about "intrinsic redshifts" is nonsense. Period! Nevertheless, Sandage and Tammann were careful to restrict their observations to such noncontroversial objects as the M 33 spiral galaxy. No quasars or peculiar galaxies were used in their determination of the Hubble constant. In addition, independent arguments concerning the age and size of the universe presented late in 1974 by J. R. Gott and J. E. Gunn at Caltech and D. N. Schramm and B. M. Tinsley at the University of Texas strongly support the value of the Hubble constant obtained by Sandage and Tammann. Finally, four decades after Hubble first discovered the cosmological redshifts of the galaxies, we can now feel confident that we know precisely how fast the universe is expanding.

14.2 The Big-Bang, Steady-State, and "Other" Cosmologies

Hardly anyone quarrels with the idea that the universe is expanding. The Hubble law is a well-established fact of modern astronomy (even though problems do exist with peculiar galaxies and quasars), and from analogies such as Grandma's raisin cake, the obvious interpretation of the Hubble law is that we live in an expanding universe. The distances between clusters of galaxies gradually are getting bigger and bigger.

But what was the universe like 10, 15, or even 20 billion years ago? Thinking backward in time, we realize that far in the past all the galaxies in the universe must have been closer together than they are today. In fact, if we think back far enough, there must have been a time when all the galaxies were piled on top of each other. Since the Hubble law tells how fast the universe is expanding, from the Hubble constant (which gives the orientation, or slope, of the line in Figure 14–1) astronomers can estimate how far into the past they must go in order to arrive at the time when all the galaxies were lumped together. As shown in the previous section, this *Hubble age* of the universe turns out to be 20 billion years, using the value of the Hubble constant as determined by Sandage and Tammann. In other words, although the galaxies today are separated by millions and billions of light years, 20 billion years ago the galaxies were not separated at all. The entire universe as well as all the matter and energy it contains was confined into a region of zero volume. The universe was a *singularity* 20 billion years ago. For some reason, a stupendous explosion must have occurred at this ancient time that caused the universe to begin expanding. This was the "creation event." This was the so-called *big bang*.

This *big-bang cosmology* is the simplest and most straightforward model of the universe astronomers can think of. For reasons that will be discussed in the next few pages, it is the currently accepted model of the universe. Some 20 billion years ago, a *primordial* explosion occurred that started the universe expanding. This expansion is still going on today, as demonstrated by the Hubble law.

The big bang is the simplest of all cosmological models, but it is possible to think of alternatives that do *not* include a creation event. For example, back in 1948, astrophysicists F. Hoyle, T. Gold, and H. Bondi proposed the *steady-state cosmology.* Everyone agrees that the universe is expanding. No argument there. But, suppose that, as the galaxies get farther and farther apart, *new* galaxies are created in the empty space left behind. These new galaxies would be created at a rate such that the universe would seem to be unchanging. Oh, surely, if you came back to the earth 10 billion years in the future you would find new people, new cities, new stars, and new galaxies in the sky. But, *on the average,* inside a gigantic cube a billion light years on a side, you would find the *same* total amount of matter. Inside a large volume of space, you would find roughly the same number of stars and galaxies then as now. For this reason, astronomers say that the universe is in a steady state. The over-all, average properties of the universe remain unchanged with time.

Just as the average number of galaxies inside a particular volume of the universe remain unchanged into the future, this average density of matter remains unchanged into the past. In the steady-state theory, as we think back into the past, we do *not* find the galaxies piled on top of each other. As we think further and further back into the past, galaxies seem to disappear; the farther we go back into the past, the fewer galaxies there are. They have not been created yet. Thus, the steady-state theory does not have a singularity. It does not have a creation event. The expanding universe always has existed and always will exist.

This points up an important difference between the big-bang and steady-state cosmologies. In the big-bang theory, the universe is evolving; it is changing. If we live in a big-bang universe, then 10 billion years ago all the galaxies in the sky must have been much nearer than they are today. Ten billion years into the future, all the galaxies in the sky will be much farther away than they are today. On the other hand, if we live in a steady-state universe, then any speculation of either 50 billion years ago *or* 50 billion years in the future keeps the universe looking roughly the *same.* The big-bang universe is evolving. The steady-state universe is *not* evolving. Figure 14-4 contrasts the big-bang and steady-state theories schematically.

Always remember that when looking into space we are really looking backward in time. Looking at galaxies with high redshifts, we are seeing how things were in the distant past. In this regard, quasars provide an important *cosmological test* of these two very different theories. Suppose that the traditional astronomers are correct: the redshifts of quasars are cosmological. The quasars are at the incredibly remote distances indicated by these huge redshifts. We then notice that *all* quasars have high redshifts; there are no quasars with very small redshifts. This means that there are lots of quasars far away, but *no* quasars nearby. Thus, long ago, there must have been lots of quasars in the universe. Today there are no quasars around. Thus, the universe *is* changing. Long ago the universe must have looked very different than it does today. *If* the cosmological interpretation of the redshifts of quasars is correct, then we must live in an evolving universe. The steady-state theory must, therefore, be wrong.

In the early 1970s, Hoyle made a noble attempt to salvage the steady-state cosmology. If the redshifts of quasars are intrinsic rather than cosmological, then the quasars can be nearby, not far away in the distant past. Furthermore, if

Chapter 14 Cosmology and the Universe   425

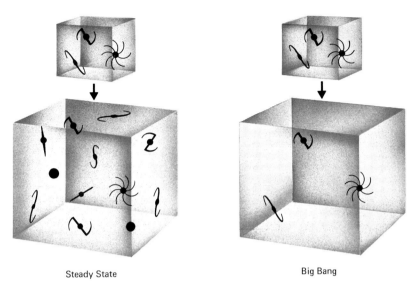

Steady State   Big Bang

Figure 14–4. *The Big Bang vs. The Steady State.* The big-bang and steady-state cosmologies are contrasted schematically in this diagram. In the big-bang cosmology, the universe simply expands from an initial singularity (the "creation event"). In the steady-state cosmology, new galaxies are created as the universe expands.

the Hoyle-Narlikar theory is correct (see Section 13.4), then matter is being created in quasars. Perhaps quasars show precisely how new galaxies are created in the empty space left behind as old galaxies get farther and farther apart! Although this is an ingenious idea, there are two other severe objections to the steady-state theory that have nothing to do with quasars at all.

Suppose a rock is thrown up into the air. Three things can happen. First of all, the rock can simply go up and come back down. The speed of the rock is *less* than the escape velocity of the earth; the strength of the gravitational field of the earth pulls the rock back down to the ground. Secondly, a rock could be thrown up into the air with the aid of a rocket ship. By putting just enough fuel in the rocket ship (so that the rock would achieve a speed equal to the escape velocity from the earth), the rock would have barely enough speed so that it would never fall back. Very far from the earth, the rock would eventually slow down and come to rest, never to return. Finally, you could throw the rock into the air with the aid of a very powerful rocketship containing a large amount of fuel. The speed of the rock would be higher than the escape velocity from the earth. Not only would the rock never return, even infinitely far from the earth, the rock always would be moving away at a high speed.

An analogous situation exists in the big-bang cosmology. There are three possibilities here. First of all, if the big bang were comparatively weak, if the big bang did not have very much energy, then the galaxies would go "up" and the galaxies would come back "down." Some day, far in the future, the expansion of the universe will stop and, after reaching a maximum size, the universe will begin collapsing in on itself. The redshifts of galaxies will turn into blueshifts. Finally, perhaps 100 billion years in the future, there will be another big bang.

The second possibility is that perhaps there was just enough energy in the primordial explosion such that the galaxies managed to achieve the escape velocity from each other. In this case, the universe will never "turn around" and start collapsing in on itself. Rather, infinitely far into the future, when the galaxies are infinitely far apart, the galaxies will finally stop rushing away from us.

The third possibility is that there was so much energy in the initial big bang that the galaxies have speeds far above the velocities needed for them to escape from each other's mutual gravitational attraction. Ever infinitely far into the future, they will always be rushing away from each other. The galaxies will always have substantial redshifts and the universe will always be expanding.

How are astronomers to distinguish between each type of big-bang universe? Do we live in a universe that will reach a maximum size and then start contracting? Do we live in a universe that will stop expanding only after the galaxies are infinitely far apart? Or do we live in a universe that will go on expanding forever, no matter how far away the galaxies get? It all depends on how much energy there was in the initial big bang. Unfortunately, astronomers have no way of measuring how violent the creation event was.

All the galaxies and matter in the universe are exerting a gravitational attraction on each other. Each galaxy is pulling on each and every other galaxy. This is true, regardless of what kind of a universe we live in. Thus, no matter what kind of universe we have, the rate of expansion *must* be slowing down. The question, therefore, becomes "How fast is the expansion of the universe slowing down?" If the expansion rate is decreasing rapidly, then the universe will stop expanding someday and begin to collapse. The gravitational attraction that all the galaxies exert on each other will have overpowered their outward motion. But, if the expansion rate is *not* decreasing very fast, then the galaxies will always be getting farther and farther apart. The exact speeds of remote galaxies in the distant future depend on precisely how fast the expansion rate is slowing down. The whole point is that, if astronomers could measure the *deceleration* of the universe, they could discover the kind of universe in which we live. We would know the ultimate fate of the universe.

Right now the universe is expanding at a certain rate. That rate is given by the Hubble constant, 10 miles per second per million light years. Since astronomers have telescopes rather than crystal balls, they cannot know how the universe will be expanding in the future. But, as we look out into space, we are looking back into the past. Thus, by measuring the redshifts and distances of the most remote galaxies, astronomers might be able to figure out how fast the expansion is slowing down. Although there is a linear relationship between speed and distance for nearby galaxies (that is, the famous Hubble law), this relationship will *not* be linear for the most remote galaxies. Depending on the kind of universe in which we live, the data could fall on one of several different curves, as shown in Figure 14–5.

Chapter 14 Cosmology and the Universe

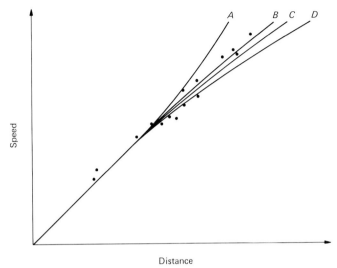

Figure 14-5. *The Distance-Redshift Relation.* The precise relationship between distance and recessional speed for the most remote galaxies depends critically on the nature of the universe. Curves A, B, and C are for big-bang cosmologies. Curve D is for a steady-state cosmology.

If astronomers find from their observations of the speeds and distances of the farthest galaxies that the data fall along curve A in Figure 14–5, then the expanding universe someday will turn into a collapsing universe. Curve A is the case in which remote galaxies are moving away at unusually high speeds. This means that the universe was expanding much faster in the past than it is today. The rate of expansion is slowing down very rapidly. The universe will, therefore, reach a maximum size, the redshifts someday will become blueshifts, and the universe will start contracting.

Curve B corresponds to the case in which the galaxies have exactly the speeds needed to manage to escape from each other's gravitational pull. If the data fall along curve B, infinitely far into the future, the galaxies will come to rest at a time when they are infinitely far apart.

If the rate of expansion of the universe is *not* slowing down very fast, the data will lie along curve C. Curve C corresponds to the case in which the galaxies *always* will be rushing away from each other, no matter how long we wait. The universe always will be expanding, even infinitely far into the future.

Finally, curve D is for the steady state. If we live in a steady state, the data will fall along curve D. Even though there are several curves, all of which correspond to big-bang universes, there is only *one* curve for the steady state. Although there is a big debate about which big-bang curve is correct, since the late 1960s one thing has become very clear: the data do *not* lie along curve D. This is the second major, devastating blow to the steady-state theory.

There are big problems in deciding which big-bang curve is the right one.

Observations must be made of the *most* remote galaxies. Astronomers then run the risk of not being sure what they are seeing. At very great distances, only the brightest galaxies in clusters can be detected. Unfortunately, these galaxies might really be Seyfert galaxies, exploding galaxies, giant cD ellipticals, or other oddities that are much brighter than normal galaxies. If these distant galaxies appear brighter than normal, astronomers run the risk of incorrectly assuming that these galaxies are closer than they really are. Thus, astronomers would be fooled into thinking that high redshift galaxies are nearby. This would favor curve A.

As recently as the early 1970s, many astronomers believed that curve A was the right one. It was, therefore, popular to talk about the *oscillating universe* that alternately expands and contracts. These conclusions were based on a very few high redshift galaxies, such as 3C 295 in the constellation of Bootes. However, in 1975, Sandage and Tammann published the final two installments (Parts 5 and 6) of their classic series of papers. (See Section 14.1.) Their conclusion is that the best data available seem to favor curve B. The big bang, therefore, had just enough energy so that the universe will always be expanding.

Up until the mid-1970s, astronomers in the field of observational cosmology worked on one central question: "Do we live in a big-bang or steady-state universe?" As more and more data piled up, the steady-state cosmology gradually began to fall out of favor. The important question became "In which kind of big-bang universe do we live?" Or, more precisely, "How fast is the expansion of the universe decelerating?" These are reasonable questions only as long as the big-bang cosmology is the only alternative to the steady state. But, by 1975, some troublemakers with impressive credentials were at it again.

The simplest, most straightforward interpretation of the Hubble law is that there was a big bang. The universe is expanding. Thinking backward in time, we can conceive of a situation in which all the galaxies were piled on top of each other. All the matter and energy in the universe must have been confined into a very small region. There must have been a *singularity*, perhaps 20 billion years in the past, when the entire universe was infinitely dense, with infinite curvature of space-time. At the moment of the big bang, the universe was a singularity in the same sense that singularities exist at the centers of black holes.

In the mid-1970s, the brilliant mathematician G. F. R. Ellis proved that a *true* singularity did not have to exist at the time of the creation event. Instead, matter could be passing into our universe from a previous universe through a *Cauchy horizon*, named after the nineteenth-century mathematician, A. L. Cauchy. This Cauchy horizon gives the illusion of being a singularity, but it is not. Even though infinite pressure, infinite density, and infinite temperatures must exist at the time of singularity, matter is "well-behaved" in crossing the Cauchy horizon. Material pouring into our universe across the Cauchy horizon does not need to have infinite pressure, density, and temperature. In other words, there was *no* big bang. There was a whimper. Ellis' cosmology is called the *whimper cosmology.* Indeed, Ellis' classic paper on this subject is entitled, "Was the Big Bang a Whimper?"

Chapter 14 Cosmology and the Universe

Finally, in the spring of 1975, F. Hoyle proposed a new and even more revolutionary alternative to the usual big-bang cosmology. In Hoyle's new theory, there was no big bang. There wasn't even a whimper. In fact, the universe is not even expanding!

Hoyle starts off by asking how we measure anything. Suppose you say that you are 6 feet tall. What you *really* mean is that your height is equal to six times the length of a wooden stick that someone has told you is 1-foot long. The lesson to be learned is that we always measure distances in reference to standard yardsticks. But, suppose the lengths of these standard yardsticks are changing with time. This means that the distances we measure also are changing with time. In particular, if the masses of particles are changing with time, according to the Hoyle-Narlikar theory (see Section 13.4), then the lengths of our standard yardsticks also are changing. If the masses of particles are increasing, the sizes of atoms must be shrinking, and the lengths of rulers must be shrinking. Thus, what we call a light year is getting shorter and shorter as the universe gets older. More light years fit between two galaxies now than did 10 billion years ago. The universe is *not* expanding. Our rulers are shrinking!

If we think back far enough, to 20 billion years ago, there must have been a time when the masses of atoms and the particles out of which they are made were *zero*. At this *zero surface* matter does not have any mass. The inch, the millimeter, the mile, and the light year are all infinitely long. It would look as though matter in the universe is all squeezed together in a singularity, but that is not so. Our rulers, by which we measure the distances between atoms, or between stars, or between galaxies, are all infinitely long.

By the summer of 1975, Hoyle had demonstrated that his new theory has some extremely attractive features. For example, if the center of the sun does not come from our universe, but rather managed to squeeze through from the previous "universe" that existed on the other side of the zero surface, then scientists can explain why they do not observe neutrinos from the sun. Hoyle's theory solves the neutrino problem discussed in Section 7.4! But the skeptics would argue that given enough time Hoyle will use these bizarre ideas to solve everything including the Arab-Israeli question and probably come up with a cure for cancer! The general feeling is that Hoyle's new theory is way out in left field.

Whether or not the new ideas of Ellis and Hoyle are correct, it is important to realize that there are possible alternatives to the classical big bang that almost everyone believes today. Both of these two new theories are similar in that both have a place in the past (a Cauchy horizon or a zero surface) that gives the illusion of being a singularity. In actuality, matter is coming across these regions from an earlier universe into our own universe without having to experience infinite pressure, infinite density, and infinite curvature of space-time. Whatever the case may be, it is, perhaps, appropriate to close this discussion with a quotation from Edwin Hubble:

Thus the explorations of space end on a note of uncertainty. And necessarily so. We are, by definition, in the very center of the observable region. We know our

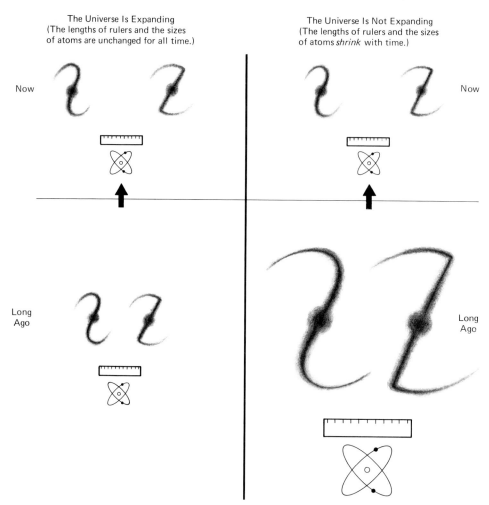

Figure 14–6. Is the Universe Really Expanding? If the masses of atoms are not changing, then the obvious interpretation of the Hubble law is that the universe is expanding. But, if the masses of atoms is increasing, then the sizes of atoms are shrinking. If this is true, then the universe is not expanding. Rather, our *rulers* are shrinking.

immediate neighborhood rather intimately. With increasing distance, our knowledge fades, and fades rapidly. Eventually, we reach the dim boundary—the utmost limits of our telescopes. There, we measure shadows, and we search among ghostly errors of measurement for landmarks that are scarcely more substantial.

The search will continue. Not until the empirical resources are exhausted, need we pass on to the dreamy realms of speculation.

> Edwin Hubble
> *The Realm of the Nebulae*
> (New Haven, Conn.: Yale University Press, 1936), p. 201–202.

14.3 The Young Universe

Back in the mid-1960s, some scientists at Bell Telephone Laboratories were in big trouble. A great deal of effort and money had gone into building a new, highly sensitive radio receiver that was to be used in conjunction with the space program. In 1965, physicists Drs. A. Penzias and R. Wilson were testing this new antenna when they discovered that they were picking up static. No matter in what direction they pointed the radio receiver, they always heard some faint background static. It seemed to be coming from everywhere. Could it be that the engineers had goofed? Could it be that the static was coming from some faulty electronic equipment inside the receiver? No. After months of careful work, Penzias and Wilson concluded that this background static was coming from outer space.

As early as 1946, physicist G. Gamow had theoretically predicted that, if the universe began with a big bang, the universe must have been very hot. It must have been filled with radiation corresponding to a temperature of at least 10 billion degrees shortly after the creation event. As the universe expanded, it cooled. Finally, almost 20 billion years after the creation, the universe has expanded so much that the background temperature should be only a few degrees above absolute zero. Using a similar line of reasoning, but with new data, Dr. R. H. Dicke at Princeton, in the mid-1960s, calculated that the background temperature of the universe today should be about 3° above absolute zero (3°K).

As learned from the discussion of ideal blackbodies in Section 5.3, anything with a temperature above absolute zero *must* emit some sort of electromagnetic radiation. The wavelength at which most of this radiation is emitted depends on temperature, according to Wien's law. Therefore, if the entire universe has a temperature of 3°K, it should be emitting radiation in the form of *microwaves* with wavelengths of around ½-inch.

Up until the time of Penzias and Wilson, everyone thought that the intensity of this background radiation would be far too weak to be detected here on Earth. It was generally believed that the technology necessary to detect such weak radiation simply did not exist. But, in 1965, it became clear that the static Penzias and Wilson had discovered was in fact this background radiation. They were hearing the echo of the big bang!

The antenna used by Penzias and Wilson (see Figure 14–7) was designed to make observations at a wavelength of about 3 inches. In the years following the announcement of this important discovery, additional observations have been made over a wide range of wavelengths. When the signal strength, or brightness, of the background radiation is plotted on a graph against wavelength, as shown in Figure 14–8, all the data fall along a curve. This curve is the theoretical curve given by Planck's formula (see Section 5.3) for a blackbody at a temperature of 2.7°K. The background temperature of the universe is, therefore, 2.7° above absolute zero. The radiation that Penzias and Wilson discovered is, therefore, called the *blackbody background radiation*.

The discovery of the 3° blackbody background radiation is the third and final devastating blow to the steady-state theory. As the universe expands it must cool down. It must have been hotter long ago; it will be colder in the distant future. Since the temperature of the universe is changing, the universe cannot be in a steady state, it must be evolving. The steady-state theory gets the ax.

With the discovery of the 3°-blackbody background radiation astronomers can piece together the probable history of our universe. Right now, we live in an expanding universe that is cool and contains matter clumped together in stars and galaxies. The age of the universe is about 20 billion years. Thinking backward into the past, there must have been a time when the galaxies were touching. This occurred when the universe was only about 500 million years old. Going back still further, before the formation of stars and galaxies, the universe must have consisted of hydrogen and helium gas spread through space. This mixture of hydrogen and helium gas must have been lumpy, so that the lumps could have condensed into stars and galaxies. If the universe had been perfectly smooth, stars and galaxies would not exist today.

Thinking back toward the creation of the universe, we realize that the further back we go, the hotter the universe gets. At 500,000 years after the big bang, it

Figure 14–7. A Microwave Antenna. Drs. Wilson (left) and Penzias (right) are shown here with the "horn-reflector antenna" at Bell Telephone Laboratories in Holmdel, New Jersey. Using this telescope, they discovered the 3° background radiation. (*Bell Labs*)

Chapter 14 Cosmology and the Universe

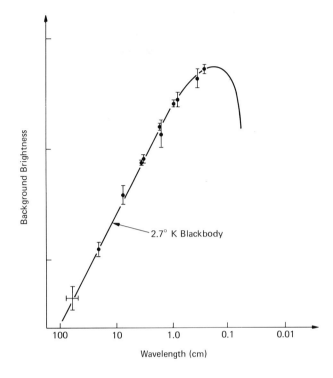

Figure 14-8. The Planck Curve for 2.7°K. All the data from measurements of the intensity of the background microwave radiation agree with a temperature of 2.7° above absolute zero. This background radiation may be thought of as the cooled-off "echo" of the big bang.

was so hot that all the electrons must have been torn off of the hydrogen and helium atoms. At earlier times, all the hydrogen and helium must have been completely ionized.

From about 2,000 years after the big bang up to the present time, the universe has been dominated by the matter that it contains. Long ago, the universe was dominated by hydrogen and helium gas. Today the most important objects in the universe are stars and galaxies. Thus, from 2,000 years after the big bang to the present (and presumably far into the future) we have been in the so-called *matter era*. The most important stuff in the universe is matter, in one form or another.

When astronomers think back to still earlier times, to when the universe was younger than 2,000 years, they find that the nature of the universe was dramatically different. During these very early stages of the universe, the background radiation was compressed to such a high state that the universe must have been very hot. The universe must have had temperatures higher than a million degrees and must have been filled with lots of X rays and γ rays. At these early stages, the energy contained in this hot radiation was greater than the energy contained in the matter in the universe. The universe must have been dominated by radiation. Thus, before the universe was 2,000 years old, we were in the so-called *radiation era*. The radiation era extended from about 1 second after the creation to about 2,000 years later.

At times earlier than 1 second after the creation, we enter the *lepton era*. During the lepton era, the temperature of the universe must have been hotter than 10 billion degrees. The γ rays that filled the universe at that time probably underwent considerable pair production (see Section 10.4), giving rise to electrons and antielectrons, muons and antimuons, neutrinos and antineutrinos. Such particles and antiparticles are called *leptons* by nuclear physicists.

Finally, during the first few microseconds of the universe (one microsecond = one millionth of a second), we arrive at the threshold of the *hadron era*. Nuclear physicists use the term *hadron* to refer to the most massive subatomic particles such as protons, neutrons, lambdas, sigmas, and so on. During these first few microseconds, when the temperature of the universe was hotter than 1 trillion degrees, the highly energetic γ rays probably underwent considerable pair production, giving rise to protons and antiprotons, neutrons and antineutrons, and so on. Therefore, the history of the universe may be summarized as in Table 14–1.

Table 14–1.

Age of Universe	Temperature of Universe	Important Events in Universe
0	Infinity	Big bang: singularity in all space-time
Couple microseconds	Trillions of degrees	Hadron era: massive particle and antiparticle pairs created
One second	Billions of degrees	Lepton era: particle and antiparticle pairs created
Up to 2,000 years	Falls to 1 million degrees	Radiation era: hydrogen and helium created
After 2,000 years	Falls below 1 million degrees	Matter era: hydrogen and helium gas condenses into stars and galaxies
20 billion years	2.7°K	Now

Although this scenario for the creation of the universe sounds reasonable at first glance, there are some severe problems. Nuclear physicists feel that all their experiments show a basic symmetry between matter and antimatter. For every proton there should be an antiproton; for every electron there should be an antielectron. But where is all this antimatter? It cannot be nearby, because whenever matter and antimatter come into contact, a violent explosion occurs in which matter and antimatter vanish in a flash of γ rays. In addition, it probably cannot be far away either. Astronomers observe many cosmic rays (see Section 10.4) coming from outer space. These cosmic rays actually consist of very high-speed electrons and protons coming from the most distant parts of the universe. If the universe consists of equal parts of matter and antimatter, on the average, for every incoming proton or electron, there should be an incoming antiproton or antielectron. Not so! The overwhelming majority of cosmic ray particles arriving at the earth are made out of matter (protons and electrons), not antimatter. Astronomers, therefore, conclude that the universe favors matter over antimatter.

The fact that the universe favors matter over antimatter is not easy to explain. Perhaps during the first few microseconds of the universe, the intense radiation field precipitated into ordinary matter in such a way that many more particles than antiparticles were produced. Crudely speaking, this is analogous to the formation of raindrops here on Earth. If the air is very moist, when the temperature falls, the water vapor condenses into tiny water droplets that form clouds and rain. Perhaps, as the temperature of the universe decreased during the first few moments after the creation, the radiation condensed into particles. This, however, appears to violate some of the basic symmetry laws of nature discovered by nuclear physicists.

Although there are some severe theoretical problems associated with understanding the first few seconds of the universe, most astronomers strongly feel that the big-bang cosmology is correct. The discovery of the 3°-blackbody background radiation is interpreted as powerful evidence in favor of a creation event, or singularity, some 20 billion years ago. The universe was born from a *primordial fireball* that must have had a temperature of at least a trillion degrees during the first microsecond. As the universe expanded, the fireball cooled, until today the radiation that permeates the entire universe is only 2.7° above absolute zero.

During the late 1960s, astronomers were quick to realize a very important feature of the 3°-blackbody background radiation. No matter where you look in the sky, the temperature of the background radiation is virtually the same. Even with the most precise radio telescopes available, astronomers find that the temperature of the blackbody background is the same in *all* directions. This discovery has some important implications. It means that violent processes during the first microsecond of the universe must have smoothed out the radiation field. If the radiation field were not perfectly smoothed out, if there were some regions of the early universe that were slightly hotter than other regions, these hot regions should be seen today. On the contrary, the background radiation is incredibly *isotropic*—it is the same temperature in all directions.

Explaining how the background radiation became smooth, or isotropic, has proven to be a very difficult problem in theoretical cosmology. A ray of hope appeared in 1969 when Charles Misner at the University of Maryland proposed the *mixmaster universe*. The mixmaster universe corresponds to a solution of the Einstein field equations of general relativity in which violent mixing occurs. The entire universe is alternately deformed into the shape of a cigar and a pancake in all directions. Although it is possible that our universe might have been a mixmaster universe during its very earliest stages, by the mid-1970s it was clear that the mixmaster universe was not a cure-all. The problem of the isotropy of the 3°-blackbody background radiation remains a mystery for cosmologists who believe there was a big bang.

It should be mentioned here that Ellis's whimper cosmology discussed in the previous section does not predict a perfectly isotropic background radiation. The whimper cosmology predicts very tiny variations in temperature for the background across the sky. If future, more precise observations of the blackbody background radiation do reveal slight temperature variations across the sky, we will have strong evidence in favor of a whimper cosmology rather than a big-bang cosmology.

On the other hand, Hoyle's new cosmology predicts a perfectly smooth blackbody background. As light from stars and galaxies in the "previous universe" reaches the zero surface where the masses of particles are zero, this light is completely scattered and smoothed out. In the new Hoyle cosmology, the 3°-blackbody background is actually the smoothed-out starlight leaking through the zero surface from the previous universe.

Although the isotropy of the 3°-blackbody background radiation is one of the current mysteries in modern astronomy, it seems reasonable to suppose that the early universe was extremely chaotic and violent. In the early 1970s, the brilliant theoretical physicist Stephan Hawking turned his attention to some of the possible consequences of chaos and violence in the first few seconds after the creation.

In Chapter 11, one of the possible final states of stellar evolution called black holes was discussed. All dying stars with masses less then $2\frac{1}{2}$ suns must become pulsars or white dwarfs. Black holes in space must come from dying stars considerably more massive than our sun. The reason for this is that a dying star must have at least 3 solar masses to overcome atomic and nuclear forces that otherwise would stop the contraction and support the star. Therefore, *classical black holes* must have masses greater than three times the mass of the sun.

In an important paper published in 1971, Dr. Hawking considered some interesting processes that could have occurred when the universe was less than 1 second old. It is entirely reasonable to suppose that the *very* early universe was not completely smoothed out. It is reasonable to suppose that there might have been many small lumps in the universe shortly after the creation. Dr. Hawking has proved that the violent and chaotic events during the early universe would have crushed these lumps into *mini black holes*. Black holes of *any* mass could have been formed by these processes. The universe today may contain many very tiny black holes that were squeezed out of lumps in the primordial fireball by crushing and violent events 20 billion years ago.

A black hole having a mass equal to a small asteroid is only about the size of an atom. Black holes with smaller masses have even smaller sizes. In order to deal properly with very tiny objects, scientists must make use of *quantum mechanics*. Quantum mechanics is a branch of physics that deals with the behavior of very tiny objects. Physicists constantly use quantum mechanics to calculate the properties of atoms and nuclei.

By the mid-1970s, Dr. Hawking had succeeded in applying quantum mechanics to mini black holes. To everyone's surprise, he discovered that mini black holes radiate particles and light. All sorts of atomic particles and light should come pouring out of mini black holes! The smaller the black hole, the more it radiates. As Dr. Hawking puts it, "Black holes are white-hot!" Ordinary classical black holes formed from dying stars are, by comparison, very large. These quantum mechanical effects are unimportant for the traditional black holes discussed in Chapter 11. But, if very tiny black holes left over from the violent birth of the universe do exist, they must be radiating lots of particles and light into space.

As a mini black hole radiates particles and light into space, its mass gets smaller and smaller. The smaller the black hole, the faster it radiates. Obviously, therefore, mini black holes *evaporate!* The more they evaporate, the smaller they get; the smaller they get, the more they evaporate. It's a vicious cycle that ends in a violent explosion. By 1974, Dr. Hawking had succeeded in proving that all mini black holes must end their lives in an explosion equivalent to the detonation of a million megaton hydrogen bombs. All mini black holes left over from the creation of the universe with masses less than a billion tons would have evaporated by now. Only mini black holes with masses greater than a billion tons (about the same as the mass of a small asteroid) would still be around today.

In the late 1960s, the United States launched a military satellite in connection with the nuclear test ban treaty. The purpose of the satellite was to detect γ rays from nuclear explosions on the earth. To everyone's surprise, the instruments aboard the satellite recorded an occasional violent burst of γ rays that obviously did not come from the earth. By 1975, our satellites had detected almost a dozen such bursts each year. No one has been able to explain these γ-ray bursts. Are they the deaths of mini black holes left over from creation of the universe? The discovery of mini black holes, some of which might even be in our own solar system, could prove to be one of the most exciting adventures in modern astronomy.

14.4 General Relativity and Cosmology

There is a lot of stuff in the universe. There are stars. There are galaxies. There are nebulae, such as HII regions and supernova remnants. There are also large quantities of gas and dust in interstellar and intergalactic space. On the "small scale," the universe looks very lumpy. Inside stars, for example, there is a lot of matter; between the stars there is almost a perfect vacuum. But, on the "large scale," the universe looks very smooth. Clusters of galaxies seem to be distributed randomly through space, and if we were somehow to stand back far enough, we would not notice the lumps at all.

This is analogous to looking at your own body. Your body is made out of atoms. Each atom has a very dense nucleus at its center and is orbited by many electrons. On the small scale, you would see lumps in the form of protons, neutrons, and electrons. An atom is mostly empty space, just as the solar system is mostly empty space. But as you run your hands over your body, you do not notice the lumps in the form of nuclear particles. On the contrary, the matter out of which your body is constructed seems to be fairly smooth, or homogenous.

If the universe were examined on a *very* large scale, individual planets, stars, or galaxies would not be seen. Rather, matter would seem to be homogeneously distributed through space. Thinking a little further about the large-scale view of the universe, it is realized that, all matter, regardless of its form, produces gravitational fields. Earth has a gravitational field; the sun has a gravitational field; galaxies have gravitational fields. Similarly, the matter in the entire universe must have a gravitational field. There must be a universal gravitational field produced by all the matter in the universe.

Additionally, Einstein's general theory of relativity (see Chapter 11) demonstrates that any gravitational field manifests itself by warping space-time. The precise details of the distribution of matter produce a particular curvature of space and time according to Einstein's field equations. The universe contains matter. This matter spread throughout the universe gives rise to a universal gravitational field. This gravitational field produces an over-all curvature to all space-time in the universe. The universe *must* have a curvature of some kind. It is, therefore, entirely reasonable to ask: *"What is the shape of the universe?"*

At first glance, the earth looks flat. Walking around or driving down a freeway, it seems reasonable to suppose that we all live on a flat earth. Of course, this is not true. But you do not have to go off into outer space like the Apollo astronauts and take a photograph to prove that the earth is round. The fact that the earth is round can be determined by experiments performed right here.

Imagine standing on the earth's equator, as shown in Figure 14–9. Beginning at two different locations on the equator, draw lines on the earth toward the North Pole. As shown in the diagram, both lines begin by looking parallel. Both lines are perpendicular to the equator. But as the lines get closer and closer to the North Pole, the distance between the lines decreases. In fact, both lines cross at the North Pole. This little example is simultaneously trivial and profound. By drawing the straightest possible lines (that is, geodesics) on the earth's surface, we can discover the exact shape of the earth.

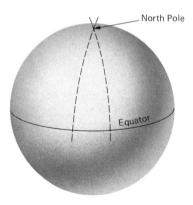

Figure 14–9. The Shape of the Earth. The shape of the earth can be deduced by drawing two "geodesics" (i.e., the straightest possible curves) on the earth's surface. Although the two lines start off "parallel," they eventually cross.

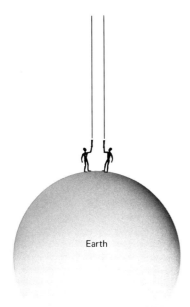

Figure 14–10. *The Shape of the Universe.* In theory, the shape of the universe could be deduced by sending two parallel beams of light out into space. The curvature of space affects how the beams of light move.

The example of drawing lines on the earth suggests an experiment to discover the shape of the universe. Imagine going outside one night with two powerful laser searchlights. Aim both searchlights straight up in the air. Further suppose that there is infinite technological precision in aiming the two searchlights. The two beams of light that leave the earth are *perfectly* parallel, as shown schematically in Figure 14–10.

For the sake of convenience, assume that nothing gets in the way of these two beams of light. The beams of light do not bump into rain clouds, planets, stars, or galaxies as they travel out into space. The two beams travel unhampered into the universe along geodesics—along the straightest possible paths.

There are three and *only* three possibilities. First of all, as these two beams of light move through space, we might find that they are always parallel. Even millions or billions of light years from earth, the two beams of light are traveling alongside each other, just as they were when they left the earth. If the two searchlights in your backyard were separated by 10 feet, then even trillions of miles from the earth the two beams of light are still exactly 10 feet apart. If this is the case, it would be concluded that the universe is *flat*. The curvature of the universe is *zero*.

There is, however, a second possibility. If we could catch up with the beams of light far from the earth, we might find that the beams of light are gradually getting closer and closer together. Instead of remaining perfectly parallel, the beams of light might converge, just as the two lines drawn from the earth's equator converge at the North Pole. In fact, billions of light years from the earth, the two beams of light might actually cross. In this case we would conclude that the universe is *spherical*. The universe has a *positive* curvature.

Finally, there is a third possibility. After catching up with the beams of light,

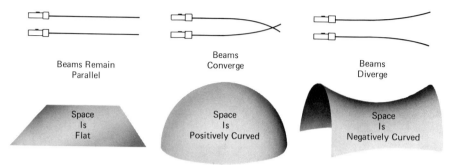

Figure 14-11. *Three Possibilities.* Depending on how space is curved, two light rays which start out "parallel" will either converge, remain parallel, or diverge.

we might find that they are gradually getting farther and farther apart. We might find that the beams of light are diverging. In this case we would conclude that the universe is *hyperbolic*. The universe has a *negative* curvature.

As is always the case in relativity, it is extremely difficult to think in four dimensions. As was pointed out in Chapter 11, it is much easier to imagine things in two dimensions and then extend the results to four-dimensional space-time. To understand concepts such as "the shape of the universe" and "the curvature of space," it is necessary to discuss what happens on two-dimensional surfaces (table tops, basketballs, saddles). This amounts to drawing *embedding diagrams* of the universe in which two of the four dimensions are suppressed.

As shown in Figure 14–11, the three possibilities of what happens to the beams of light correspond to three basic geometrical surfaces. If the light rays remain perfectly parallel forever and ever, then we live in a flat universe. Three-dimensional space is "flat" in exactly the same sense that a table top is flat. If the light rays converge, then we live in a positively curved universe. Three-dimensional space is "spherical" in exactly the same sense that the surface of a basketball is spherical. Finally, if the light rays diverge, then we live in a negatively curved universe. Three-dimensional space is "hyperbolic" in exactly the same sense that the surface of a saddle is hyperbolic. The universe is either flat (zero curvature), spherical (positive curvature), or hyperbolic (negative curvature).

Of course, this experiment with laser searchlights in your backyard is absurd. We cannot follow beams of light out into space. We cannot catch up with a couple of light rays 5 billion light years away. That's nonsense. But is there something we *can* do? Astronomers are literally glued to the earth. Do we have any hope of making some sort of observations from the earth that might tell us about the geometry of the universe?

Imagine three surfaces that correspond to the three possible shapes of the universe, as shown in Figure 14–11: a flat surface (a piece of a table top), a positively curved surface (a piece of a basketball) and a negatively curved surface (a piece of a saddle). Smear glue over each surface. Then sprinkle salt randomly over each surface. Finally, take three very intelligent bugs and stick them down, one on each surface. The crosses on the surfaces shown in Figure 14–12 mark the locations of the bugs. Then ask each bug to measure the distances between himself and all the grains of salt. From such measurements, the bugs can discover the shape of the surfaces to which they are glued!

Let's think about the bug on the flat surface. Of course, he does not know the shape of his "universe" and he starts measuring the distances between himself and all the grains of salt. He draws a map of his results on a flat piece of paper. His map is shown at the bottom of Figure 14–12. The cross marks his location, and each dot marks the location of a grain of salt. He finds that the dots on his map are distributed randomly. Since he was told (or assumes) that the grains of salt are randomly distributed in his universe, he correctly concludes that he is on the *flat* surface.

The other two bugs get very different maps. When the bug on the spherical surface makes a map on a flat piece of paper, he finds that there seem to be more grains of salt nearby than far away. This is because making a flat map of a spherical surface is like squashing down the spherical surface. Cracks and empty spaces appear near the edges. The numbers of grains of salt thin out at large distances from the bug's location. On the other hand, when the bug on the hyperbolic surface makes his map on a flat pad of paper, he finds that there

Figure 14–12. "Counting" Galaxies. In theory, by measuring the distances to all the remote galaxies and by making a map of the resulting observations, the general appearance of the map should reveal the shape of the universe.

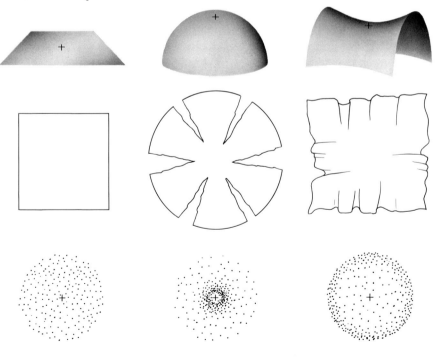

seem to be more grains of salt far away than nearby. This is because squashing a saddle down on a flat piece of paper will produce wrinkles around the edges. His universe piles up around the edges; the numbers of grains of salt seems to pile up around the edges. Assuming that the grains of salt really are distributed randomly on each of the surfaces, the maps (on flat pieces of paper) indicate the kinds of surfaces to which the bugs are glued.

The whole point of this little story is that astronomers are glued to the earth. The best they can hope to do is measure the distances to the remote galaxies that are sprinkled around the universe. If the resulting maps of the locations of the galaxies show a uniform distribution throughout space, then we live in a flat universe. If the maps indicate that there seem to be more galaxies nearby than far away, then we live in a positively curved (spherical) universe. Or, if the maps indicate that there are more galaxies farther away than nearby, then we live in a negatively curved (hyperbolic) universe.

Although this method sounds fine at first glance, there are some big problems. First of all, the over-all curvature of the universe is very slight. Space is almost perfectly flat. The reason for this is that, on the average, the universe is almost empty. If we look at the universe on the large scale, we find that, on the average, there are only about 100 pounds of matter in every million trillion cubic miles! With such a low density of matter, the curvature of space is so slight that astronomers must look to very great distances to notice any departures from flatness, if there are any. In order to decide the issue of curvature, astronomers must observe objects with redshifts (according to the Hubble law) that correspond to speeds greater than 80 per cent or 90 per cent of the speed of light. As learned in the previous chapter, these distances are beyond the limits of the galaxies. No one has ever seen a galaxy with such high redshifts. If anything is out there, it is the quasars. But what is a quasar?

Astronomers have, nevertheless, tried to take a stab at the issue of the shape of the universe. Instead of actually making a map of the universe, they count how many extragalactic objects they see in the sky with a certain brightness. In this regard, radar astronomy is more helpful than optical astronomy. There are a lot of radio sources that do not seem to correspond to visible objects. These sources, therefore, probably lie beyond the limits of the 200-inch telescope. Since the issue of curvature will be decided by the most distant objects, astronomers prefer to turn to the data from radio surveys of the sky. For convenience, let's assume that all extragalactic radio sources have the *same* intrinsic brightness— they all emit the same amount of energy. Obviously, this is a *very* crude assumption, but let's see where it leads. *If* all the radio galaxies (or quasars, or whatever is out there) have the same real brightness, then the intensity of the radio signals that astronomers detect at their telescopes is a direct measure of the distance to the radio source. The fainter the signal, the farther the source. Looking to greater and greater distances, we should discover more and more radio sources since we are taking in a larger and larger region of the universe. Therefore, there should be lots of dim (distant) radio sources, compared to very few bright (nearby) radio sources.

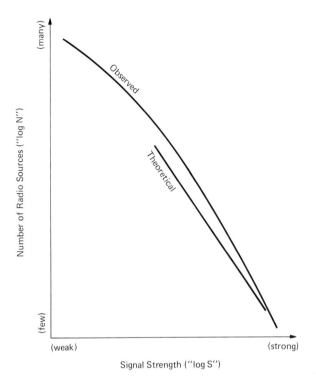

Figure 14–13. The "Log N-Log S" Plot. The number of radio sources (N) is plotted against the signal strength (S) of the radio sources. This graph is similar to making an actual map, as shown in Figure 14–12. Unfortunately, astronomers do not know enough about distant radio galaxies to draw any firm conclusions from this graph. (*Adapted from Drs. Pooley and Ryle*)

Astrophysicists can calculate exactly how many radio sources should be expected that have a certain brightness or signal strength. The results of such calculations can be expressed on a *log N-log S plot*, as shown in Figure 14–13. This graph is nothing more than a plot of the number N, of radio sources against the signal strength, S, of the radio sources. The straight line labeled *theoretical* corresponds to what we would expect from a flat universe. There is a large number of weak (distant) sources and a small number of strong (nearby) sources.

In the late 1960s, G. G. Pooley and M. Ryle published their radio observations of the numbers of radio sources having various signal strengths. The curve they obtained is labeled *observed* in Figure 14–13. They found that there was an unusually high number of very weak radio sources. Unfortunately, this does not tell us very much. One interpretation of the Pooley-Ryle work is that there are more radio sources at very great distances than would be expected from a flat universe. But, before jumping to any conclusions, always recall that looking out into space means looking backward into the past. Suppose radio sources were brighter in the past than they are today. Suppose radio galaxies *evolve* in such a way that billions of years ago they were emitting lots of energy, but today they are rather weak. This means that as we look to great distances we should also see large numbers of radio sources. Thus, unfortunately, counting the numbers of radio sources of different radio brightnesses has not proven fruitful in disclosing the shape of the universe.

In conclusion, we realize that the universe must have a shape. The universe is either flat, positively curved, or negatively curved. In principle, counting the numbers of galaxies or radio sources at various distances from the earth should result in the ability to discover the shape of the universe. In practice, things turn out to be much more complicated. Since the over-all curvature of the universe is very slight, astronomers would like to turn to the most distant objects for decisive answers. But these distant objects can be seen only as they were billions of years ago in the distant past. In all probability, these remote objects were very different long ago than they are now. Things are not as simple as in the story with the bugs and the grains of salt. If the grains of salt are *not* all the same, if the grains of salt change in some way as the bugs look to greater and greater distances, then the bugs could become confused. They would not be all that sure which surface they are on. Only the most remote galaxies and radio sources are useful in discovering the shape of the universe. Since they are so far away, they are the most difficult objects to observe. Indeed, these remote galaxies that are important in deciding the issue of curvature are the *same* galaxies that should tell how fast the expansion of the universe is slowing down. These are the very same galaxies that must be examined to learn about the deceleration of the expanding universe discussed in Section 14.2.

In the final pages of this book we will show how the curvature of the universe and the rate of deceleration are intimately related. By learning precisely how fast the expansion of the universe is slowing down, astronomers can conclude what the shape of the universe must be. From this knowledge, they can make some important statements about the ultimate fate of the entire universe.

14.5 The Fate of the Universe

There are three possible shapes or geometries of the universe: flat, spherical, and hyperbolic. One way of demonstrating the differences between these three curvatures involves some ideas about triangles. In high school geometry classes students are taught that "The sum of the angles of a triangle is 180°." But this is only true if we are dealing with so-called Euclidean geometry, in which space is flat. If a triangle is drawn on a curved surface, the sum of the angles could be very different from 180°, depending on the curvature of the surface.

As in the previous section, imagine three surfaces (this time without glue, bugs, or salt) that represent, in two dimensions, the three possible geometries of the universe. As shown in Figure 14–14, if a triangle is drawn on the flat surface, the sum of the angles is exactly 180°, just as we learned in high school. But, if a triangle is drawn on a spherical surface, such as a basketball, the sides of the triangle are bowed out. The sum of the three angles of this triangle is *greater* than

Figure 14-14. Triangles in Flat and Curved Space. On a flat surface, the sum of the angles of a triangle is exactly 180°. On a spherical (positively curved) surface, the sum is greater than 180°. On a hyperbolic (negatively curved) surface the sum is less than 180°.

180°. And, finally, if a triangle is drawn on a negatively curved, saddle-shaped surface, the sides of the triangle are squeezed in. The sum of the three angles of this triangle is *less* than 180°.

Therefore, in principle, the geometry of the universe could be discovered by measuring the sums of the angles in huge triangles in space. Since the curvature of the universe is very slight, a gigantic triangle would have to be employed. The universe is *locally flat,* just as the earth looks flat if you confine your observations to your home town. Little triangles drawn in the universe, only a couple of million light years on a side, will not indicate anything. Over a few million light years space looks flat, and the sum of the angles in these little triangles would always be 180°. To discover departures from flatness, *big* triangles are needed—a couple of billion light years on a side. The assistance of some little green men living on quasars at the two distant corners of this big triangle is also required to measure the sizes of the two distant angles (presumably one corner of the triangle is located here at the earth). Therefore, this is a useless and absurd method of detecting the curvature of the universe. In fact, no one has ever thought of any good methods of directly measuring the shape of the universe. Is there perhaps some way we can *indirectly* infer the shape of the universe?

When we discussed the expanding universe in Section 14.2, we realized that the rate of expansion *must* be slowing down. All the galaxies in the universe are pulling on each other due to their mutual gravitational attraction. The critical question for astronomers is "How fast is the expansion slowing down?" There are three and only three possibilities: (1) If the deceleration is very slight—if the expansion of the universe is *not* rapidly slowing down—then the universe will go on expanding forever. (2) If the expansion of the universe is slowing down very rapidly, then the universe will reach a maximum size and begin collapsing. (3) In between these two possible cases, if the expansion of the universe is slowing down at just the "right" rate, it insures that, some day, infinitely far into the future, when the galaxies are infinitely far apart, the expansion will stop.

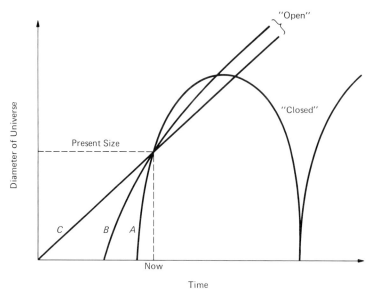

Figure 14–15. *The Future of the Universe.* How fast the universe is expanding is directly related to how much matter there is in space. This, in turn, is directly related to the geometry of space. Depending on the shape of the universe, the universe will either expand forever or eventually collapse in upon itself.

The answer to the question "How fast is the expansion of the universe slowing down?" is, in turn, directly related to the answer to the question "How much matter is there in the universe?" If there is very little matter in the universe, then the gravitational attraction between this matter will *not* be sufficient to halt the expansion of the universe; it will go on expanding forever. If there is a great deal of matter in the universe, then the gravitational attraction between all this matter will overcome the expansion of the universe, and the universe will some day start collapsing. Finally, if there is just the right amount of matter in the universe, then expansion will stop only after all the galaxies are infinitely far apart.

The easiest way to express how much matter there is in the universe is to assume that the matter is all smoothed out (large-scale view) and ask what this comes out to in pounds per cubic mile or tons per cubic light year, and so forth. The critical value seems to be 100 pounds per million trillion cubic miles. If, on the average, there are less than 100 pounds of matter in each million trillion cubic miles of space, then the universe will expand forever and ever. If, on the average, there are more than 100 pounds of matter in each million trillion cubic miles of space, then the universe will reach a maximum size and start collapsing. But, if it turns out that, on the average, there are exactly 100 pounds of matter in each million billion cubic miles, then the universe will stop expanding when all the galaxies are infinitely far apart.

From the general theory of relativity, recall that the gravitational field of

Chapter 14 Cosmology and the Universe

matter manifests itself by warping space-time. More precisely, the whole idea behind the Einstein field equations is that matter tells space how to curve and curved space tells matter how to behave. Thus, astronomers conclude that the three possible geometries of the universe are directly related to the three possible rates at which the expansion of the universe is slowing down!

If there is lots of matter in the universe (average density *greater* than 100 pounds per million trillion cubic miles), then the expansion of the universe is slowing down so fast that the universe will reach a maximum size and begin collapsing. This means that there is enough matter in space to "close" the universe. This corresponds to a spherical, or positively curved, geometry.

If the amount of matter in the universe is exactly equal to the critical value (average density *equal* to 100 pounds per million trillion cubic miles), then the expansion will gradually slow down and stop only after the galaxies are infinitely far apart. In this case the universe is flat. The over-all curvature of space is zero.

Finally, if there is very little matter in the universe (average density *less* than 100 pounds per million trillion cubic miles), then the expansion never stops. Even infinitely far into the future, the galaxies will be moving away from each other at high speeds. In this case, the universe is open. This corresponds to a hyperbolic, or negatively curved, geometry.

By way of summary we see that there are three possibilities as presented in Table 14–2.

Table 14–2.

Geometry of Universe	Curvature of Universe	Average Density of Matter	Fate of Universe
Spherical	Positive	Greater than 100 pounds per million trillion cubic miles	Expansion stops and collapse begins
Flat	Zero	Equal to 100 pounds per million trillion cubic miles	Expansion stops when galaxies are infinitely far apart
Hyperbolic	Negative	Less than 100 pounds per million trillion cubic miles	Expansion never stops

If the universe is either flat or hyperbolic, then the universe is infinite in extent. The universe is said to be *open*. It extends in all directions forever and ever. There is no edge and there is no center. Questions such as "What is the universe expanding into?" What is beyond the edge of the universe?" or "Where is the center of the universe?" make no sense at all. They are meaningless questions. Something that is infinite has no edge or center; there simply cannot be anything "beyond" something that extends forever.

But what if the universe is *closed*? What if there is so much matter in the universe that the curvature of space is positive, and thus the shape of the

Figure 14–16. *An Expanding Balloon.* By analogy with the surface of a balloon, a "closed" universe does not have an edge nor a center. A litle bug could walk around the balloon forever and never come to "the edge" or "the center."

universe is spherical? Would these same questions about the edge and center of the universe then have meaning?

Imagine a little girl blowing up a balloon, as shown in Figure 14–16. Suppose, in addition, that there are many little dots all over the balloon. As the spherical balloon gets bigger and bigger, the dots get farther and farther apart. Standing on one of these dots, you would find that the nearby dots are moving away from you slowly, while the more distant dots are moving away much more rapidly. That's the Hubble law!

Now imagine some (more) intelligent bugs on this expanding spherical balloon. These bugs could walk around and around this balloon forever, but they would *never* get to an "edge." None of these bugs could get to a place where he could honestly say, "Here I am at the edge of the balloon." The *surface* of the balloon has no edge! Similarly, they could walk around on the balloon and never get to the center. The *surface* of the balloon has no center!

By analogy, even if the universe is closed and spherical, it does *not* have an edge or a center. It is, therefore, *always* stupid and meaningless to ask such questions as "What's beyond the edge of the universe?" or "Where is the center of the universe?"

Continuing the story of the little girl's balloon, if the universe is closed, there is enough matter to stop the expansion. The universe (or the balloon) some day far into the future will reach a maximum size and then begin collapsing. The redshifts of galaxies will turn into blueshifts as the universe gets smaller and smaller. Eventually, all the matter in the universe will be crushed into a *singularity* of infinite density, infinite pressure, and infinite curvature of space-time! But this universal singularity in the distant future will have some important differences from the kinds of singularities believed to exist at the centers of black holes. If matter is dropped down an ordinary black hole, *all* information about this matter is lost *except* for its mass, charge, and angular momentum. If we drop a ton of bricks or a ton of steel down a black hole, the final result is the same. This is because, as shown schematically in Figure 14–17, the black hole is attached to the rest of the universe. The curvature of space far from the black hole gets flatter and flatter as it connects onto the nearly flat space-time of the rest of the universe. But if the *entire* universe becomes a singularity, then there is nothing on which to connect space-time. This means that even the most fundamental properties of matter, such as mass, charge, and angular momentum, are lost! The next universe emerging from this singularity may have properties that are *fundamentally* different from our universe. *If* the universe is closed, there will

Figure 14–17. *Matter Falling into a Black Hole.* Regardless of whatever is dropped down a black hole, the final result is always the same. Only the mass, charge, and angular momentum of the in-falling object are preserved. But *if* the entire universe were to collapse down its own black hole, even these most fundamental properties of matter would be lost. (*Adapted from Dr. Wheeler*)

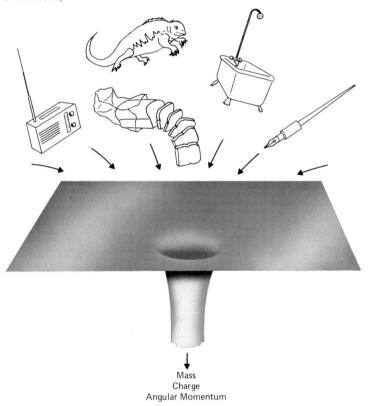

be a universal singularity in the distant future. There will be *another* big bang. The next universe that will emerge from the ashes of our universe will be made of matter whose atomic and nuclear properties are totally unlike anything ever seen.

So far only theory has been discussed. What do the observations tell us? In which kind of a universe do we really live? Unfortunately, as emphasized many times, only the most distant galaxies can possibly provide the answers. These are the most difficult galaxies to observe. Consequently, the observations are not totally conclusive.

Up until the early 1970s, it was popular to suppose that we live in a closed, spherical universe. The most distant galaxies seemed to be slowing down fast enough to indicate that the expansion would someday stop. We then would be in an *oscillating universe* that has a big bang every couple of hundred billion years. However, in 1975, Dr. Alan Sandage published a follow-up paper to the series he wrote with Tammann (mentioned in Section 14.1) concerning the redshifts of distant galaxies. His observations indicate that the universe will *never* stop expanding! This is in agreement with independent arguments presented by Gott, Gunn, Schramm, and Tinsley (also mentioned in Section 14.1) based on observations of the average density of matter in the universe. There does not seem to be enough matter in the universe to ever stop the expansion. Our universe is then a one-shot deal.

It would be premature to say that we have heard the "final word." One of the most important projects in astronomy during the late 1970s will be to improve on Sandage's current observations (that the universe is infinite and possibly hyperbolic). In this regard it should be noted that some astronomers are not completely certain that a big bang ever occurred in the first place. Nevertheless, we can feel quite sure that, regardless of what observational astronomers do, the theoretical physicists will be hard at work thinking up new ideas that will probably shed more heat than light on the subject of cosmology. For example, if tiny white holes exist in the universe, as Dr. Stephen Hawking has recently suggested (see Section 14.3), then *everything* seems hopeless. Recall that when things fall into *black* holes, information is *lost* from the universe. Black holes may be thought of as "information sinks." But if *white* holes exist, then new information is constantly being *added* to the universe. White holes may be thought of as "information sources." If new information is constantly being pumped into the universe, we could never predict anything!

As we stand at the frontiers of human knowledge, groping our way through the most remote reaches of the universe for understanding that might not even exist, we recall the words of Mark Twain:

> *There is something fascinating about science. One gets such a wholesale return of conjecture out of such a trifling investment of fact.*
>
> *Life on the Mississippi* (1874)

Chapter 14 Cosmology and the Universe

Questions and Exercises

1. Use an analogy, such as an expanding raisin cake or balloon, to argue that the universe is expanding.
2. What is the Hubble constant?
3. Briefly describe the step-by-step observations that Drs. Sandage and Tammann used to determine the Hubble constant.
4. Contrast and compare the big-bang and steady-state cosmologies.
5. Explain why the existence of quasars can be interpreted as being in direct conflict with a steady-state cosmology.
6. Briefly discuss how the deceleration of the universe is related to the ultimate fate of the universe.
7. What possible alternatives are there to the big-bang and steady-state cosmologies?
8. What is meant by the 3°-blackbody background radiation?
9. Where and by whom was the 3°-background radiation first discovered?
10. Contrast and compare the radiation era and the matter era.
11. What happened as the universe went from the radiation to the matter era?
12. What evidence is there for the idea that the universe contains more matter than antimatter?
13. What is meant by the primordial fireball?
14. What is meant by the isotropy of the background radiation?
15. Under what condition could mini black holes have been formed?
16. Contrast and compare ordinary black holes and mini black holes.
17. Why is it reasonable to suppose that any mini black holes around today must have masses greater than a billion tons?
18. By discussing the behavior of light rays traveling through space, present an argument demonstrating that there are only three possible "shapes" for the universe.
19. Discuss why making a map of the universe should (in theory) reveal the shape of the universe.
20. Why is it difficult to draw any firm conclusions about the shape of the universe from a log N-log S plot?
21. How is the geometry of the universe related to the rate of deceleration of the universe?
22. How is the average density of matter in the universe related to the deceleration of the universe?

Glossary

Absolute magnitude: A measure of the real brightness of celestial objects. Technically, the apparent magnitude a star would have at a distance of 10 parsecs.

Absolute zero: The lowest possible temperature; a temperature of 0°K or −273°C.

Absorption spectrum: Dark lines superimposed on a continuous spectrum of electromagnetic radiation.

Acceleration: A change in velocity; speeding up, slowing down, or changing direction.

Albedo: The percentage of sunlight reflected by a planet, a satellite, or an asteroid.

Almanac: A book or table listing astronomical events.

Altitude: Angular distance above or below the horizon, measured along a vertical circle, to a celestial object.

Amplitude: A range of variability, as in the light from a variable star.

Angstrom (Å): A unit of length equal to one hundred millionth of a centimeter.

Angular size: The angle subtended by an object.

Annular eclipse: An eclipse of the sun in which the moon is too distant to cover the sun completely, so that a ring of sunlight shows around the moon.

Aphelion: The point in a planet's orbit at which it is farthest from the sun.

Apogee: The point in the orbit of a satellite at which it is farthest from the earth.

Apparent magnitude: A measure of the brightness of a star or another celestial object viewed from the earth.

Apparent solar day: The interval between two successive transits of the sun's center across the meridian.

Apparent solar time: Time reckoned by the actual position of the sun in the sky.

Ascending node: The point in the orbit of a celestial body where it crosses from the south to the north of a reference plane, usually the plane of the celestial equator or the ecliptic.

Aspect: The position of the sun, satellites, or planets with respect to one another.

Association: A loose cluster of stars whose motions, spectral types, or positions in the sky indicate a common origin.

Asteroid: A minor planet.

Astrometric binary: A binary star in which one component is not observable but can be deduced from the motions of the visible star.

Astronautics: The science of laws and methods of space flight.

Glossary

Astronomical unit (AU): The average distance between the earth and the sun; a unit of length equal to approximately 93 million miles.

Astronomy: The study of the universe beyond the earth's atmosphere.

Astrophysics: The study of the physical properties and phenomena associated with planets, stars, and galaxies.

Atom: The smallest particle of an element; it retains the properties that characterize that element.

Aurora: Light radiated by atoms and ions high in the earth's atmosphere, sometimes called the northern lights.

Autumnal equinox: An intersection of the ecliptic and the celestial equator; the point at which the sun crosses the celestial equator moving from north to south.

Azimuth: The angle along the celestial horizon, measured eastward from the north point, to the intersection of the horizon with the vertical circle passing through an object.

Barred spiral: A spiral galaxy in which the arms project from the ends of a bar that runs through the nucleus.

Big-bang theory: A cosmological theory in which the expansion of the universe is presumed to have begun with a primeval explosion.

Binary star: A double star; two stars revolving about each other.

Blackbody: A hypothetical, perfect radiator; a body that absorbs and reemits all the radiation falling on it.

Black hole: A region of highly warped space-time caused by an intense gravitational field and surrounded by an event horizon.

Bode's law: A technique for obtaining a sequence of numbers that gives the approximate distances of the planets from the sun in astronomical units.

Bohr atom: A model of the atom developed by Niels Bohr in which the electrons revolve about the nucleus in circular orbits.

Bolometric correction: The difference between the visual and bolometric magnitude of a star.

Bolometric magnitude: A measure of the total amount of radiation from a celestial object received just outside the earth's atmosphere and detected by a device sensitive to *all* forms of electromagnetic energy.

cD galaxy: A large, bright elliptical galaxy.

Celestial equator: A great circle on the celestial sphere, 90° from the celestial poles.

Celestial mechanics: A branch of astronomy dealing with the motions of the members of the solar system.

Celestial navigation: The art of navigating at sea or in the air from sightings of the sun, moon, planets, and stars.

Celestial poles: The points about which the celestial sphere appears to rotate; the points of intersections of the celestial sphere with the extension of the earth's polar axis.

Celestial sphere: The apparent sphere of the sky; a sphere of large radius centered on the earth.

Cepheid variable: A star that belongs to

one of two classes (type I and type II) of yellow, supergiant, pulsating stars.

Ceres: The largest of the minor planets and the first to be discovered.

Chromatic aberration: A defect in an optical system that causes light of different colors to be focused at different points.

Chromosphere: The part of the solar atmosphere immediately above the photosphere.

Cluster of galaxies: A system of galaxies with at least several, and in some cases thousands, of members.

Cluster variable: A member of a certain class of pulsating variable stars, all with periods less than 1 day; an RR Lyrae variable.

Color excess: The amount by which the color index of a star increases when its light is reddened in passing through interstellar absorbing material.

Color index: The difference between the magnitudes of a star measured in light from two different regions of the spectrum.

Color magnitude diagram: A plot of the apparent or absolute magnitudes of the stars in a cluster against their color indexes.

Comet: A swarm of solid particles and gases that revolves about the sun, usually in a highly eccentric orbit.

Comet head: The main part of a comet, consisting of a nucleus and a coma.

Cometary coma: The diffuse, gaseous component of the head of a comet.

Conic section: The curve of intersection between a circular cone and a plane—an ellipse, circle, parabola, or hyperbola.

Conjunction: The configuration in which a planet appears nearest to the sun or some other planet.

Constellation: A configuration of stars named for a particular object, person, or animal.

Continuous spectrum: A spectrum of light comprised of radiation of a continuous range of wavelengths, or colors.

Corona: The outer atmosphere of the sun.

Coronagraph: An instrument for photographing the chromosphere and corona of the sun when it is not eclipsed.

Cosmic rays: Atomic nuclei (mostly protons) that strike the earth's atmosphere at exceedingly high speeds.

Cosmological model: A specific theory of the organization and evolution of the universe.

Cosmological principle: The assumption that, on the average, the universe is the same everywhere at any given time.

Cosmology: The study of the organization and evolution of the universe.

Crater (lunar): A circular depression in the surface of the moon.

Crater (meteoritic): A terrestrial crater caused by the collision of a meteoroid with the earth.

Crescent moon: A phase of the moon during which its elongation from the sun is less than 90° and it appears less than half full.

Glossary

Dark nebula: A dark cloud of interstellar dust that obscures the light of more distant stars.

Daylight savings time: A time 1 hour ahead of standard time, usually adopted in spring and summer to take advantage of long evening twilights.

Declination: Angular distance north or south of the celestial equator.

Deferent: A stationary circle in the Ptolemaic system along which the center of another circle (an epicycle) moves.

Density: The ratio of the mass of an object to its volume.

Descending node: The point in the orbit of a celestial body where it crosses from the north to the south of a reference plane, usually the plane of the celestial equator or the ecliptic.

Differential gravitational force: The difference between the gravitational force exerted on two bodies near each other by a third, more distant body.

Differential rotation: Rotation in which all the parts of an object do not behave like a solid. The sun and galaxies exhibit differential rotation.

Diffraction grating: A system of closely spaced equidistant slits through which light is passed to produce a spectrum.

Diffuse nebula: A reflection, or emission, nebula produced by interstellar matter.

Disk (galactic): The central, wheel-like portion of a spiral galaxy.

Diurnal motion: Motion during 1 day.

Doppler shift: An apparent change in wave length of the radiation from a given source due to its relative motion in the line of sight.

Dwarf (star): A main sequence star with a low mass compared to that of a giant or supergiant star.

Eclipse: The cutting off of all or part of the light of one body by another body passing in front of it.

Eclipse path: The track along the earth's surface swept out by the tip of the shadow of the moon during a solar eclipse.

Eclipse season: A period during the year when an eclipse of the sun or moon is possible.

Eclipsing binary star: A binary star in which the plane of revolution of the two components is nearly edge on to our line of sight, so that one star periodically passes in front of the other.

Ecliptic: The apparent annual path of the sun on the celestial sphere.

Electromagnetic radiation: Radiation consisting of waves propagated with the speed of light. Radio waves; infrared, visible, and ultraviolet light; X rays; and gamma rays are all forms of electromagnetic radiation.

Electromagnetic spectrum: The entire range of electromagnetic waves.

Electron: A negatively charged subatomic particle that normally moves about the nucleus of an atom.

Ellipse: A conic section; the curve of intersection of a circular cone and a plane cutting completely through it.

Elliptical galaxy: A galaxy whose apparent contours are ellipses and that contains no conspicuous interstellar material.

Emission line: A discrete, bright spectral line.

Emission nebula: A gaseous nebula that derives its visible light from the fluorescence of ultraviolet light from a star.

Emission spectrum: A spectrum of emission lines.

Energy level: A particular level, or amount, of energy possessed by an atom or ion above that which it possesses in its ground state.

Ephemeris: A table giving the positions of a celestial body at various times, or other astronomical data.

Ephemeris time: Time that passes at a strictly uniform rate, used to compute the instant of various astronomical events.

Epicycle: A circular orbit of a body in the Ptolemaic system, the center of which revolves about another circle (the deferent).

Equation of time: The difference between apparent and mean solar time.

Equator: A great circle on the earth, 90° from the North and South Poles.

Equatorial mount: A mounting for a telescope, one axis of which is parallel to the earth's axis, so that a motion of the telescope about that axis compensates for the earth's rotation.

Equinox: One of the intersections of the ecliptic and celestial equator.

Eruptive variable: A variable star whose changes in light are erratic or explosive.

Escape velocity: The speed needed to escape from the gravitational field of an object.

Event horizon: A region of highly warped space-time where the escape velocity equals the speed of light; the "surface" of a black hole.

Evolutionary cosmology: A cosmology that assumes that all parts of the universe have a common age and evolve together.

Excitation: The imparting of energy to an atom or an ion.

Extinction: The attenuation of light from a celestial body produced by the earth's atmosphere or by interstellar absorption.

Extragalactic: Beyond the galaxy.

Eyepiece: A magnifying lens used to view the image produced by the objective lens of a telescope.

Faculus (faculae): Bright region near the limb of the sun.

Filtergram: A photograph of the sun taken through a special narrow band-pass filter.

Fireball: A spectacular meteor.

Fission: The breakup of a heavy atomic nucleus into two or more lighter ones.

Flare (solar): A sudden and temporary outburst of light from an extended region of the solar surface.

Flare star: A member of a class of stars that show occasional, sudden, and unpredicted increases in light.

Flocculus (flocculi): A bright region of the solar surface observed in the monochro-

Glossary

matic light of a spectral line; usually called a plage.

Focal length: The distance from a lens, or mirror, to the point where light converged by it is focused.

Focus: The point where the rays of light converged by a mirror or lens meet.

Forbidden lines: Spectral lines not usually observed under laboratory conditions because they result from highly improbable atomic transitions.

Fraunhofer line: An absorption line in the spectrum of the sun or a star.

Fraunhofer spectrum: The array of absorption lines in the spectrum of the sun or of a star.

Frequency: The number of vibrations in a unit of time; the number of waves that cross a given point in a unit of time.

Full moon: A phase of the moon during which it is in opposition (180° from the sun), and its full daylight hemisphere is visible from the earth.

Fusion: The building up of heavier atomic nuclei from lighter ones.

Galactic cluster: An open cluster of stars in the spiral arms or disk of the galaxy.

Galactic equator: The intersection of the principal plane of the Milky Way with the celestial sphere.

Galactic latitude: Angular distance north or south of the galactic equator measured along a great circle passing through the galactic poles.

Galactic longitude: Angular distance measured eastward along the galactic equator from the galactic center to the intersection of the galactic equator with a great circle passing through the galactic poles.

Galactic poles: The poles of the galactic equator; the intersections with the celestial sphere of a line perpendicular to the plane of the galactic equator.

Galactic rotation: The rotation of the galaxy.

Galaxy: A large assemblage of stars. A typical galaxy contains millions to hundreds of billions of stars.

Gamma rays: Photons of electromagnetic radiation with wavelengths shorter than those of X rays. Gamma rays have the highest frequency of any form of electromagnetic radiation.

General relativity: A theory of gravitation proposed by Dr. Einstein in which gravity is expressed as the warping of space-time.

Giant (star): A highly luminous star with a large radius.

Gibbous moon: A phase of the moon during which more than half, but not all, of the moon's sunlit hemisphere is visible from the earth.

Globular cluster: A large, spherical cluster of stars.

Granulation (solar): The ricelike pattern evident in photographs of the solar photosphere.

Gravitation: The attraction of matter for matter.

Greenwich meridian: The meridian of longitude passing through the site of the old Royal Greenwich Observatory, near

London; the great circle from which terrestrial longitude is measured.

Gregorian calendar: A calendar (now in common use) introduced by Pope Gregory XIII in 1582.

H I region: A region of neutral hydrogen in interstellar space.

H II region: A region of ionized hydrogen in interstellar space.

Hertzsprung-Russell (H-R) diagram: A diagram showing the relationship of the absolute magnitudes of a group of stars to their temperature, spectral class, or color index.

High-velocity object: An object with a high space velocity; generally, an object that does not share the high orbital velocity of the sun about the galactic nucleus.

Horizon (astronomical): A great circle on the celestial sphere, 90° from the zenith.

Horizon system: A system of celestial coordinates (altitude and azimuth) based on the astronomical horizon and a north point.

Hour angle: An angle measured westward along the celestial equator from a local meridian to an hour circle passing through an object.

Hour circle: A great circle on the celestial sphere passing through the celestial poles.

Hubble constant: The constant of proportionality in the relationship between the velocities and distances of remote galaxies.

Hyperbola: A curve of intersection between a circular cone and a plane that is at too small an angle with the axis of the cone to cut all the way through it and is not parallel to a line in the face of the cone.

Image: The optical representation of an object produced by the refraction or reflection of light rays from the object by a lens or mirror.

Image tube: A device in which electrons emitted from a photocathode surface exposed to light are focused electronically.

Inclination (orbital): The angle between the orbital plane of a revolving body and a fundamental plane—usually the plane of the celestial equator or the ecliptic.

Index Catalogue (IC): A supplement to Dryer's *New General Catalogue* of star clusters and nebulae.

Inertia: The property of matter that makes the action of a force necessary to change the state of motion of an object.

Inferior conjunction: The configuration of an inferior planet when it has the same longitude as the sun and is between the sun and the earth.

Inferior planet: A planet whose distance from the sun is less than the earth's.

Infrared radiation: Electromagnetic radiation with a wavelength longer than the longest visible wavelength (red) but shorter than wavelengths in the radio range.

Intercalate: To insert, as a day, in a calendar.

Interferometer (stellar): An optical device that uses light-interference phenomena to measure small angles.

International Date Line: An arbitrary line on the surface of the earth, near a longitude of 180°, on either side of which the date changes by one day.

Interplanetary medium: Gas and solid particles in interplanetary space.

Glossary

Interstellar dust: Microscopic solid grains in interstellar space.

Interstellar gas: Sparse gas in interstellar space.

Interstellar lines: Absorption lines superimposed on stellar spectra, produced by interstellar gas.

Interstellar matter: Interstellar gas and dust.

Ion: An atom that has become electrically charged through the addition or loss of one or more electrons.

Ionization: The process by which an atom gains or loses electrons.

Ionosphere: The upper region of the earth's atmosphere, in which many of the atoms are ionized.

Irregular galaxy: A galaxy without rotational symmetry (that is, neither spiral nor elliptical).

Irregular variable: A variable star whose period of light variation is not regular.

Jovian planet: Jupiter, Saturn, Uranus, or Neptune.

Julian calendar: A calendar introduced by Julius Caesar in 45 B.C.

Jupiter: The fifth planet from the sun in the solar system.

Kepler's laws: Three laws developed by Johannes Kepler to describe the motions of the planets.

Kerr black hole: A rotating black hole.

Kerr solution: A mathematical solution of the equations of general relativity describing the curvature of space-time about a rotating black hole.

Kirkwood's gaps: Gaps in the spacing of the periods of the minor planets due to perturbations produced by the major planets.

Latitude: A north-south coordinate on the surface of the earth; the angular distance north or south of the equator measured along a meridian.

Leap year: A year with 366 days, intercalated approximately every 4 years to make the average length of the calendar year as nearly equal as possible to the tropical year.

Libration: A change in the visible hemisphere of the moon viewed from earth.

Light: Electromagnetic radiation visible to the human eye.

Light curve: A graph showing the variation in light, or magnitude, of a variable or an eclipsing binary star.

Light year: The distance light travels in a vacuum in 1 year; 6 trillion miles.

Limb: The apparent edge of a celestial body.

Limiting magnitude: The faintest magnitude that can be observed with a given instrument or under given conditions.

Line of nodes: The line connecting the nodes of an orbit.

Local Group: The cluster of galaxies to which our galaxy belongs.

Longitude: An east-west coordinate on the earth's surface; the angular distance along the equator east or west of the Greenwich meridian to another meridian.

Low-velocity object: An object that has a low space velocity; generally, an object that shares the sun's high orbital speed about the galactic center.

Luminosity: The rate of radiation of energy into space by a celestial object.

Luminosity class: A subclassification of a star of a given spectral class according to its luminosity.

Lunar eclipse: An eclipse of the moon.

Magellanic Clouds: Two neighboring galaxies visible to the naked eye in southern latitudes.

Magnetic field: The region of space near a magnetized body within which magnetic forces can be detected.

Magnifying power: A measure of the strength of a telescope based on the increase in angular diameter of an object viewed through it.

Magnitude: A measure of the amount of light received from a star or another luminous object.

Main sequence: The largest sequence of stars on the Hg-R diagram.

Major axis (of an ellipse): The maximum diameter of an ellipse.

Mare: Latin for sea; the name applies to many sealike features on the moon and Mars.

Mars: The fourth planet from the sun in the solar system.

Mass: A measure of the total amount of material in a body.

Mass-luminosity relationship: An empirical relationship between the masses and luminosities of many stars.

Mass-radius relationship (of white dwarfs): A theoretical relationship between the masses and radii of white dwarf stars.

Maximum elongation: The point in the orbit of an inferior planet at which the difference between its celestial longitude and that of the sun is greatest.

Mean solar day: The interval between successive passages of the mean sun across the meridian; the average length of the apparent solar day.

Mean solar time: Time reckoned from the position of the mean sun.

Mean sun: A fictitious body that moves eastward with uniform angular velocity along the celestial equator.

Mechanics: The branch of physics that deals with the behavior of material bodies.

Mercury: The nearest planet to the sun in the solar system.

Meridian (celestial): A great circle on the celestial sphere that passes through an observer's zenith and the north (or south) celestial pole.

Meridian (terrestrial): A great circle on the surface of the earth that passes through a particular place and the North and South Poles of the earth.

Messier Catalogue: A catalogue of nonstellar objects compiled by Charles Messier in 1787.

Meteor: The luminous phenomenon observed when a meteoroid enters the earth's atmosphere and burns up; sometimes called a shooting star.

Meteor shower: The apparent descent of a large number of meteors radiating from a common point in the sky, caused by the

Glossary

collision of the earth with a swarm of meteoritic particles.

Meteorite: A portion of a meteoroid that survives passage through the atmosphere and strikes the ground.

Meteoroid: A meteoritic particle in space.

Milky Way: A band of light encircling the sky, caused by the many stars lying near the plane of the galaxy; also used as synonym for the galaxy to which the sun belongs.

Minor axis (of an ellipse): The smallest, or least, diameter of an ellipse.

Minor planet: One of several tens of thousands of small planets, ranging from a few hundred miles to less than 1 mile in diameter.

Mira-type variable: Any of a large class of red giant, long-period, or irregularly pulsating variables, of which the star Mira is a prototype.

Molecule: Two or more atoms bound together.

Monochromatic: Of one wave length, or color.

Nadir: A point on the celestial sphere 180° from the zenith.

Neap tides: The lowest tides in the month, which occur when the moon is near the first or third quarter phase.

Nebula (nebulae): A cloud of interstellar gas or dust.

Neptune: The eighth planet from the sun in the solar system.

Neutron: A subatomic particle with no charge and with a mass approximately equal to that of the proton.

Neutron star: A star of extremely high density composed entirely of neutrons.

New General Catalogue (NGC): A catalogue of star clusters, nebulae, and galaxies compiled by J. L. E. Dreyer in 1888.

New moon: A phase of the moon during which its longitude is the same as that of the sun.

Newton's laws: Laws of mechanics and gravitation formulated by Isaac Newton.

Newtonian focus: An optical arrangement in a reflecting telescope whereby light is reflected by a flat mirror to a focus at the side of the telescope tube just before it reaches the focus of the objective lens.

Node: An intersection of the orbit of a celestial body with a fundamental plane, usually the plane of the celestial equator or the ecliptic.

North point: That intersection of the celestial meridian and the astronomical horizon lying nearest the north celestial sphere.

Nova: A star that experiences a sudden outburst of radiant energy, temporarily increasing its luminosity hundreds or thousands of times.

Nuclear transformation: The transformation of one atomic nucleus into another.

Nucleus (atomic): The heavy part of an atom, composed mostly of protons and neutrons, about which the electrons revolve.

Nucleus (cometary): A swarm of solid particles in the head of a comet.

Nucleus (galactic): A concentration of stars, and possibly gas, at the center of a galaxy.

Nutation: The nodding motion of the earth's axis.

Objective lens: The principal image-forming component of a telescope or another optical instrument.

Objective prism: A prismatic lens placed in front of the objective lens of a telescope to transform each star image into an image of the stellar spectrum.

Oblate spheroid: A solid formed by rotating an ellipse about its minor axis.

Oblateness: A measure of the flattening of an oblate spheroid.

Obliquity (of the ecliptic): The angle between the planes of the celestial equator and the ecliptic; about $23\frac{1}{2}°$.

Obscuration (interstellar): The absorption of starlight by interstellar dust.

Occultation: An eclipse of a star or planet by the moon or another planet.

Open cluster: A comparatively loose, or "open," cluster of a few dozen to a few thousand stars in the spiral arms or disk of the galaxy; a galactic cluster.

Opposition: The configuration of a planet when its elongation is 180°.

Optical binary: Two stars at different distances that, when viewed in projection, are so nearly lined up that they appear close together, although they are not dynamically associated.

Optics: The branch of physics that deals with the properties of light.

Orbit: The path of a body revolving about another body or point.

Parabola: The curve of intersection between a circular cone and a plane parallel to one side of the cone.

Paraboloid: A parabola of revolution; a curved surface whose cross section is parabolic. The surface of the primary mirror in a standard reflecting telescope is a paraboloid.

Parallax: An apparent displacement of an object due to a motion of the observer.

Parallax (stellar): An apparent displacement of a nearby star that results from the motion of the earth around the sun; numerically, the angle subtended by 1 AU at the distance of a particular star.

Parsec: The distance of an object with a stellar parallax of 1 second of arc. A parsec equals 3.26 light years.

Partial eclipse: An eclipse in which the concealed body is not completely obscured.

Penumbra: The portion of a shadow from which only part of the light source is occulted by an opaque body.

Penumbral eclipse: A lunar eclipse in which the moon passes through the penumbra, but not the umbra, of the earth's shadow.

Periastron: The place in the orbit of a star in a binary star system where it is closest to its companion star.

Perigee: The place in the orbit of an earth satellite where it is closest to the center of the earth.

Perihelion: The place in the orbit of an object revolving about the sun where it is closest to the center of the sun.

Period: Generally the interval of time required for a celestial body to rotate once

Glossary

on its axis, revolve once about a primary, or return to its original state after an increase in luminosity.

Period-luminosity relationship: An empirical relationship between the periods and luminosities of Cepheid variable stars.

Perturbation: A small disturbance in the motion of a celestial body produced by the proximity of another body.

Phases (lunar): The changes in the moon's appearance as different portions of its illuminated hemisphere become visible from the earth.

Photocell (photoelectric cell): An electron tube in which electrons are dislodged from the cathode when it is exposed to light and accelerated, thus producing a current in the tube whose strength serves as a measure of the light striking the cathode.

Photographic magnitude: The magnitude of an object, as measured on traditional blue- and violet-sensitive photographic emulsions.

Photometry: The measurement of light intensities.

Photomultiplier: A photoelectric cell in which the electric current generated is amplified at several stages within the tube.

Photon: A discrete unit of electromagnetic energy.

Photosphere: The region of the solar (or stellar) atmosphere from which radiation escapes into space.

Photovisual magnitude: A magnitude corresponding to the spectral region to which the human eye is most sensitive, but measured photographically with green- and yellow-sensitive emulsions and filters.

Plage: A bright region of the solar surface observed in the monochromatic light of a particular spectral line; a flocculus.

Planck's law of radiation: A formula for calculating the intensity of radiation at various wave lengths emitted by a blackbody.

Planet: Any of nine solid, nonluminous bodies revolving about the sun.

Planetary nebula: A shell of gas ejected from, and enlarging about, an extremely hot star.

Planetoid: A minor planet.

Pluto: The ninth planet from the sun in the solar system.

Polar axis: The axis of rotation of a planet; also, an axis in the mounting of a telescope that is parallel to the earth's axis.

Position angle: The direction in the sky of one celestial object with respect to another; for example, the angle, measured to the east from the north, showing the position of the fainter component of a visual binary star with respect to the brighter component.

Precession (of the equinoxes): The slow westward motion of the equinoxes along the ecliptic due to precession.

Precession (terrestrial): A slow, conical motion of the earth's axis of rotation, caused principally by the gravitational torque of the moon and sun on the earth's equatorial bulge.

Primary minimum (in the light curve of an eclipsing binary): The middle of the eclipse, during which the most light is lost.

Prime focus: The point in a telescope

where the objective lens focuses the light.

Prime meridian: The meridian of longitude passing through the site of the old Royal Greenwich Observatory, near London; the great circle from which terrestrial longitude is measured.

Primordial fireball: The extremely hot opaque gas presumed to have contained the entire mass of the universe at the time of, or immediately following, the explosion of the primeval atom.

Prism: A wedge-shaped piece of glass used to disperse white light into a spectrum.

Prominence: A flamelike phenomenon in the solar corona.

Proper motion: The angular change in direction of a star during 1 year.

Proton: A heavy subatomic particle that carries a positive charge; one of the two principal constituents of the atomic nucleus.

Pulsar: A small but powerful celestial radio source with very regular, short periods.

Pulsating variable: A variable star that periodically changes in size and luminosity.

Quadrature: A configuration of a planet in which its elongation is 90°.

Quarter moon: Either of the two phases of the moon when its longitude differs by 90° from that of the sun and it appears half full.

Quasar: A quasi-stellar source or object.

Quasi-stellar galaxy: An apparently stellar object with a very large redshift, presumed to be extragalactic and highly luminous.

Quasi-stellar object: A stellar object with a very large redshift.

Quasi-stellar source: A stellar object with a very large redshift that is a strong source of radio waves.

Radial velocity: The component of relative velocity that lies in the line of sight.

Radial-velocity curve: A plot of the variation in radial velocity over time of a binary or variable star.

Radiant (of a meteor shower): The point in the sky from which the meteors belonging to a shower seem to radiate.

Radiation: The emission and transmission of energy in the form of waves or particles.

Radio astronomy: The use of radio wavelengths to make astronomical observations.

Radio telescope: A telescope designed to make observations in radio wavelengths.

Ray (lunar): Any of a system of bright, elongated streaks, sometimes associated with a crater.

Recurrent nova: A nova that has erupted more than once.

Red giant: A large, cool star of high luminosity in the upper right portion of the Hg-R diagram.

Reddening (interstellar): A change in the color, or wavelength, of starlight passing through interstellar dust, which scatters blue light more effectively than red.

Redshift: A shift to longer wavelengths

Glossary

of the light from remote galaxies, presumably caused by their velocity of recession.

Reflecting telescope: A telescope in which the principal optical component is a concave mirror.

Reflection nebula: A relatively dense interstellar dust cloud that is illuminated by starlight.

Refracting telescope: A telescope in which the principal optical component (objective) is a lens or system of lenses.

Refraction: The bending of light rays passing from one transparent medium to another.

Relative orbit: The orbit of one of two mutually revolving bodies about the other.

Resolution: The degree to which fine details in an image are separated, or resolved.

Resolving power: A measure of the ability of an optical system to resolve, or separate, fine details in the image it produces.

Retrograde motion: An apparent westward motion of a planet on the celestial sphere or with respect to the stars.

Revolution: The motion of one body around another.

Right ascension: A coordinate for measuring the east-west positions of celestial bodies; the angle, measured eastward along the celestial equator, from the vernal equinox to the hour circle passing through a body.

Rill (lunar): A crevasse, or trenchlike depression, in the moon's surface.

Rotation: The turning of a body about an axis running through it.

RR Lyrae variable: One of a class of giant pulsating stars with periods of less than 1 day; a cluster variable.

Satellite: A body, such as the earth's moon, that revolves about a larger body.

Saturn: The sixth planet from the sun in the solar system.

Schmidt telescope: A reflecting telescope in which aberrations produced by a spherical concave mirror are compensated for by a thin objective correcting lens.

Schwarzschild black hole: A nonrotating, uncharged black hole.

Schwarzschild solution: A mathematical solution of the equations of general relativity describing the curvature of space-time about a static black hole.

Secondary minimum (in an eclipsing binary's light curve): The middle of the eclipse of the fainter star by the brighter, at which time the light of the system diminishes less than it does during the eclipse of the brighter star by the fainter.

Seleno-: A prefix referring to the moon.

Semimajor axis: Half the major axis of a conic section.

Separation (in a visual binary): The angular separation of the two components of a visual binary star.

Seyfert galaxy: A galaxy with an unusually bright nucleus.

Shadow cone: The umbra of the shadow of a spherical body (such as the earth) in sunlight.

Shell star: A type of star, usually a class B, A, or F star, surrounded by a gaseous ring, or shell.

Sidereal day: The interval between two successive meridian passages of the vernal equinox.

Sidereal month: The period of the moon's revolution about the earth with respect to the stars.

Sidereal period: The period of revolution of one body about another with respect to the stars.

Sidereal time: The local hour angle of the vernal equinox.

Sidereal year: The period of the earth's revolution about the sun with respect to the stars.

Singularity: A place in space-time of infinite curvature; the "center" of a black hole.

Small circle: Any circle on the surface of a sphere that is not a great circle.

Solar activity: Phenomena in the solar atmosphere associated with sunspots, plages, and the like.

Solar antapex: The direction away from which the sun is moving with respect to the local standard of rest.

Solar apex: The direction toward which the sun is moving with respect to the local standard of rest.

Solar motion: The motion, or velocity, of the sun with respect to the local standard of rest.

Solar parallax: The angle subtended by the equatorial radius of the earth at a distance of 1 AU.

Solar system: The system of the sun and the planets, satellites, asteroids, comets, meteoroids, and other objects revolving around it.

Solar time: Time measured according to the sun.

Solar wind: A radial flow of corpuscular radiation leaving the sun.

Solstice: Either of two points on the celestial sphere where the sun reaches its maximum distances north or south of the celestial equator.

South point: The intersection of the celestial meridian and the astronomical horizon 180° from the north point.

Spectral class (or type): A classification of a star according to the characteristics of its spectrum.

Spectral sequence: The sequence of spectral classes of stars arranged in order of decreasing temperature.

Spectrogram: A photograph of a spectrum.

Spectrograph: An instrument for photographing a spectrum; usually attached to a telescope to photograph the spectrum of a star.

Spectroheliogram: A photograph of the sun obtained with a spectroheliograph.

Spectroheliograph: An instrument for photographing the sun, or part of the sun, in the monochromatic light of a particular spectral line.

Spectroscope: An instrument for viewing the spectrum of a light source directly.

Spectroscopic binary star: A binary star in which the components are not resolvable optically, but whose binary nature is indicated by periodic variations in its radial velocity indicating orbital motion.

Spectroscopic parallax: A parallax derived by comparing the apparent magni-

Glossary

tude of the star with its absolute magnitude as deduced from its spectral characteristics.

Spectroscopy: The study of spectra.

Spectrum: The array of colors, or wavelengths, obtained when light from a source is dispersed by passing it through a prism or grating.

Spectrum analysis: The study and analysis of spectra, especially stellar spectra.

Spectrum binary: A star whose binary nature is revealed by spectral characteristics that can result only from a composite of the spectra of two different stars.

Spicule: A narrow jet of material rising in the solar chromosphere.

Spiral arms: Armlike areas of interstellar material and young stars that wind out in a plane from the central nucleus of a spiral galaxy.

Spiral galaxy: A flattened, rotating galaxy with wheel-like arms of interstellar material and young stars winding out from its nucleus.

Spring tide: The highest tide of the month, produced when the longitudes of the sun and moon differ from each other by nearly 0° or 180°.

Standard time: Local mean solar time of a standard meridian, adopted over a large region to avoid the inconvenience of continuous time changes around the earth.

Star: A self-luminous sphere of gas.

Star cluster: An assemblage of stars held together by their mutual gravitation.

Steady-state theory: A cosmological theory based on the cosmological principle and the continuous creation of matter.

Stefan-Boltzmann law: A formula for computing the rate at which a blackbody radiates energy.

Stellar evolution: The changes that take place in the size, luminosity, structure, and other characteristics of a star as it ages.

Stellar parallax: The angle subtended by 1 AU at the distance of a star, usually measured in seconds of arc.

Subdwarf: A star less luminous than main sequence stars of the same spectral type.

Subgiant: A star more luminous than main sequence stars of the same spectral type but less so than normal giants of that type.

Summer solstice: The point on the celestial sphere where the sun is farthest north of the celestial equator.

Sun: The star about which the earth and other planets revolve.

Sunspot: A temporary cool region in the solar photosphere that appears dark in contrast to the surrounding hotter photosphere.

Sunspot cycle: A semiregular 11-year period during which the frequency of sunspots fluctuates.

Supergiant: A star of very high luminosity.

Superior conjunction: The configuration of a planet in which it has the same longitude as the sun but is more distant than the sun.

Superior planet: A planet farther from the sun than the earth.

Supernova: A stellar outburst, or explosion, in which a star suddenly increases in luminosity by hundreds of thousands or hundreds of millions of times.

Surface gravity: The acceleration of gravity at the surface of a planetary body.

Synodic month: The period of revolution of the moon with respect to the sun; or the period of the cycle of lunar phases.

Synodic period: The interval between successive occurrences of the same configuration of a planet (for example, between successive oppositions or successive superior conjunctions).

Tail (cometary): Gases and solid particles ejected from the head of a comet and forced away from the sun by radiation pressure or corpuscular radiation.

Tektites: Rounded glassy bodies suspected to be of meteoritic origin.

Telescope: An optical instrument used to view, measure, or photograph distant objects.

Telluric: Of terrestrial origin.

Temperature (absolute): Temperature measured in centigrade degrees from absolute zero.

Temperature (centigrade): Temperature measured on a scale calibrated so that water freezes at 0° and boils at 100°.

Temperature (color): Temperature of a star as estimated from the intensity of the stellar radiation in two or more colors, or wavelengths.

Temperature (effective): Temperature of a blackbody that would radiate the same total amount of energy as a particular body.

Temperature (Fahrenheit): Temperature measured on a scale calibrated so that water freezes at 32° and boils at 212°.

Temperature (Kelvin): Absolute temperature measured in centigrade degrees (Kelvin temperature = centigrade temperature $+273°$).

Temperature (radiation): The temperature of a blackbody that radiates the same amount of energy in a given spectral region as a particular body.

Terminator: The line of sunrise or sunset on a celestial body such as the moon.

Terrestrial planet: Mercury, Venus, Earth, Mars, and sometimes Pluto.

Thermonuclear energy: Energy associated with thermonuclear reactions.

Thermonuclear reaction: A nuclear reaction or transformation that results from encounters between high-velocity nuclear particles.

Tidal force: A differential gravitational force that tends to deform a body.

Tide: A deformation of a body caused by the differential gravitational force exerted on it by another body.

Total eclipse: An eclipse of the sun in which the photosphere is hidden entirely by the moon; a lunar eclipse in which the moon passes completely into the umbra of the earth's shadow.

Train (of a meteor): A temporarily luminous trail in the wake of a meteor.

Triangulation: The measurement of some of the elements of a triangle so that other elements can be calculated by trigonometric operations; a method of determining distances without taking direct measurements.

Glossary

Trojan minor planet: One of several minor planets that share Jupiter's orbit around the sun but are located approximately 60° around the orbit from Jupiter.

Tropical year: The period of revolution of the earth about the sun with respect to the vernal equinox.

UBV system: A system of measuring stellar magnitudes in the ultraviolet, blue, and green-yellow regions of the spectrum.

Uhuru: The nickname for Explorer 42, an astronomical satellite designed to detect X rays.

Ultraviolet radiation: Electromagnetic radiation whose wavelength is shorter than the shortest wavelengths to which the eye is sensitive (violet); radiation whose wavelength ranges from approximately 100 to 4,000 Å.

Umbra: The central, completely dark part of a shadow.

Universal time: The local mean time of the prime meridian.

Universe: The totality of matter, radiation, and space.

Uranus: The seventh planet from the sun in the solar system.

Venus: The second planet from the sun in the solar system.

Vernal equinox: The point on the celestial sphere where the sun crosses from the south to the north of the celestial equator.

Vertical circle: Any great circle passing through the zenith.

Visual binary star: A binary star in which the two components can be resolved telescopically.

Visual photometer: An instrument used with a telescope to measure, visually, the light flux from a star.

Volume: A measure of the total space occupied by a body.

Wavelength: The spacing of the crests or troughs in a wave train.

Weight: A measure of the force of gravitational attraction.

West point: The point on the horizon 270° from the north point, measured clockwise from the zenith.

White dwarf: A star that has exhausted most or all of its nuclear fuel and has collapsed to a very small size.

Widmanstätten figures: Crystalline structures observable in cut and polished meteorites.

Wien's law: A formula relating the temperature of a blackbody to the exact wavelength at which it emits the greatest intensity of radiation.

Winter solstice: The point on the celestial sphere where the sun is farthest south of the celestial equator.

Wolf-Rayet star: One of a class of very hot stars that eject shells of gas at a very high velocity.

X-ray stars: Stars that emit observable amounts of radiation at X-ray frequencies.

X rays: Photons whose wavelengths are shorter than ultraviolet wavelengths and longer than gamma wavelengths.

Zeeman effect: A splitting or broadening of spectral lines due to magnetic fields.

Zenith: The point on the celestial sphere

opposite the direction of gravity; the direction opposite to that indicated by a plumb bob.

Zenith distance: The distance, in degrees of arc, of a point on the celestial sphere from the zenith.

Zodiac: A belt around the sky centered on the ecliptic.

Zodiacal light: A faint illumination along the Zodiac, believed to be due to sunlight reflected and scattered by interplanetary dust.

Zone of avoidance: A region near the Milky Way where the interstellar dust is so thick that few or no exterior galaxies can be seen.

Zone time: The time, kept in a zone 15° wide, equal to the local mean time of the central meridian of the zone.

Appendix

Table A-1. The Planets (Physical Data)

Planet	Diameter (in miles)	Diameter (Earth = 1)	Mass (Earth = 1)	Surface Gravity (Earth = 1)	Period of Rotation	Number of Moons
Mercury	3,025	0.38	0.06	0.38	58.65 days	0
Venus	7,526	0.95	0.82	0.90	243 days	0
Earth	7,927	1.00	1.00	1.00	23 hrs. 56 min.	1
Mars	4,218	0.53	0.11	0.38	24 hrs. 37 min.	2
Jupiter	88,700	11.19	318.0	2.64	9 hrs. 50 min.	14
Saturn	75,100	9.47	95.2	1.13	10 hrs. 14 min.	10
Uranus	29,200	3.69	14.6	1.07	10 hrs. 49 min.	5
Neptune	31,650	3.50	17.3	1.08	16 hrs.	2
Pluto	3,500?	0.5?	0.1?	0.3?	6.39 days	0

Table A-2. The Planets (Orbital Data)

Planet	Average Distance from the Sun (in AUs)	Average Distance from the Sun (in millions of miles)	Orbital Period (in years)	Orbital Period (in days)	Orbital Speed (in miles per second)	Orbital Inclination (in degrees)
Mercury	0.387	36.0	0.241	88.0	29.7	7.0
Venus	0.723	67.2	0.615	224.7	21.8	3.4
Earth	1.000	92.9	1.000	365.3	18.5	0.0
Mars	1.524	141.5	1.881	687.0	15.0	1.8
Jupiter	5.203	483.4	11.862		8.1	1.3
Saturn	9.539	886.0	29.458		6.0	2.5
Uranus	19.18	1782.0	84.013		4.2	0.8
Neptune	30.06	2792.0	164.793		3.4	1.8
Pluto	39.44	3664.0	247.686		2.9	17.2

Table A-3. Satellites of Planets

Name	Maximum Magnitude	Diameter (in miles)	Average Distance from Planet (in miles)	Period of Revolution	Discoverer
Satellite of Earth					
Moon	−12.7	2160	238,900	27^d 07^h 43^m	
Satellites of Mars					
Phobos	11.6	12	5,800	0^d 07^h 39^m	Hall, 1877
Deimos	12.8	(< 10)*	14,600	1 06 18	Hall, 1877
Satellites of Jupiter†					
V	13.0	(100)	112,000	0^d 11^h 57^m	Barnard, 1892
Io	4.8	2020	262,000	1 18 28	Galileo, 1610
Europa	5.2	1790	417,000	3 13 14	Galileo, 1610
Ganymede	4.5	3120	665,000	7 03 43	Galileo, 1610
Callisto	5.5	2770	1,171,000	16 16 32	Galileo, 1610
XIII	(20)	(< 10)	2,707,000	239 06	Kowal, 1974
VI	13.7	(50)	7,133,000	250 14	Perrine, 1904
VII	16	(20)	7,295,000	259 16	Perrine, 1905
X	18.6	(< 10)	7,369,000	263 13	Nicholson, 1938
XII	18.8	(< 10)	13,200,000	631 02	Nicholson, 1951
XI	18.1	(< 10)	14,000,000	692 12	Nicholson, 1938
VIII	18.8	(< 10)	14,600,000	738 22	Melotte, 1908
IX	18.3	(< 10)	14,700,000	758	Nicholson, 1914
Satellites of Saturn					
Janus	(14)	< 300	100,000	0^d 17^h 59^m	Dollfus, 1966
Mimas	12.1	300	116,000	0 22 37	Herschel, 1789
Enceladus	11.8	400	148,000	1 08 53	Herschel, 1789
Tethys	10.3	600	183,000	1 21 18	Cassini, 1684
Dione	10.4	600	235,000	2 17 41	Cassini, 1684
Rhea	9.8	810	327,000	4 12 25	Cassini, 1672
Titan	8.4	2980	759,000	15 22 41	Huygens, 1655
Hyperion	14.2	(100)	920,000	21 06 38	Bond, 1848
Iapetus	11.0	(500)	2,213,000	79 07 56	Cassini, 1671
Phoebe	(14)	(100)	8,053,000	550 11	Pickering, 1898
Satellites of Uranus					
Miranda	16.5	(200)	77,000	1^d 09^h 56^m	Kuiper, 1948
Ariel	14.4	(500)	119,000	2 12 29	Lassell, 1851
Umbriel	15.3	(300)	166,000	4 03 38	Lassell, 1851
Titania	14.0	(600)	272,000	8 16 56	Herschel, 1787
Oberon	14.2	(500)	365,000	13 11 07	Herschel, 1787
Satellites of Neptune					
Triton	13.6	2300	220,000	5^d 21^h 03^m	Lassell, 1846
Nereid	18.7	(200)	3,461,000	359 10	Kuiper, 1949

*Parentheses indicate that numbers are estimated.
†While this book was in press, a fourteenth satellite was discovered orbiting Jupiter. Precise data concerning this satellite are not as yet available.

Table A-4. The Brightest Stars

Name	Common Name	Apparent Magnitude	Right Ascension	Declination	Distance (in light years)
1. α Canis Majoris	Sirius	−1.58	$6^h 44^m$	−16°41′	8.7
2. α Carinae	Canopus	−0.86	6 23	−52 41	98
3. α Centauri	Rigil Kentaurus	0.06	14 38	−60 43	4.3
4. α Lyrae	Vega	0.14	18 36	+38 45	26.5
5. α Aurigae	Capella	0.21	5 14	+45 58	45
6. α Bootis	Arcturus	0.24	14 14	+19 20	36
7. β Orionis	Rigel	0.34	5 13	−08 14	900
8. α Canis Minoris	Procyon	0.48	7 38	+05 18	11.3
9. α Eridani	Achernar	0.60	1 37	−57 23	118
10. β Centauri	Agena	0.86	14 02	−60 13	490
11. α Aquilae	Altair	0.89	19 49	+08 47	16.5
12. α Orionis	Betelgeuse	0.92 (var.)	5 54	+07 24	520
13. α Crucis	Acrux	1.05	12 25	−62 56	370
14. α Tauri	Aldebaran	1.06	4 34	+16 27	68
15. β Geminorum	Pollux	1.21	7 44	+28 06	35
16. α Virginis	Spica	1.21	13 24	−11 00	220
17. α Scorpii	Antares	1.22	16 28	−26 22	520
18. α Piscis Austrini	Fomalhaut	1.29	22 56	−29 47	22.6
19. α Cygni	Deneb	1.33	20 40	+45 10	1600
20. α Leonis	Regulus	1.34	10 07	+12 07	84
21. β Crucis		1.50	12 46	−59 32	490
22. α Geminorum	Castor	1.58	7 33	+31 57	45
23. γ Crucis	Gacrux	1.61	12 30	−56 57	220
24. ε Canis Majoris	Adhara	1.63	6 57	−28 56	680
25. ε Ursae Majoris	Alioth	1.68	12 53	+56 07	68
26. γ Orionis	Bellatrix	1.70	5 24	+06 19	470
27. λ Scorpii	Shaula	1.71	17 32	−37 05	310
28. ε Carinae	Avior	1.74	8 22	−59 24	340
29. ε Orionis	Alnilam	1.75	5 35	−01 13	1600
30. β Tauri	El Nath	1.78	5 24	+28 35	300
31. β Carinae	Miaplacidus	1.80	9 13	−69 36	86
32. α Trianquli Australis	Atria	1.88	16 46	−68 59	82
33. α Persei	Marfak	1.90	3 22	+49 45	570
34. η Ursae Majoris	Alkaid	1.91	13 46	+49 28	210
35. γ Velorum	Regor	1.92	8 09	−47 16	520
36. γ Geminorum	Alhena	1.93	6 36	+16 26	105
37. α Ursae Majoris	Dubhe	1.95	11 02	+61 55	105
38. ε Sagittarii	Kaus Australis	1.95	18 22	−34 24	124
39. δ Canis Majoris	Wezen	1.98	7 07	−26 21	2100
40. β Canis Majoris	Mirzam	1.99	6 21	−17 56	750
41. δ Velorum		2.01	8 44	−54 36	76
42. θ Scorpii	Sargas	2.04	17 35	−42 59	650

Table A–4. *The Brightest Stars (Continued)*

			Position (1970)		
Name	Common Name	Apparent Magnitude	Right Ascension	Declination	Distance (in light years)
43. ζ Orionis	Alnitak	2.05	5 39	−01 57	1600
44. β Aurigae	Menkalinan	2.07	5 57	+44 57	88
45. α Cassiopeiae	Schedar	2.1 (var.)	00 39	+56 22	150
46. α Pavonis	Peacock	2.12	20 23	−56 50	310
47. α Ursae Minoris	Polaris	2.12	2 02	+89 08	680
48. σ Sagittarii	Nunki	2.14	18 53	−26 20	300
49. α Ophiuchi	Rasalhague	2.14	17 34	+12 35	58
50. α Andromedae	Alpheratz	2.15	00 07	+28 55	90

Table A–5. *The Nearest Stars*

	Position (1970)			
Name	Right Ascension	Declination	Distance (in light years)	Apparent Magnitude
α Centauri*	14^h 37^m	−60°43′	4.3	0.1
Barnard's Star	17 56	+04 36	5.9	9.5
Wolf 359	10 55	+07 13	7.6	13.5
Lalande 21185	11 02	+36 10	8.1	7.5
Sirius*	6 44	−16 41	8.6	−1.5
Luyten 726–8*	1 37	−18 07	8.9	12.5
Ross 154	18 48	−23 51	9.4	10.6
Ross 248	23 40	+44 01	10.3	12.2
ε Eridani	3 32	−09 34	10.7	3.7
Luyten 789–6	22 37	−15 31	10.8	12.2
Ross 128	11 46	+01 01	10.8	11.1
61 Cygm*	21 06	+38 36	11.2	5.2
ε Indi	22 02	−56 55	11.2	4.7
Procyon*	7 38	+05 18	11.4	0.3
Σ 2398*	18 42	+59 35	11.5	8.9
Groom. 34*	00 17	+43 51	11.6	8.1
Lacaille 9352	23 04	−36 02	11.7	7.4
τ Ceti	1 43	−16 06	11.9	3.5
BD + 5° 1668	7 26	+05 28	12.2	9.8
Lacaille 8760	21 15	−39 00	12.5	6.7
Kapteyn's Star	5 11	−45 00	12.7	8.8
Kruger 60*	22 27	+57 33	12.8	9.7
Ross 614*	6 28	−02 48	13.1	11.3

Table A-5. The Nearest Stars (Continued)

Name	Position (1970) Right Ascension	Declination	Distance (in light years)	Apparent Magnitude
BD −12° 4523	16 29	−12 35	13.1	10.0
van Maanen's Star	00 47	+05 16	13.9	12.4
Wolf 424*	12 32	+09 12	14.2	12.6
CD −37° 15492	00 03	−37 30	14.5	8.6
Groom. 1618	10 09	+49 36	15.0	6.6

An asterisk indicates that this star is actually double or multiple. In such cases the magnitude of the brightest component is given.

Table A-6. The Brightest Galaxies

Name	Position (1970) Right Ascension	Declination	Apparent Magnitude	Distance (in millions of light years)
1. LMC	5h 23m8	−69°47′	0.9	0.2
2. SMC	00 51.7	−72 59	2.9	0.2
3. M31	00 41.1	+41 07	4.3	2.1
4. M33	1 32.2	+30 30	6.2	2.4
5. M83	13 35.4	−29 43	7.0	8.0
6. NGC 253	00 46.1	−25 27	7.0	7.5
7. M81	9 53.1	+69 12	7.8	6.5
8. NGC 5128	13 23.6	−42 51	7.9	
9. NGC 55	00 13.5	−39 23	7.9	7.5
10. NGC 4945	13 03.5	−49 19	8.0	
11. M101	14 02.1	+54 29	8.2	14.0
12. NGC 300	00 53.5	−37 51	8.7	7.5
13. NGC 2403	7 33.9	+65 40	8.8	6.5
14. M51	13 28.6	+47 21	8.9	14.0
15. NGC 205	00 38.7	+41 32	8.9	2.1
16. NGC 4258	12 17.5	+47 28	8.9	14.0
17. M94	12 49.5	+41 16	8.9	14.0
18. M32	00 41.1	+40 43	9.1	2.1
19. Fornax	2 38.3	−34 39	9.1	0.4
20. M104	12 38.3	−11 28	9.2	37.0
21. M82	9 53.6	+69 50	9.2	6.5
22. NGC 6822	19 43.2	−14 50	9.2	1.7
23. M63	13 14.4	+42 11	9.3	14.0
24. M64	12 55.3	+21 51	9.3	12.0
25. M49	12 28.3	+08 09	9.3	37.0
26. NGC 247	00 45.6	−20 54	9.5	7.5
27. NGC 2903	9 30.4	+21 39	9.5	19.0

Table A-7. *The Nearest Galaxies*

Name	Position (1970) Right Ascension	Position (1970) Declination	Apparent Magnitude	Distance (in thousands of light years)
LMC	$5^h\ 23^m8$	$-69°47'$	0.9	160
SMC	00 51 .7	$-72\ 59$	2.9	190
Ursa Minor	15 08 .4	$+67\ 13$		250
Draco	17 19 .7	$+57\ 57$		260
Sculptor	00 58 .4	$-33\ 52$	10.5	280
Fornax	2 38 .3	$-34\ 39$	9.1	430
Leo I	10 06 .9	$+12\ 27$	11.3	750
Leo II	11 11 .9	$+22\ 19$	12.8	750
NGC 6822	19 43 .2	$-14\ 50$	9.2	1700
M31	00 41 .1	$+41\ 07$	4.3	2100
NGC 205	00 38 .7	$+41\ 32$	8.9	2100
M32	00 41 .1	$+40\ 43$	9.1	2100
NGC 185	00 37 .2	$+48\ 11$	10.3	2100
NGC 147	00 31 .5	$+48\ 11$	10.6	2100
M33	1 32 .2	$+30\ 30$	6.2	2400
IC 1613	1 03 .5	$+01\ 58$	10.0	2400

Table A-8. *Messier Catalogue*

M	NGC	Position (1970) Right Ascension	Position (1970) Declination	Constellation	Apparent Magnitude	Distance (in light years)	Description
1	1952	$5^h\ 32^m7$	$+22°01'$	Tau	10	3,500	Crab nebula, (supernova of 1054)
2	7089	21 31 .9	$-00\ 57$	Aqr	7	45,000	Globular cluster
3	5272	13 40 .8	$+28\ 32$	CVn	6	40,000	Globular cluster
4	6121	16 21 .8	$-26\ 26$	Sco	6	10,000	Globular cluster
5	5904	15 17 .0	$+02\ 13$	Ser	6	30,000	Globular cluster
6	6405	17 38 .1	$-32\ 11$	Sco	6	1,800	Open cluster
7	6475	17 51 .9	$-34\ 48$	Sco	5	800	Bright open cluster
8	6523	18 01 .8	$-24\ 23$	Sgr		5,000	Lagoon nebula
9	6333	17 17 .5	$-18\ 29$	Oph	7	26,000	Globular cluster
10	6254	16 55 .5	$-04\ 04$	Oph	7	23,000	Globular cluster
11	6705	18 49 .5	$-06\ 19$	Sct	6	5,500	Open cluster
12	6218	16 45 .6	$-01\ 54$	Oph	7	23,000	Globular cluster
13	6205	16 40 .6	$+36\ 31$	Her	6	26,000	The "great" globular cluster
14	6402	17 36 .0	$-03\ 14$	Oph	8	23,000	Globular cluster
15	7078	21 28 .6	$+12\ 02$	Peg	6	40,000	Globular cluster

Table A-8. Messier Catalogue (Continued)

M	NGC	Position (1970) Right Ascension	Declination	Constellation	Apparent Magnitude	Distance (in light years)	Description
16	6611	18 17.2	−13 48	Ser	7	8,000	Open cluster
17	6618	18 19.1	−16 12	Sgr		5,000	Omega or Horseshoe nebula
18	6613	18 18.2	−17 09	Sgr	7	5,000	Open cluster
19	6273	17 00.2	−26 13	Oph	7	23,000	Globular cluster
20	6514	18 00.6	−23 02	Sgr		4,000	Trifid nebula
21	6531	18 02.8	−22 30	Sgr	7	4,000	Open cluster
22	6656	18 34.6	−23 56	Sgr	6	10,000	Globular cluster
23	6494	17 55.1	−19 00	Sgr	7	2,100	Open cluster
24	6603	18 16.7	−18 27	Sgr	6	9,000	Star cloud
25	IC 4725	18 29.9	−19 16	Sgr		2,000	Open cluster
26	6694	18 43.6	−09 26	Sct	8	5,000	Open cluster
27	6853	19 58.4	+22 38	Vul	8	700	Dumbbell nebula (planetary)
28	6626	18 22.6	−24 52	Sgr	8	16,000	Globular cluster
29	6913	20 22.9	+38 25	Cyg	7	4,000	Open cluster
30	7099	21 38.6	−23 18	Cap	8	40,000	Globular Cluster
31	224	0 41.1	+41 06	And	4	2,000,000	Great galaxy in Andromeda
32	221	0 41.1	+40 42	And	9	2,000,000	Elliptical companion to M31
33	598	1 32.2	+30 30	Tri	7	3,000,000	Spiral galaxy
34	1039	2 40.1	+42 40	Per	6	1,400	Open cluster
35	2168	6 07.0	+24 21	Gem	6	2,800	Open cluster
36	1960	5 34.3	+34 05	Aur	6	4,200	Open cluster
37	2099	5 50.4	+32 33	Aur	6	4,100	Open cluster
38	1912	5 26.6	+35 48	Aur	7	4,200	Open cluster
39	7092	21 31.1	+48 18	Cyg	6	900	Open cluster
40				UMa			No cluster or nebulas? (2 stars)
41	2287	6 45.8	−20 42	CMa	6	2,300	Open cluster
42	1976	5 33.9	−05 24	Ori		1,500	Great Orion nebula
43	1982	5 34.1	−05 18	Ori		1,500	Associated with M42
44	2632	8 38.2	+20 06	Cnc	4	520	Praesepe (open cluster)
45		3 45.7	+24 01	Tau		400	Pleiades (open cluster)
46	2437	7 40.4	−14 45	Pup	9	2,300	Open cluster
47	2422	7 35.1	−14 26	Pup	5	1,600	Open cluster
48	2548	8 12.0	−05 41	Hya	6	1,800	Open cluster
49	4472	12 28.3	+08 10	Vir	9	40,000,000	Elliptical galaxy
50	2323	7 01.5	−08 18	Mon	6	3,000	Open cluster
51	5194	13 28.6	+47 21	CVn	9	15,000,000	Whirlpool galaxy (spiral)
52	7654	23 22.9	+61 26	Cas	7	6,000	Open cluster
53	5024	13 11.5	+18 20	Com	8	65,000	Globular cluster
54	6715	18 53.2	−30 31	Sgr	8	55,000	Globular cluster
55	6809	19 38.1	−31 01	Sgr	5	20,000	Globular cluster
56	6779	19 15.4	+30 07	Lyr	8	40,000	Globular cluster

Table A-8. Messier Catalogue (Continued)

M	NGC	Right Ascension	Declination	Constellation	Apparent Magnitude	Distance (in light years)	Description
57	6720	18 52.5	+33 00	Lyr	9	1,800	Ring nebula (planetary)
58	4579	12 36.2	+11 59	Vir	10	40,000,000	Barred spiral galaxy
59	4621	12 40.5	+11 50	Vir	11	40,000,000	Elliptical galaxy
60	4649	12 42.1	+11 44	Vir	10	40,000,000	Elliptical galaxy
61	4303	12 20.3	+04 39	Vir	10	40,000,000	Spiral galaxy
62	6266	16 59.3	−30 04	Sco	7	26,000	Globular cluster
63	5055	13 14.4	+42 11	CVn	10	16,000,000	Spiral galaxy
64	4826	12 55.2	+21 51	Com	8	12,000,000	Spiral galaxy
65	3623	11 17.3	+13 16	Leo	10	20,000,000	Spiral galaxy
66	3627	11 18.6	+13 10	Leo	9	20,000,000	Spiral galaxy
67	2682	8 49.5	+11 56	Cnc	7	27,000	Old open cluster
68	4590	12 37.8	−26 35	Hya	8	40,000	Globular cluster
69	6637	18 29.8	−32 23	Sgr	8	23,000	Globular cluster
70	6681	18 41.3	−32 19	Sgr	9	65,000	Globular cluster
71	6838	19 52.4	+18 42	Sge	9	16,000	Globular cluster
72	6981	20 51.8	−12 41	Aqr	9	68,000	Globular cluster
73	6994	20 57.3	−12 46	Aqr			Open cluster
74	628	1 35.1	+15 38	Psc	11	25,000,000	Pinwheel spiral galaxy
75	6864	20 04.3	−22 01	Sgr	8	94,000	Globular cluster
76	650	1 40.3	+51 25	Per	11	15,000	Planetary nebula
77	1068	2 41.1	−00 07	Cet	9	40,000,000	Seyfert galaxy
78	2068	5 45.3	+00 02	Ori		1,600	Small gaseous nebula
79	1904	5 22.9	−24 33	Lep	8	50,000	Globular cluster
80	6093	16 15.2	−22 55	Sco	7	36,000	Globular cluster
81	3031	9 53.4	+69 12	UMa	8	7,000,000	Spiral galaxy
82	3034	9 53.6	+69 50	UMa	9	7,000,000	Exploding galaxy
83	5236	13 35.3	−29 43	Hya	9	15,000,000	Spiral galaxy
84	4374	12 23.6	+13 03	Vir	10	40,000,000	Elliptical galaxy
85	4382	12 23.8	+18 21	Com	10	40,000,000	Elliptical galaxy
86	4406	12 24.6	+13 06	Vir	10	40,000,000	Elliptical galaxy
87	4486	12 29.2	+12 33	Vir	10	40,000,000	Elliptical galaxy with jet
88	4501	12 30.4	+14 35	Com	10	40,000,000	Spiral galaxy
89	4552	12 34.1	+12 43	Vir	11	40,000,000	Elliptical galaxy
90	4569	12 35.3	+13 19	Vir	11	40,000,000	Spiral galaxy
91							(M58?)
92	6341	17 16.2	+43 11	Her	7	32,000	Globular cluster
93	2447	7 43.2	−23 48	Pup	6	36,000	Open cluster
94	4736	12 49.6	+41 17	CVn	9	15,000,000	Spiral galaxy
95	3351	10 42.3	+11 52	Leo	10	25,000,000	Barred spiral galaxy
96	3368	10 45.1	+11 59	Leo	10	25,000,000	Spiral galaxy
97	3587	11 13.1	+55 11	UMa	11	2,600	Owl nebula (planetary)
98	4192	12 12.2	+15 04	Com	10	40,000,000	Spiral galaxy
99	4254	12 17.3	+14 35	Com	10	40,000,000	Spiral galaxy
100	4321	12 21.4	+15 59	Com	10	40,000,000	Spiral galaxy
101	5457	14 02.1	+54 30	UMa	8	15,000,000	Pinwheel spiral galaxy
102							(M101?)
103	581	1 31.2	+60 32	Cas	7	8,000	Open cluster

Index

Numbers in **boldface** indicate a major discussion of the subject.

A

Abell Catalogue, 368
Abell, G. O., 368, 419
Aberration (chromatic), 154–156, 159
Aberration (spherical), 155–156, 159
Aberration (of starlight), 300–305
Absolute zero, 81, 139
Absorption lines, 135, 191
Accretion disk, 341–342, 403
"Age of Aquarius," 67
Aldebaran, 13–14, 16, 198, 246, 286
Almagest, 30, 32
Alpha Centauri, 187–188
Altair, 14, 16
Altitude (of stars), 18
American Ephemeris and Nautical Almanac, 20, 71–72
Anaxagoras, 24–25
Ancient astronomy, 24–52
Anderson, C. D., 303, 306–307
Angel, J. R. P., 256
Angstrom, 124
Annihilation, 307
Antarctic Circle, 62
Antares, 168, 198–199, 246
Antimatter, 303, 307, 435
Aphelion, 56
  of earth, 59
Apogee, 71
Apparent solar day, 54–58
Apparent solar time, 54–55, 57–58
Arctic Circle, 62
Arcturus, 198, 246
Arecibo Ionospheric Observatory, 170
Aristarchus, 28–29, 31–32
Arp, H. C., 391, 406, 409–413
Asteroid, 84, **114–115**, 121
Asteroid belt, 115, 121

Astigmatism, 155
Astroarchaeology, 5
Astrology, 5, 7, 22
Astronomical unit (AU), 39
Astronomy, 1
Atlas of Stellar Spectra, 199–200
Atmosphere (of a planet), 81
Atom, **131–138**, 142, 148, 416, 437
  ground state of, 133
  ionization of, 191, 252
  nucleus, 132, 213
Atomic structure, 132–134
Autumnal equinox, 62
Azimuth, 18

B

B magnitude, 167
B. D. Catalogue, 13
B. D. numbers, 13
Baade, W., 266
Babcock, H., 208
Barred spiral galaxies, 366–367
Bayer, J., 13
Bennett, S. A., 380
Bessel, F., 187
Betelgeuse, 13, 142, 198, 246, 286
"Big-bang" theory, 8, 423–428, 450
Binary stars, **216–221**, 254, 337–341
  position angle, 217
  separation, 217
  types of, 217–219
Black body, 139, 141
Black body background radiation, 431–433, 435–436
Black body radiation, 139–142
Black hole, 48, 52, 257, 277, **325–342**, 449–450
  electrically charged, 337

479

Black hole (*cont.*)
  evaporation of, 436–437
  exploding, 437
  falling into a, 329–330
  Kerr, 331–337, 341, 398
  mini, 436–437
  Schwarzschild, 325–333, 337
  supermassive, 350, 398–399, 403
Black hole bomb, 335–336
Blandford, R. D., 400
Bode's law, 114
Bohm-Vitense, E., 270–271
Bohr, N., 132–135
Bolometer, 174–175
Bolometric correction, 189
Bolton, J. G., 379
Bondi, H., 423
Bonner Durchmusterung Catalogue, 13
Brahe, T., 36, 42, 225, 249
Breccia, 87
Burbidge, G. R., 405–406, 410

C

cD-galaxies, 369, 393
CNO cycle, 236
CP 0834, 257, 262
CP 0950, 257, 261–262
CP 1133, 257, 261–262
CP 1919, 257, 262–263
Caesar, J. 58–59
Calcium, 143, 165, 208–209, 374–375
  H and K lines, 143, 165
Caloris basin, 107
Camera lens, 164–165
Carbon burning, 248
Carruthers, G., 177–180
Cassegrain reflector, 159, 167
Castor, 216
Cauchy, A. L., 428
Cauchy horizon, 428
Causality, 310–312
Celestial equator, 19–21, 60, 67
Celestial north pole, 19, 67–69
Celestial south pole, 19
Celestial sphere, 17–22
Centaurus A, 379, 400, 402
Centigrade temperature scale, 81
Centrifugal force, 276, 415–416

Cepheid variable stars, 222–224, 239, 258, 357–361
Ceres, 114–115
Cerro Tololo Inter-American Observatory, 161
Chandrasekher limit, 254, 265, 274–275
Chandrasekher, S., 253–254, 265, 274–275
Chichen Itza, 6
Chromatic aberration, 154–156, 159
Chromosphere, 204, 209
Clusters of galaxies, 368–373
  globular (regular), 369–370
  open (irregular), 370–371
Cluster-type variables, 359
Collapsing star (appearance of), 325–329, 332
Collimating lens, 164–165
Collisional broadening, 148
Color index, 168, 197, 202
Color-magnitude diagram, 202–204
Coma (aberration), 155
Comet(s), 49, 51, 84, **118–121,** 353
  Halley's, 119, 121
  Kohoutek, 178, 180
  Mrkos, 120
Compact galaxy, 403
Conic sections, 49–50
Conjunction, 33–34
Conservation of angular momentum, 256, 266
Constellations, **10–13,** 22
  names of, 10, **21**
Coordinate systems, 17–20
Copernicus, N., 7, **32–35,** 36, 42, 78, 103
Coriolis force, 415–416
Corona, 74–76, 204, **211**
Correcting lens, 163
Cosmic ray, 303, 306–307, 311–312, 434
  shower, 306, 311–312
Cosmology, 1, 24, **418–430**
  big-bang, 423–428, 435
  geocentric, 7, 29, 31–32, 36
  heliocentric, 7, 31–33, 36
  Hoyle-Narlikar, 429–430, 436
  observational, 371
  steady-state, 423–425, 432
  theoretical, 373
  whimper, 428, 435–436
Coudé focus, 160

Index

Coudé room, 160
Crab Nebula, 226, 249, 269, 354
Crab pulsar, 264–265, 267, 269–273
Curie, M., 127
Curtis, H. D., 355–356, 415
Curvature (aberration), 155
Curvature of space, 438–443
   flat, 439–441, 443
   negative, 440–442
   positive, 439–441
Cygnus A, 379, 399–400
Cygnus X-1, 339–342, 403

D

Darwin, C., 85
Davis, R., 214–216
Day, 54–58
   apparent solar, 54–58
   mean solar, 56
   sidereal, 55–56, 58
Daylight savings time, 57
De Revolutionibus Orbium Celestium, 33, 35, 38
Declination, 20
Deferent, 29–30
Deflection of light, 322–323
Degenerate gas, 253, 266
Deimos, 96
Diagonal mirror, 157–158
Dicke, R. H., 431
Differential gravitational force, 64
Differential rotation, 206, 208
Diffraction grating, 165
Diffraction patterns, 383
Dilation of time, 291–292, 298–300
Dirac, P. A. M., 303
Dirac equation, 303, 307
Direct motion (of planets), 29
Discrepant redshifts, 410
Distance determination, 199–204
Diurnal motion, 54
Doppler effect, **143–145,** 148, 373
   relative to spectral lines, 165
   shifts, 218–219, 223
Double radio sources, 399–400
Double stars, 152, 216–221, 254, 337–341
   HDE 226868, 219, 340–341
   70 Opiuchi, 217

Dragging of inertial frames, 334
Dreyer, J. L. E., 355
Dwarf elliptical galaxies, 370

E

Earth, 85–86
   atmosphere, 85, 127–129
   axis, tilt of, 59–60
   revolution of, 281
   rocks, 91
   rotation, 19, 54–56, 59, 66
Earth's axis (tilt of), 59–60
Eclipse(s), **69–76**
   annular, 72–73
   duration of totality, 71
   lunar, 26–27, 69–70, 73–75
   numbers of, 72–73
   occurrence, 71
   partial lunar, 73–74
   partial solar, 71
   path, 71–72
   penumbral, 73–75
   seasons, 70
   solar, 69–72
   total lunar, 73
   total solar, 71, 76, 322
Eclipsing binary, **219–220,** 222
Ecliptic, 20–21, 54
Einstein, A., 3, 52, 126, 307, 343, 377, 414, 438
   assumption of the speed of light as an absolute constant, 284, 290–291
   conversion of mass into energy, 213
   general theory of relativity, 315–316, 320–323
   special theory of relativity, 284
Einstein-Rosen bridge, 331
Electromagnetic radiation, **123–129,** 138, 267, 281, 431
Electromagnetic spectrum, 125, 127–128
Electromagnetic wave, 125, 280–281, 283–284, 290
Electrons, 131, 133–135, 252–253, 303
   spin of, 348–349
Elementary particles, 131, 214, 434
Ellipse, 35, 37–39, 42
   focus, 37–39
   major axis, 37–38

Ellipse (*cont.*)
  minor axis, 37–38
  precessing, 319
  semimajor axis, 38–40
  semiminor axis, 38
Elliptical galaxies, 365–367
Ellis, G. F. R., 428
Elsmore, B., 379
Embedding diagram, 316, 322–323, 330–331, 440
Ephemeris, 20, 71–72
Epicycle, 29–30, 32, 35
Epoch, 68
Equant point, 30
Equation of energy generation, 251
Equation of hydrostatic equilibrium, 251
Equation of radiative transport, 251
Equation of state, 251
Equatorial bulge, 65–66
Equatorial system, 20
Equinoxes, 62
Eratosthenes, 27–28
Ergosphere, 334–336, 398
Eruptive variables, **224–227**, 233
Escape velocity, 50, 80
"Ether," 281
Euclidean geometry, 444
Event horizon, 327–331, 332–336
Exit cone, 327, 329
Eyepiece, 151

F

Fahrenheit temperature scale, 81
Faraday, M., 4, 125, 280
Faraday rotation, 262
Field equations (of general relativity), **316**, 328, 331–332, 377, 435
Filaments (solar), 209–211
First Cambridge Catalogue (1C), 379
Fitzgerald contraction, 292–293
Focal length, 154, 157
Focal point, 154, 157, 169
Focus (of ellipse), 37–39
Fowler, W., 215
Franklin, B., 125, 280
Fraunhofer, J., 129
Fundamental particles, 131, 214, 434

G

Galactic astronomy, 347
Galactic clusters, 201
Galactic halo, 347–348
Galactic nucleus, 348–351
Galactic plane, 347
Galactic year, 347
Galaxies, 7–8, 172, 175, **345–377**, 394–405
  magnetic fields, 380
  Our Galaxy, 345–353, 384
  as radio sources, 172, 350–351, 399–403
  spectra of, 374–375
  types of, 365–367, 394–405
Galaxy-quasar pairs, 406–407
Galilean satellites, 41–43, 84, **108–110**
Galileo, **41–43**, 44, 78, 84, 151, 153, 204, 206, 225, 314–315, 345
Gamma rays, 127, 129, 433
  bursts, 437
Gamow, G., 431
General theory of relativity, 52, 277, **314–343**, 352, 414, 438, 446–447
  classical tests of, 319–325
  field equations, **316**, 328
  weak field limit, 321
Geocentric cosmology, 7, 29, 36, 42
Geodesic, 317, 320, 322; equations, 317
Glitch, 263
Globular clusters, 201, 347–348, 359
  integrated magnitude of, 361
Gold, T., 266, 423
Gott, J. R., 414–415, 422
"Grand Canyon" of Mars, 98
"Grand Tour," 113
Gravitation, 44–49, 52, 314–343
  perturbations, 63–64
  Universal Law of, 47–49
Gravitational field, 316, 320–329, 338
Gravitational lens effect, 414–415
Gravitational redshift, 323–324, 329
Gravitational wave antenna, 352
Gravitational waves, 352–353
Gravity, 22, 46–49, 52, 66, 79–80, 132, 314–343
Great Red Spot, 108–110
Greatest eastern elongation, 33–34
Greatest western elongation, 33–34

Index

Greenhouse effect, 102
Greenstein, J. L., 386
Greenwich meridian, 18–19
Gregorian calendar, 59
Gregory, XIII, Pope, 58–59
Gum, C. S., 265
Gum nebula, 265
Gunn, J. E., 414–415, 423
Guseynov, O. H., 337–338

H

H II regions, 363, 411–412
H-alpha filtergrams, 209
H-alpha photographs, 209
H-bomb, 4, 214
H. D. Catalogue, 14
H. D. numbers, 14
HDE 226868, 219, 340–341
Hadron, 434
Hadron era, 434
Hale, G. E., 208
Halley, E., 51
Halley's comet, 51, **119,** 121
Harmony of the Worlds, The, 40
Hawking, S. W., 353, 436–437, 450
Hayashi, C., 232
Hayashi contraction, 232–233, 244
Hazard, C., 383
Heliocentric cosmology, 7, 36, 42
Helium burning, **237–238,** 247
  shell, 247
Henbest, S. N., 399–400
Henderson, T., 187
Henry Draper Catalogue, 14
Herschel, J., 216, 355
Herschel, W., 51, 216, 345–346, 355, 418
Hertz, 127
Hertzsprung, E., 196
Hertzsprung-Russell Diagram, **195–204,**
  221–222, 235–236, 247–248
  composite, 240–241
  evolutionary tracks, 238
  protostar to main sequence, 232–233
Hewish, A., 257
Hey, J. S., 379
Hi and Ho, 6
High tides, 64–65

High-velocity stars, 384
Hipparchus, 14, 29–30, 69
History of the Sung Dynasty, 273
Homer, 6
Horizon system, 17–18
Hoyle, F., 406, 410–411, 416, 423–424, 429
Hoyle-Narlikar cosmology, 429–430, 436
Hoyle-Narlikar theory, 416, 425
Hubble, E., 7–8, 359, 365, 373–376, 394, 415, 418, 420, 429–430
Hubble age (of universe), 423
Hubble classification (of galaxies), 365–368
Hubble constant, **376,** 420–423
Hubble law, 374–377, 391–393, 405, 418–423, 448
  as distance indicator, 376–377, 393, 418–420
Humason, M. L., 374, 376, 418, 420
Hyades, 240–241, 246
Hydrogen, 21-centimeter line, 348–349
Hydrogen burning, 213, **235–239,** 247, 252
  shell, 237–238
  in sun, 213
Hyperbola, 49–50
Hyperfine transition (in hydrogen), 348

I

Iben, I., 238
Igneous rocks (of maria), 87
Index Catalogue (IC), 355
Inferior conjunction, 33–34
Inferior planets, 33–34, 102–108
Infrared astronomy, 174–176
Infrared radiation, **126,** 129, 139, 174–176, 350–351
Inner Lagrangian point, 340–341
Interference fringes, 283
Interplanetary matter, 113–121
Interstellar absorption, 347
Interstellar medium, 249, 261–262, 346–347
  enrichment of, 249
Intrinsic redshifts, 406, 408
Io, 110
Ionization, 191
Ions, 191

Irregular galaxies, 366–367
Irwin, J. B., 358

J

Jansky, K., 168–169, 379
Jodrell Bank Observatory, 170, 381
Jovian planets, 83, 108–113
Juno, 114–115
Jupiter, 41–43, 79–84, **108–110**

K

Kant, I., 355
Kapteyn, J. C., 346
Kelvin temperature scale, 80–81
Kepler, J., 7, 35, **36–41**, 42, 47–48, 78, 225
Kepler's laws of planetary motion, **36–41**, 42, 47, 319–321
 first law, 38, 319–321
 second law, 38–39, 56
 as a special case of general relativity, 321
 third law, 39–42
Kerr, R. P., 332
Kerr solution, 332–336
Kippenhahn, R., 241, 244
Kirchoff, G., 135
Kitt Peak National Observatory, 161, 269
Kovacs, S., 353
Kruger 60, 216–217
Krupp, E. C., 399

L

Landau, L., 257
Large Magellanic Cloud (LMC), 178–179 357, 367, 371
Latitude, 18, 20
Law of causality, 310–312
Leap years, 58
Leavitt, H., 357
Lenses, **154–157**, 164–165
Lepton(s), 434
Lepton era, 434
"LGM theory," 258–259
Lick Observatory, 154, 157, 161

Light, **123–149**
 as a limiting velocity, 295–296
 quantization of, 141
 speed of, 125, 283–284, 290
Light-gathering power, 153
Light year, 188, 286
Lightlike geodesic, 322
Lightlike trip, 290
Limiting magnitude, 15–16
Line blanketing effect, 142
Line of nodes, 69–70
Line profiles, 149
Local Group, 370–371, 420–422
Local supercluster, 371
Log N-log S plot, 443
Long-period variable stars, 223
Longitude, 18, 20
Lorentz transformations, 291–298, 302, 310
Low, F., 175
Low tides, 65
Low-velocity stars, 384
Luminosity class, 200
Lunar bedrock, 87
Lunar craters, 87–91
Lunar mare, 87
Lunar mountains, 90–92
Lunar orbiters, 87–88
Luxon, 308
Lyman alpha, 387–389
Lynden-Bell, D., 350, 398
Lynds, C. R., 388, 396

M

"M007," 241–245
M3, 201–204, 240
M11, 240–241, 245–246
M31, 354–355, 360, 364, 371
M33, 360, 364, 371, 422
M67, 240–241, 246
M81, 394–396, 421
M82, 394–396
M87, 361–362, 397–399
MKK Atlas, 199–200
Mach, E., 416
Mach's principle, 416
Mackey, M. B., 383

Index

Magnetic fields, 95, 123, 125–126, 145
Magnifying power, 151
Magnitudes (of stars), **188–189**
  absolute, **188–189,** 196–197, 225
  absolute bolometric, 189
  apparent visual, 14–16, 189, 204
  bolometric, 189
  photographic, 16
  UBV magnitude, 167–168
Main sequence fitting, 201–205
Main sequence stars, 198, 220–221, 232–233, 235–236, 240–241
Major axis (ellipse), 37–38
Mariner 2, 102
Mariner 4, 95–96
Mariner 5, 102
Mariner 6, 95–96
Mariner 7, 95–96
Mariner 9, 84, 96, 100
Mariner 10, 104–108, 132
Markarian 205, 409
Mars, 2, 38, 40–41, **92–101**
Mascons, 88
Mass, 45–46
  proper mass, 308
  relativistic behavior of, 293–294
  rest mass, 293–294, 308
Mass equation, 251
Mass-luminosity relation, 220–221, 236
Mass vs. weight, 45–46
Matter era, 433–434
Matthews, T. A., 381, 386
Maxwell, J., 125–127, 132, 280
Maxwell's electromagnetic field equations, 125–126, 280, 284, 290, 298
Maxwell's electromagnetic theory, **125–126,** 280
McDonald Observatory, 161
Mean solar day, 56, 58
Mean solar time, 56–58
Mean sun, 56
Mercury, 33, 79, **102–108**
  relativistic precession, 319–321
Meridian, 54–56
Messier, C., 353–354
Messier Catalogue, 353–354, 394
Meteor(s), 117–118
Meteor shower, 117–188
  Leonids, 118
  Perseids, 117
Meteorites, 88, **116–117**
Meteoroids, 84, 86, 88, **115–118**
Michelson, A. A., 281–283
Michelson interferometer, 282–283
Michelson-Morley experiment, 283
Microdensitometer, 146–147, 165–166
Micrometeorites, 115
Milky Way Galaxy, 259–260, **345–353,** 371
  Carina-Cygnus arm, 349
  central bulge, 349–350
  diameter of, 347–348
  Sagittarius arm, 349
  size of, 347–348
  spiral arms, 349
  structure of, 349
Mini black holes, 436–437
Minor axis (ellipse), 37–38
Minor planets, 115
Mira, 223
Mira-type variable stars, 223, 258
Misner, C. W., 336, 435
Mitton, S., 399
Mixmaster universe, 435
Mizar, 218
Molecules, 136–138
Molonglo Radio Observatory, 259
Monochrometer, 208–209
Month, 54
Moon, 24–29, **85–92**
  craters, 87–91
  diameter, 151
  eclipses of, 27
  features, 87–91
  maria, 87–88
  mass, 115
  mountains, 90–91
  phases of, 25–29, 54, 69
  rocks, 90–92
  temperature, 86, 91
Moon rocks, 90–92
Morley, E., 282–283
Mössbauer effect, 324
Mössbauer, R., 324
Mt. Wilson Observatory, 153, 161, 355
Mullard Radio Astronomy Observatory, 170, 400

N

N-type galaxies, 403, 405
NGC 188, 240–241, 246
NGC 1275, 396–398
NGC 2264, 244–245
NGC 2362, 240
NGC 4319, 409
NGC 5128, 400
NP 0527, 259
NP 0532, 259, 263, 265, 267, 269–273
Naked singularity, 353
Narlikar, J. V., 391, 416
National Radio Astronomy Observatory, 170, 259
Neap tides, 65
Nebula(e), 7, 172, 231, 353–354, 359–360, 380
Neptune, 52, 84, **112–113**, 114
Neutrino(s), 214–216
Neutrino problem, 215–216, 429
Neutron(s), 131–132
Neutron star, 257, **265–273**, 274–277, 325, 339
  core of, 274
  crust, 268, 274
  density of, 267
  diameter, 267
  magnetic field of, 267–268
  mass vs. central density, 274–275
  mass vs. diameter, 276
  rotation rate, 266–267, 271
Neutron star geology, 273
New Astronomy, The, 38
New General Catalogue (*NGC*), 355, 405
Newton, I., 2, 7, 35, **44–52**, 78, 123, 125, 129, 157, 319, 418
Newtonian mechanics, 4, 50–52, 78–79, 319–320, 325
Newtonian reflector, 157, 159
Newton's laws of motion, 44–48
  first law, 44
  second law, 45–46
  third law, 46–47
Nonvelocity redshifts, 408
North Star, 67
Nova(e), 224–226, 363–364
Nova Persei, 225–226
Nuclear forces, 213, 236–237

O

Objective lens, 155
Oblique-rotator model, 268–269
Observational cosmology, 371
Observatories
  Arecibo Ionospheric, 170
  Cerro Tololo Inter-American, 161
  Jodrell Bank, 170, 381
  Kitt Peak National, 161
  Lick, 154, 157, 161
  McDonald, 161
  Molonglo Radio, 259
  Mt. Wilson, 153, 161, 355
  Mullard Radio Astronomy, 400
  National Radio Astronomy, 170, 259
  orbiting astronomical, 8–9
  Palomar Mt., 6, 15–16, 152–153, 160–163
  Steward, 269
  Yerkes, 154–155, 157
Observing cage, 161
Oersted, H. C., 125, 280
Oke, J. B., 389
Olympus Mons, 98
Open clusters (of stars), 201
Oppenheimer, J. R., 257, 266, 277
Opposition, 33–34
Orion, 11–12
Orion nebula, 354, 363, 411
Osterbrock, D. E., 387
Oxygen burning, 248

P

PSR 0833–45, 265
Pair production, 307, 434
Pallas, 114–115
Palomar Mountain Observatory, 6, 15–16, 152–153, 160–163
Parabola, 49–50
Parallactic angle, 187
Parallax, 186–187
Parsec, 187
Parsons, S. J., 379
Pauli exclusion principle, 252
Penrose, R., 334, 399
Penumbra, 71
Penzias, A., 431

Periapsis, 108
Perigee, 71
Perihelion, 56; of earth, 59
Period-luminosity relation, 358–359
Perturbations, 51
Phillips, J. W., 379
Phobos, 96–97
Photoelectric photometer, 166–167, 197, 202
Photoelectric photometry, **166–168**, 202
Photographic limiting magnitude, 16
Photomultiplier, 167
Photon, 141
Photon sphere, 326, 328
Photosphere, 204–209
Piazzi, G., 114
Pioneer flights, 108–111, 121
Plages, 209, 211
Planck, M., 141
Planck black body radiation law, 141–142
Planet(s), 20–21, 28–31, 33–35, 38, **79–113**
 atmospheres, 80–82
 conjunctions, 33–34
 distances from sun, 113
 gravity of, 79–80
 inferior, 33–34
 masses of, 79–81, 92
 motions, 28–31, 33
 orbits of, 38–40, 45–47, **79–80**
 relative sizes, 79
 speeds of, 38
 superior, 33–34
 temperatures, 81
Planetary nebulae, 247–248, 354
Planetoids, 115
Plasma, 336
Pleiades, 201–204, 240, 245–246
Pluto, 79, **83–84**, 114
Pogson, N., 14, 16
Polaris, 67, 188–189, 222, 302
Pooley, G. C., 399, 443
Pore, 206
Positrons, 303
Post-red giant stages, 246–249
Pound, R. V., 324
Precession, 22, **65–69**; relativistic, 319–321
Press, W. H., 335
Pressure broadening, 148

Prime focus, 161
Primordial fireball, 249, 435
Pringle, J. E., 339
Prism, 123–124, 129
Prominences, 75, **209–211**
Proper mass, 308
Proportional counters, 180–181
Proton(s), 131–132
Proton-proton chain, 236
Protostar, 231–234, 237
Ptolemaic system, 29–32
Ptolemy, C., 7, 22, 29–31, 42, 69
Pulsars, 173, **257–273,** 339
 CP 0834, 257, 262
 CP 0950, 257, 261–262
 CP 1133, 257, 261–262
 CP 1919, 257, 262–263
 distribution of, 259–261
 periods of pulsation, 259, 261–262
 as radio sources, 173, 257–265
 slow-down of, 263, 268
 in supernova remnants, 273
 TV observations of, 269–271
 X-ray observations of, 270–273
Pulsating variable stars, 223–224, 239, 258, 357–361
Pulse profile, 261

Q

Quantum mechanics, **132–137,** 252–253, 436–437
 relativistic effects, 303
Quasars, 3–4, 145, 172–173, 357, **379–394,** 404–415, 424–425
 magnitude variations, 382
 Markarian 205, 409
 as radio sources, 379–389, 413–414
 redshifts, 384–394, 406–415, 424
 3C 9, 386–387
 3C 48, 381–382, 384–385
 3C 147, 385
 3C 196, 385
 3C 216, 386
 3C 245, 386
 3C 273, 382–385, 398
 3C 287, 387
 4C 05.34, 388
 5C 2.56, 388

Quasars (*cont.*)
  OH 471, 389, 393
  OQ 172, 389, 393
  PHL 957, 388

R

RR Lyrae variable stars, **223**, 239, 258, 347
Radiation era, 433–434
Radiation laws, 138–142
Radio astronomy, 169–173, 379–389, 431–432, 442–444
Radio sky, 169
Radio surveys, 171–173, 379–381, 388–389, 442
Radio waves, 127–128, 348
Raisin cake analogy, 419–420
Rangers, 89–90
Reber, G., 169, 379
Red giant stars, 198, 221, **237–239**, 241, 246
Redshift, 145, 373–375
  cosmological, 393, 424
  discrepant, 410, 413
  gravitational, 323–324, 329
  intrinsic, 406
  law of, 374–375
  nonvelocity, 408
Rees, M. J., 339, 350, 400, 403
Reflectors, 157–164
Refractors, 153–157
Regolith, 87
Reissner-Nordstrom solution, 337
Relativity, general theory of, 52, 277, **314–343**, 352, 414, 438, 446–447
  special theory of, 280–312
Resolving power, 152
Rest mass, 293–294, 308
Retrograde motion (of planets), 28–31, 33, 102
Richter scale, 269
Riemann, G. F. B., 318
Riemannian geometry, 318
Rigel, 13, 142, 286
Right ascension, 20, 55
Roche limit, 339
Roche lobes, 340–342
Roentgen, W., 127

Rotational broadening, 149
Russell, H. N., 196
Rutherford, E., 132–133, 212
Ryle, M., 379, 386, 398–399, 443

S

Sagittarius A, 347, 350–351
Sandage, A., 364–365, 376, 381, 386, 388, 420–423, 428, 450
Saslaw, W. C., 403
Saturn, 20, **110–112**
Schiaparelli, G., 94
Schmidt, M., 384–388, 390
Schmidt telescope, 161–163
Schramm, D. N., 423
Schwarzschild, K., 328
Schwarzschild solution, 328–329, 331
Seasons, 21, **59–63**
Second Cambridge Catalogue (*2C*), 379
Semimajor axis (ellipse), 37–38
Semiminor axis (ellipse), 37–38
Seyfert, C., 403
Seyfert galaxies, 403–405
  3C 120, 403
  M77, 403–404
  NGC 1068, 403
  NGC 4151, 403
Seyfert's sextet, 410–411
Shakura, N. I., 339
Shapley, H., 347, 355–356, 358, 415
Shapley-Curtis debate, 355–356, 377, 415
Shimmins, A. J., 383
Shock wave, 249, 396
"Shooting star," 115
Sidereal day, 55–56, 58
Sidereal period, 34–35, 40
Sidereal time, 55–56, 58
Siderites, 117
Siderolites, 117
Silicon burning, 248
Simultaneity, 296–297
Single-line spectroscopic binary, 219, 337–338
Singularity, 328, 333–334
Sirius, 13–16, 168, 195, 247, 254, 286
Sirius B, 247, 254
Slipher, V. M., 373
Small Magellanic Cloud (SMC), 357

Index

Smith, F. G., 379
Snyder, H., 277
Solar atmosphere, 204–205
Solar eclipse, 26–27
Solar limb, 209
Solar luminosity, 197
Solar system, 78–121
Solar wind, 120–121, 340
Solstices, 61–63
Space-time, 287–297, 309–310, 314–318, 449
  curvature of, 316–317, 320, 322–323, 326–327, 338, 447
  diagrams of, 287–290
  as a fourth dimension, 287
Space-time diagrams, 287–290
  elsewhere, 289
  future, 289–290
  here and now, 288–289
  past, 289
Spacelike trip, 290
Special theory of relativity, 280–312
Spectra, 129–131, 135–137
  absorption line spectrum, 135
  band spectra, 136
  comparison spectrum, 165
  continuous spectrum, 135
  emission line spectrum, 136
  molecular spectra, 137
  stellar, 143–145, 149, **190–195**
  tracing of, 147, 165
Spectral analysis, 130–131, 135
Spectral classification, 192–195, 199
Spectral line(s), **129–131**, 134–136, 142–149, 190, 192, 218, 373
  broadening of, 146–149, 199
  natural shape of, 147
  natural width of, 147
  strengths of, 190, 192
Spectral line broadening, 146–149, 199
Spectrograph, 129, **164**, 190
Spectroscope, 129, 135, **164**
Spectroscopic binary stars, 218–219
Spectroscopic parallax, 200–201
Spectroscopy, **164–166**, 190
Spectrum binary, 219
Speed of light, 125, 283–284, 290
  as a limiting velocity, 295–296
Spherical aberration, 155–156, 159

Spiral galaxies, 365–367
Spring tides, 65
Star(s), 13–16, 142, 148–149, **186–227**
  absolute magnitude, 188–189, 196–197, 225
  apparent visual magnitudes, 14–16, 204
  atmosphere, 237
  catalogues of, 13–14, 199, 338–339, 353–354
  chemical composition of, 190–193
  distance determination of, 199–204
  distances, 186–188
  double, 152, 216–221, 337
  energy distribution, 142
  formation of, 230–235
  luminosity of, 196, 200, 225, 236
  masses, 216, **219–221**, 226, 246
  metal–poor, 358
  metal–rich, 358
  names of, 13–14
  Population I, 358
  Population II, 358
  spectral classes of, 192–194, 196, 199, 202, 236, 244
  temperatures, 191–193, 195, 199, 202, 232
  theoretical models, 230–233, 236, 241, 246, 251–252
Star charts, 11–12
Star clusters, **239–246**, 346
  double cluster (H and chi Persei), 240–241, 245
Star gauging, 345–346
Starquakes, 268–269
Static limit, 333–334
Steady-state theory, 423–425
Stefan, J., 140
Stefan's law, 140–141
Stellar distances, 186–188
Stellar evolution, **230–249**, 251–257, 265–267, 274, 277
Stellar wind, 340–341
Stephan's quintet, 411–412
Steward Observatory, 269
Stonehenge, 5–6
Strittmatter, P. A., 389
Struve, F., 187
Subatomic particles, 303, 306
Summer solstice, 61–63

Sun, 2–3, 41, 45, 54–56, 60–63, 78, 148, 197, **204–216**
 activity center, 210–212
 atmosphere, 204–211
 diameter, 78
 energy of, 212–216
 future of, 239
 in the galaxy, 347
 magnetic field, 208, 210, 256
 mass, 78
 prominences, 75
 rotation, 204, 206
 spectra, 129–131
Sunspots, 204, 206–208
 cycle, 207
 magnetic field of, 208
 maximum, 207
 minimum, 207
 parts of, 207
Sunyaev, R. A., 339
Superfluidity, 274
Supergiant stars, 361
Superior conjunction, 33–34
Superior planets, 33–34
Supermassive black hole, 350, 398–399, 403
Supernova(e), 225–227, 249, 273, 277, 306, 363
 3C 10, 249
 radio, 249
 remnant, 226, 249
Superradiant scattering, 336
Surface gravity, 79–80
Synchrotron, 380
Synchrotron radiation, 380
Synodic period, 34–35

T

T Tauri stars, 233
Tachyon, 308–312
Tammann, G. A., 420–423, 428, 450
Tardyon, 308
Telescopes, 15–16, **151–164, 174–184**
 infrared, 175
 limiting magnitude of, 15–16
 neutrino, 214–215
 photographic limiting magnitude, 16
 radio, 168–173, 257, 431–432
 reflecting, 157–164, 169, 269
 refracting, 151–157
 ultraviolet, 177–180
 X-ray, 180–184
Temperature scales, 80–81
Terminator, 107
Terrell, J., 391
Terrestrial planets, **83,** 92–108
Teukolsky, S. A., 335
Thermal Doppler broadening, 148
Thermonuclear fusion, 213, 248
Thermonuclear reactions, 213–214, 232, 235–239, 247–249
Third Cambridge Catalogue (3C), 379–381, 397
Third Uhuru Catalogue (3U), 338–339, 396
Thorne, K. S., 336, 338, 342, 353
Tidal force, 64
Tide(s), **63–65;** types of, 64–65
Time, **54–59,** 314
 apparent solar, 54–55, 57–58
 daylight savings, 57–58
 fourth dimension, 287
 mean solar, 56–58
 sidereal, 55–56, 58
 standard, 57–58
 zones, 57–58
Time machine, 332
Time zones, 57
Timelike trip, 290
Tinsley, B. M., 423
Transits, 33, 54, 56
Trimble, V., 338
Tropic of Cancer, 62
Tropic of Capricorn, 62
Tropical year, 58
Trumpler, R. J., 346
Turbulence broadening, 148
Twin paradox, 298–300
Two-element objective, 155

U

U magnitude, 167
U, B, V system, **167–168,** 202, 386
Uhuru, 181–183, 338
Ultraviolet astronomy, 177–180
Ultraviolet excess, 386, 388
Ultraviolet radiation, 9, 126–129, 156–157

Index 491

Umbra, 71, 73–74
Universal Law of Gravitation, 47–49, 251, 277, 319–320, 322, 325
　as a special case of general relativity, 321
Universe, 8, 418–423, **431–450**
　age of, 422–423
　critical density of, 446–447, 450
　deceleration of, 426, 446–447
　expansion of, 8, 418–423, 445
　fate of, 444–450
　history of, 432–434
　mixmaster, 435
　oscillating, 428, 450
　shape of, 438–445, 447–448
Uranus, 51, **112–113**, 114, 345

V

V magnitude, **167–168**, 202–203
VV 172, 410–411
Van Allen radiation belts, 95
van de Hulst, H. C., 347–348
Variable stars, 222–227, 233
　Cepheids, 222–224, 239, 258, 357–361
　eruptive, 224–227, 233
　light curve of eruptive, 224–225
　pulsating, 222–224
Vega, 187
Veil nebula, 249
Vela pulsar, 265, 269, 272
Velocity dispersion, 261
Venus, 33, 41–42, **102–108**, 137
Verena flights, 102
Vernal equinox, 19–22, 55, 62, 67–69
Vesta, 114–115
Viking Lander, 100–101
Virgo A, 397
Virgo cluster of galaxies, 370
Visual binary stars, 217–218
Volkoff, G., 266

W

Wampler, E. J., 389
Waning crescent, 25
Waning gibbous, 25

Wave phenomenon, 123
Wavelengths, 123–126, 128, 134–136, 139–140, 144, 166–167, 170, 175, 180, 431
Waxing crescent, 25
Waxing gibbous, 25
Weber, J., 352
Wheeler, J. A., 336
Whimper cosmology, 428, 435–436
White dwarf stars, 145, 198–199, 221, 247, **251–257**, 265, 275–277, 325
　cooling curves, 255
　magnetic field of, 256
　mass-diameter relation, 253–254
　rotation, 256
　Sirius B, 247, 254
White hole, 331, 368, 416, 450
Wien, W., 139
Wien's law, 139–141
Wilson, R., 431
Winter solstice, 62–63
Wollaston, W., 129
Wormhole, 331

X

X rays, 9, 127–129, 139, 180–184, 211, 433
　from black holes, 338–342
X-ray astronomy, **180–184**, 338–342

Y

Year(s), 54, 58
　leap, 58
　tropical, 58
Yerkes Observatory, 154–155, 157

Z

Zap craters, 90–91
Zeeman, P., 145
Zeeman effect, 145, 208, 256
Zel'dovich, Ya. B., 337–338
Zenith, 18, 55–56
Zodiac, 5, 21–22
Zwicky, F., 266

Star Charts

The Night Sky in January

Latitude of chart is 34° N, but it is practical throughout the continental United States.

To use: Hold chart vertically and turn it so the direction you are facing shows at the bottom.

Chart time (Local Standard):
10 P.M. First of month
9 P.M. Middle of month
8 P.M. Last of month

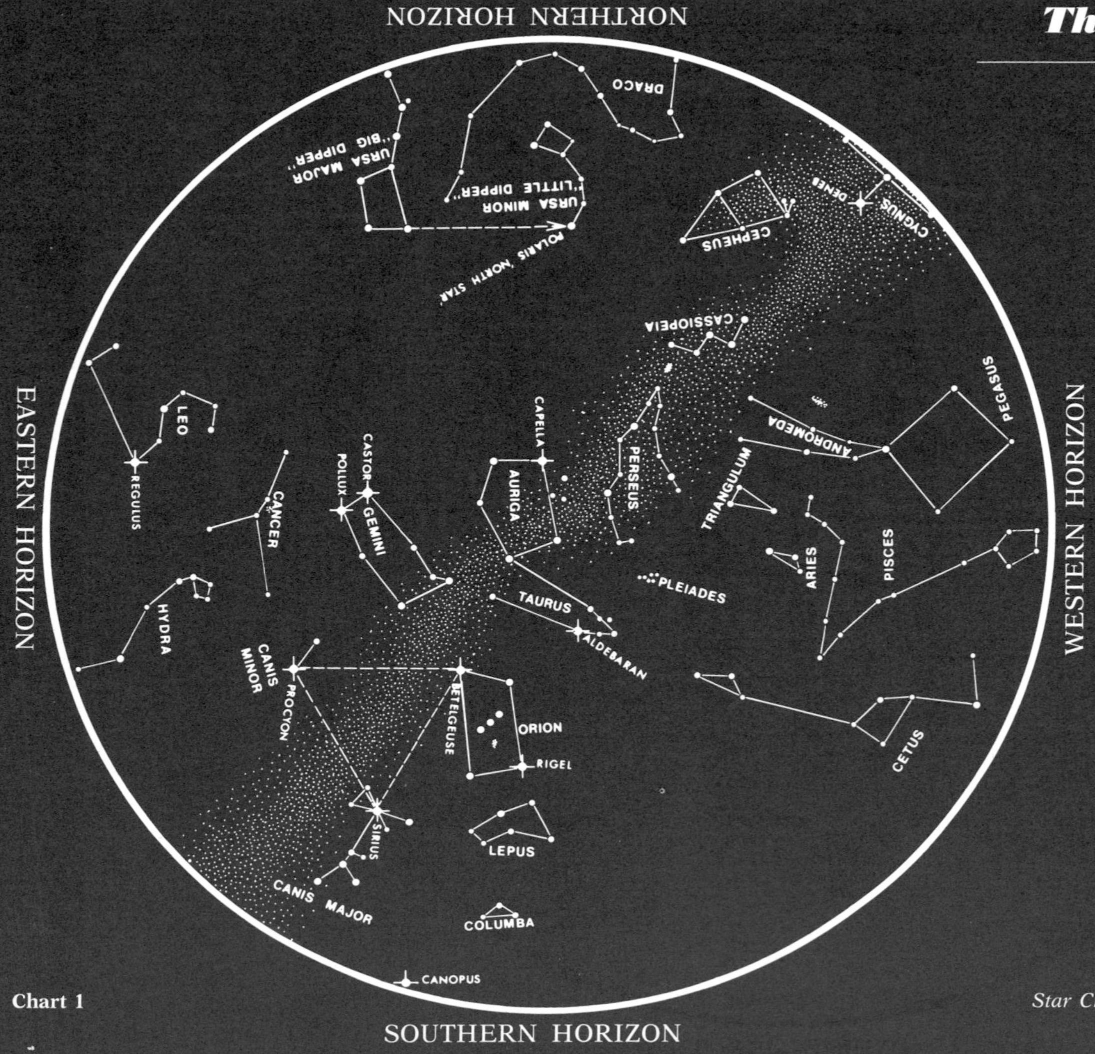

Chart 1

Star Chart from GRIFFITH OBSERVER monthly magazine

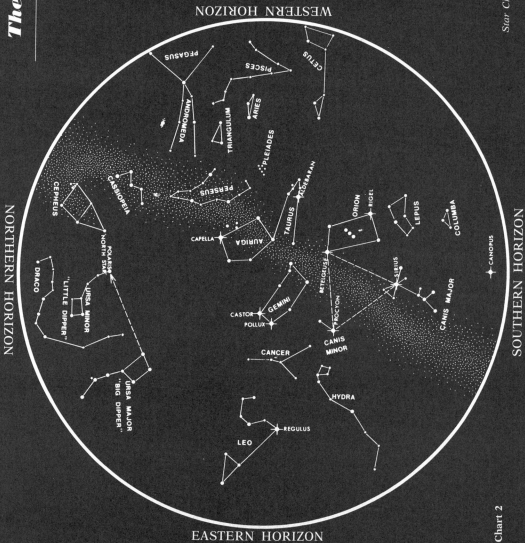

The Night Sky in March

Latitude of chart is 34°N, but it is practical throughout the continental United States.

To use: Hold chart vertically and turn it so the direction you are facing shows at the bottom.

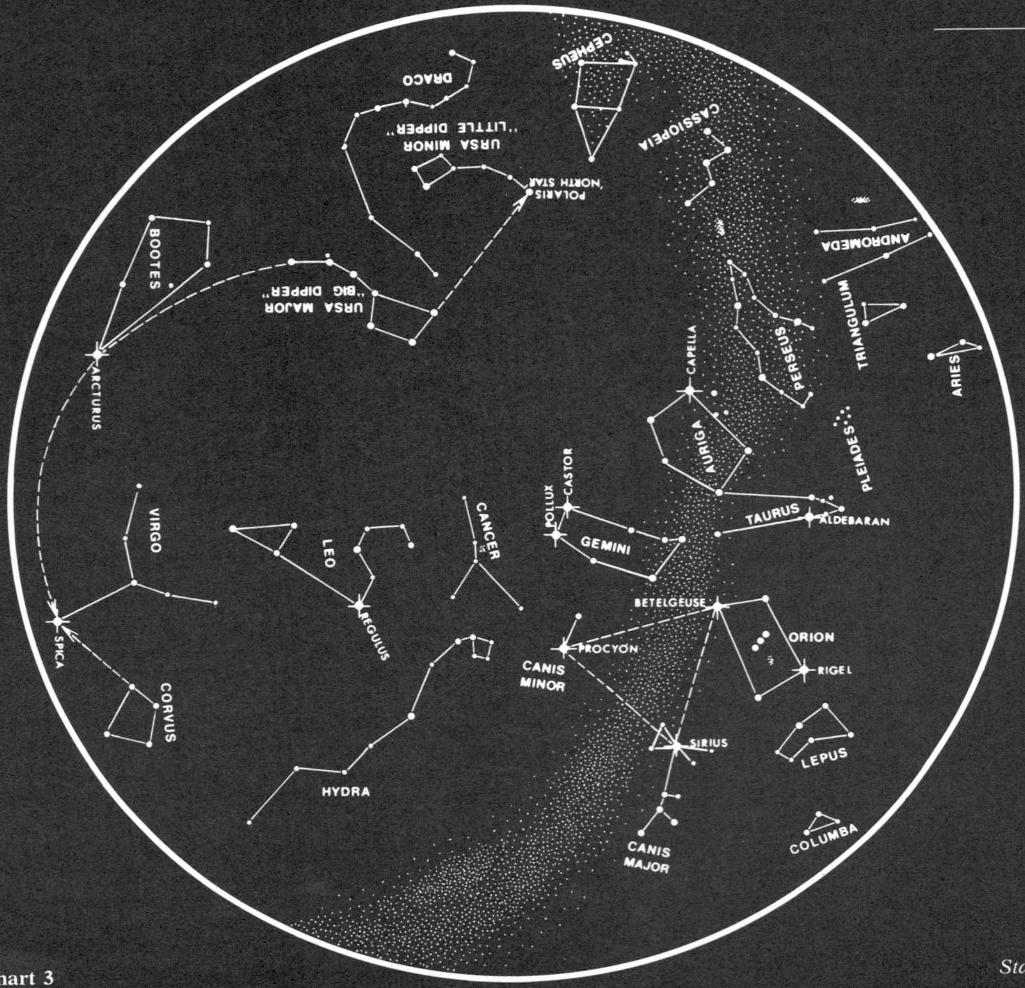

Chart time (Local Standard):
10 P.M. First of month
9 P.M. Middle of month
8 P.M. Last of month

Chart 3

Star Chart from GRIFFITH OBSERVER monthly magazine

The Night Sky in April

Latitude of chart is 34° N, but it is practical throughout the continental United States.

To use: Hold chart vertically and turn it so the direction you are facing shows at the bottom.

Chart time (Local Standard):

10 P.M. First of month
9 P.M. Middle of month
8 P.M. Last of month

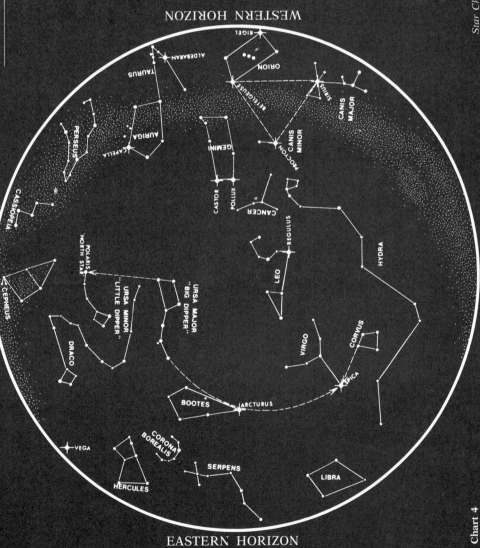

Star Chart from GRIFFITH OBSERVER monthly magazine

Chart 4

Chart 5

EASTERN HORIZON

NORTHERN HORIZON

SOUTHERN HORIZON

WESTERN HORIZON

The Night Sky in May

Latitude of chart is 34°N, but it is practical throughout the continental United States.

To use: Hold chart vertically and turn it so the direction you are facing shows at the bottom.

Chart time (Local Standard):
10 P.M. First of month
9 P.M. Middle of month
8 P.M. Last of month

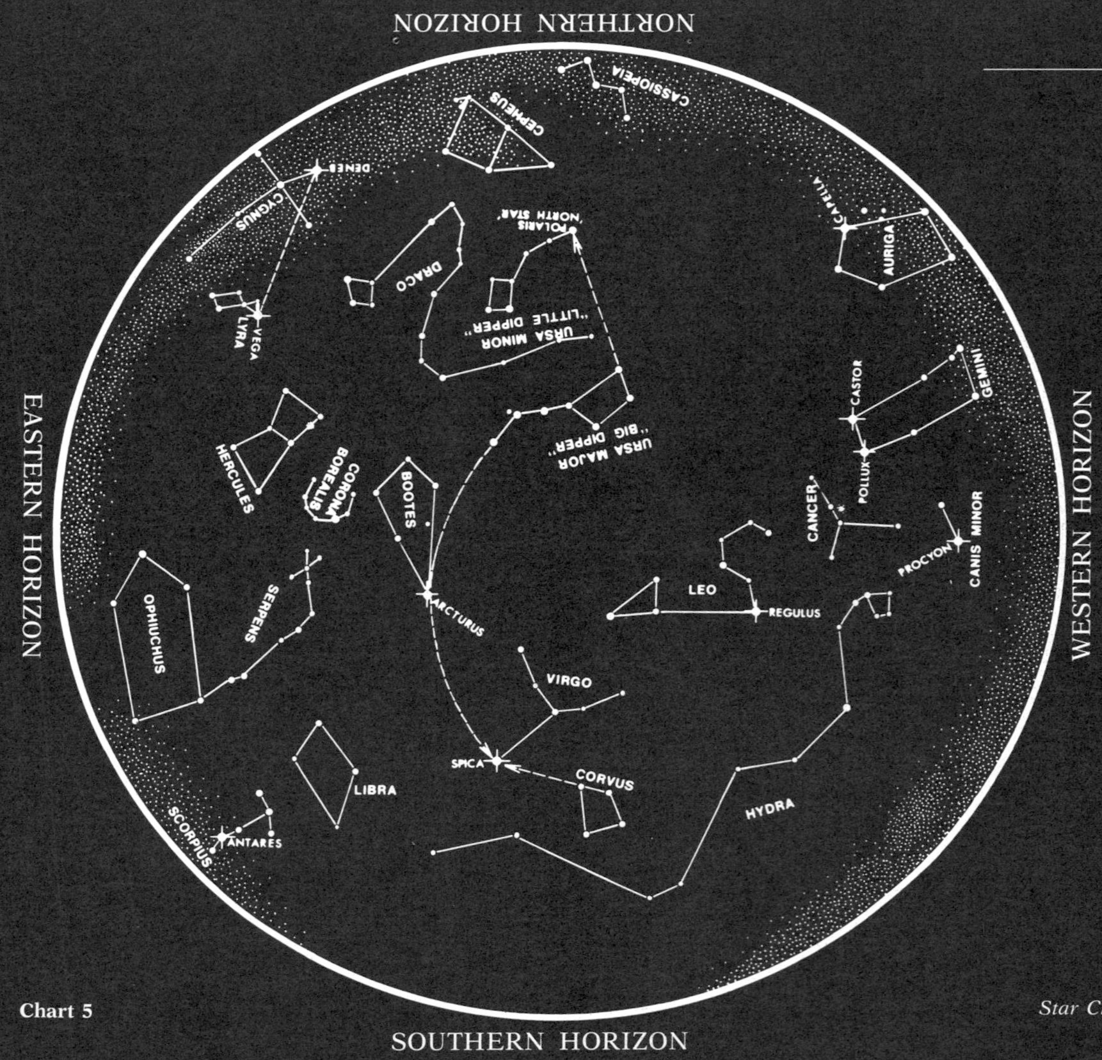

Star Chart from GRIFIFTH OBSERVER monthly magazine

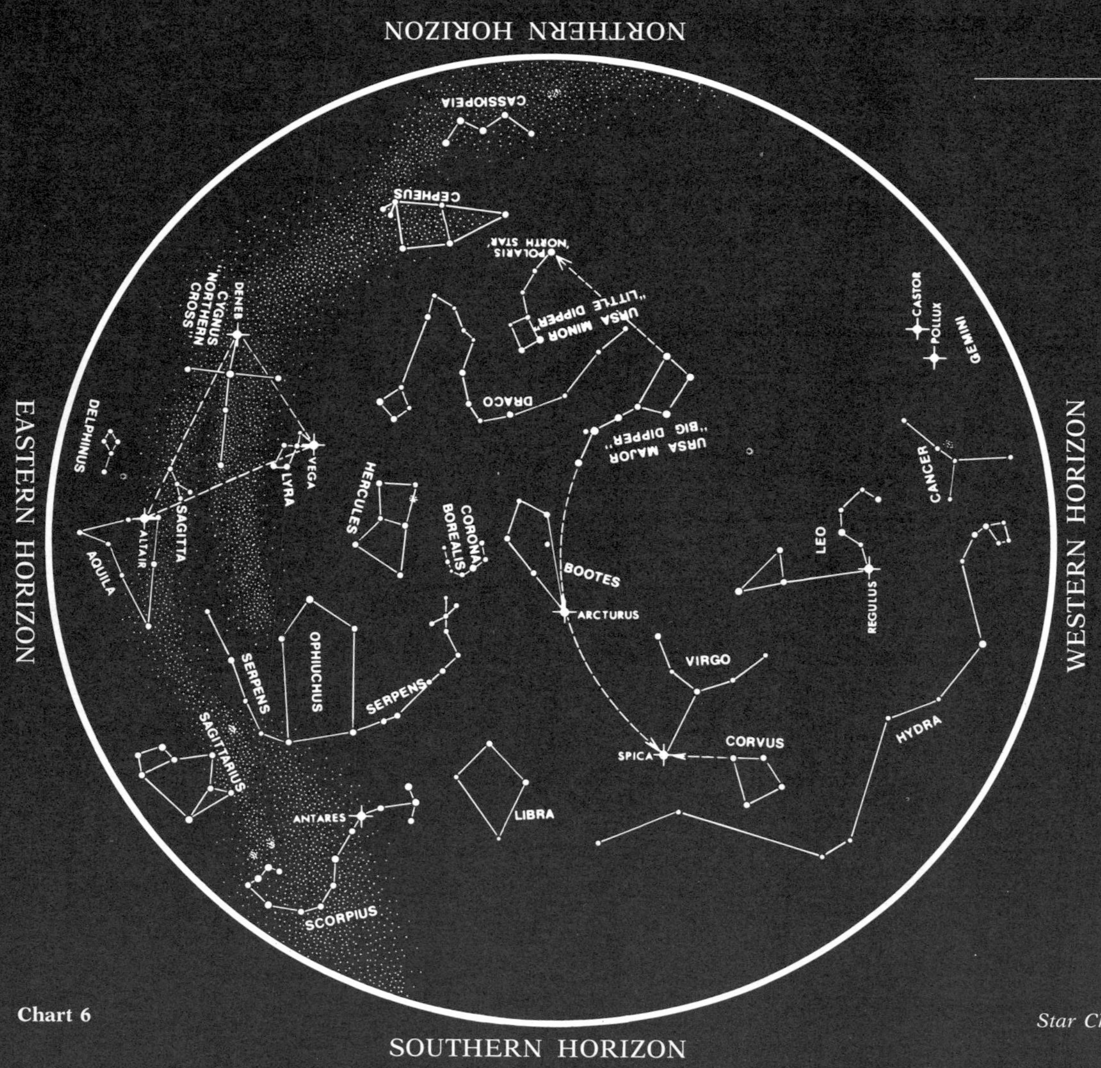

The Night Sky in June

Latitude of chart is 34°N, but it is practical throughout the continental United States.

To use: Hold chart vertically and turn it so the direction you are facing shows at the bottom.

Chart time (Local Standard):
10 P.M. First of month
9 P.M. Middle of month
8 P.M. Last of month

Star Chart from GRIFFITH OBSERVER monthly magazine

Chart 6

Chart 7

EASTERN HORIZON

NORTHERN HORIZON

SOUTHERN HORIZON

WESTERN HORIZON

Star Chart from GRIFFITH OBSERVER monthly magazine

The Night Sky in July

Latitude of chart is 34° N, but it is practical throughout the continental United States.

To use: Hold chart vertically and turn it so the direction you are facing shows at the bottom.

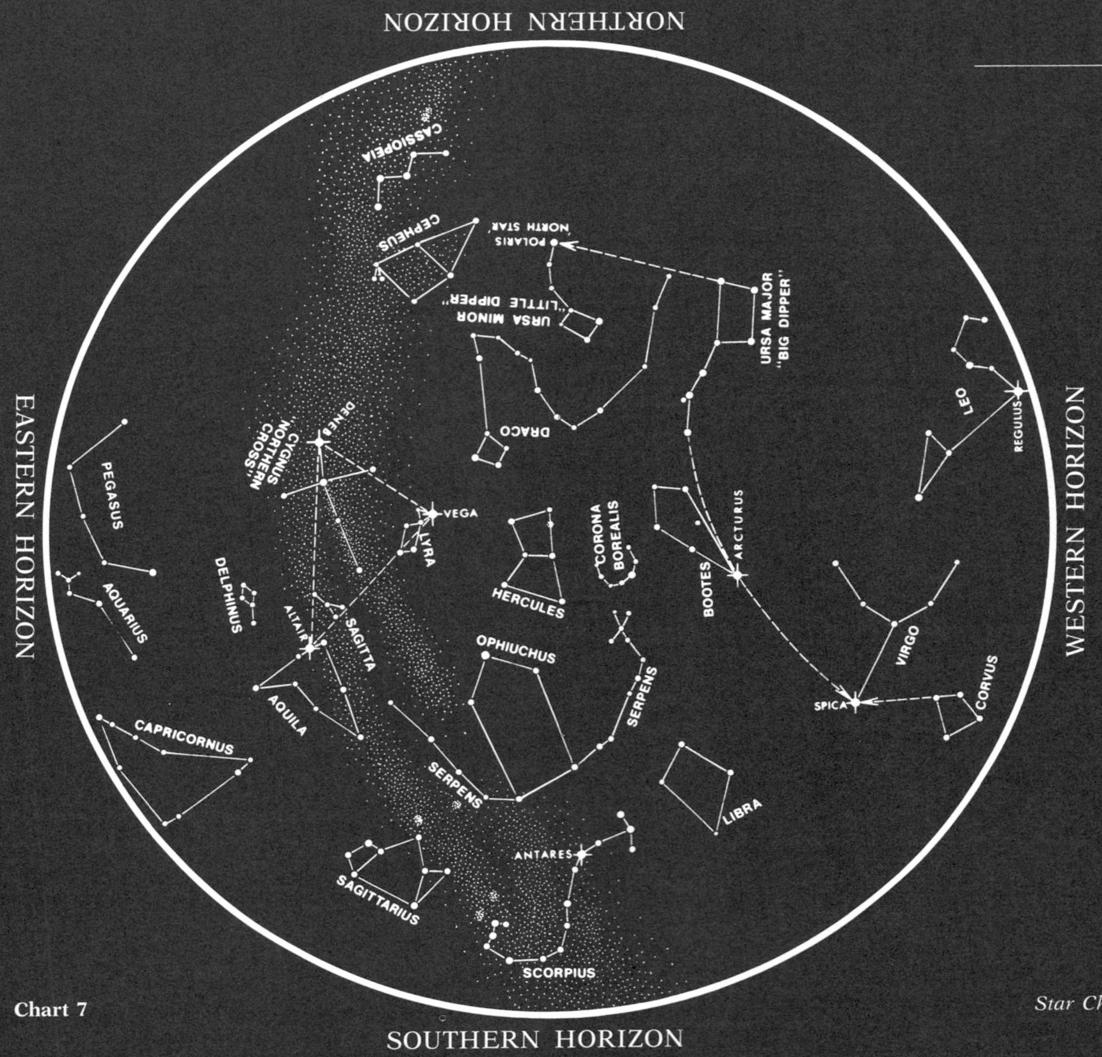

Chart time (Local Standard):
10 P.M. First of month
9 P.M. Middle of month
8 P.M. Last of month

The Night Sky in August

Latitude of chart is 34°N, but it is practical throughout the continental United States.

To use: Hold chart vertically and turn it so the direction you are facing shows at the bottom.

Chart time (Local Standard):

10 P.M. First of month
9 P.M. Middle of month
8 P.M. Last of month

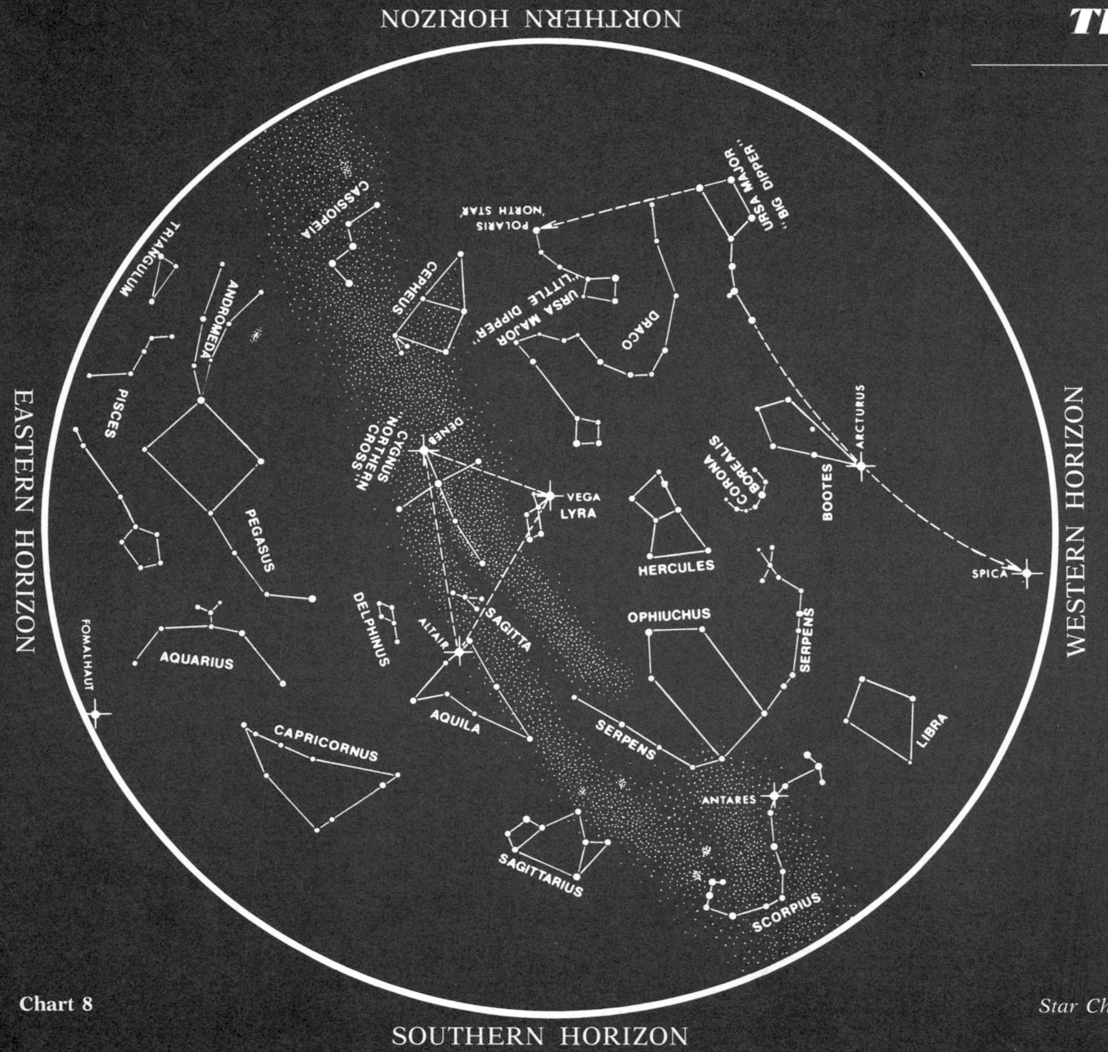

Star Chart from GRIFFITH OBSERVER monthly magazine

Chart 8

The Night Sky in September

Latitude of chart is 34°N, but it is practical throughout the continental United States.

To use: Hold chart vertically and turn it so the direction you are facing shows at the bottom.

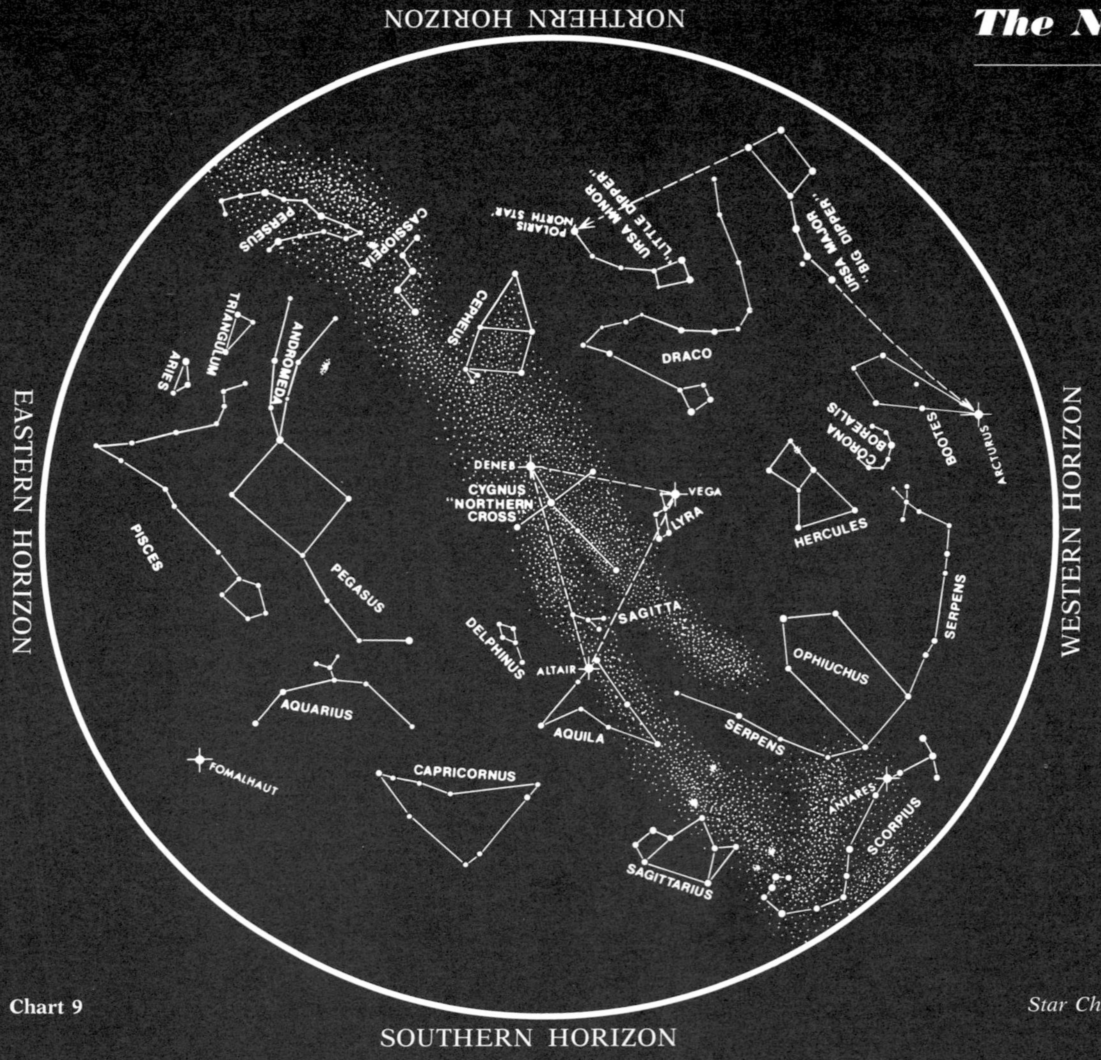

Chart time (Local Standard):
10 P.M. First of month
9 P.M. Middle of month
8 P.M. Last of month

Chart 9

Star Chart from GRIFFITH OBSERVER monthly magazine

The Night Sky in October

Latitude of chart is 34°N, but it is practical throughout the continental United States.

To use: Hold chart vertically and turn it so the direction you are facing shows at the bottom.

Chart time (Local Standard):

10 P.M. First of month
9 P.M. Middle of month
8 P.M. Last of month

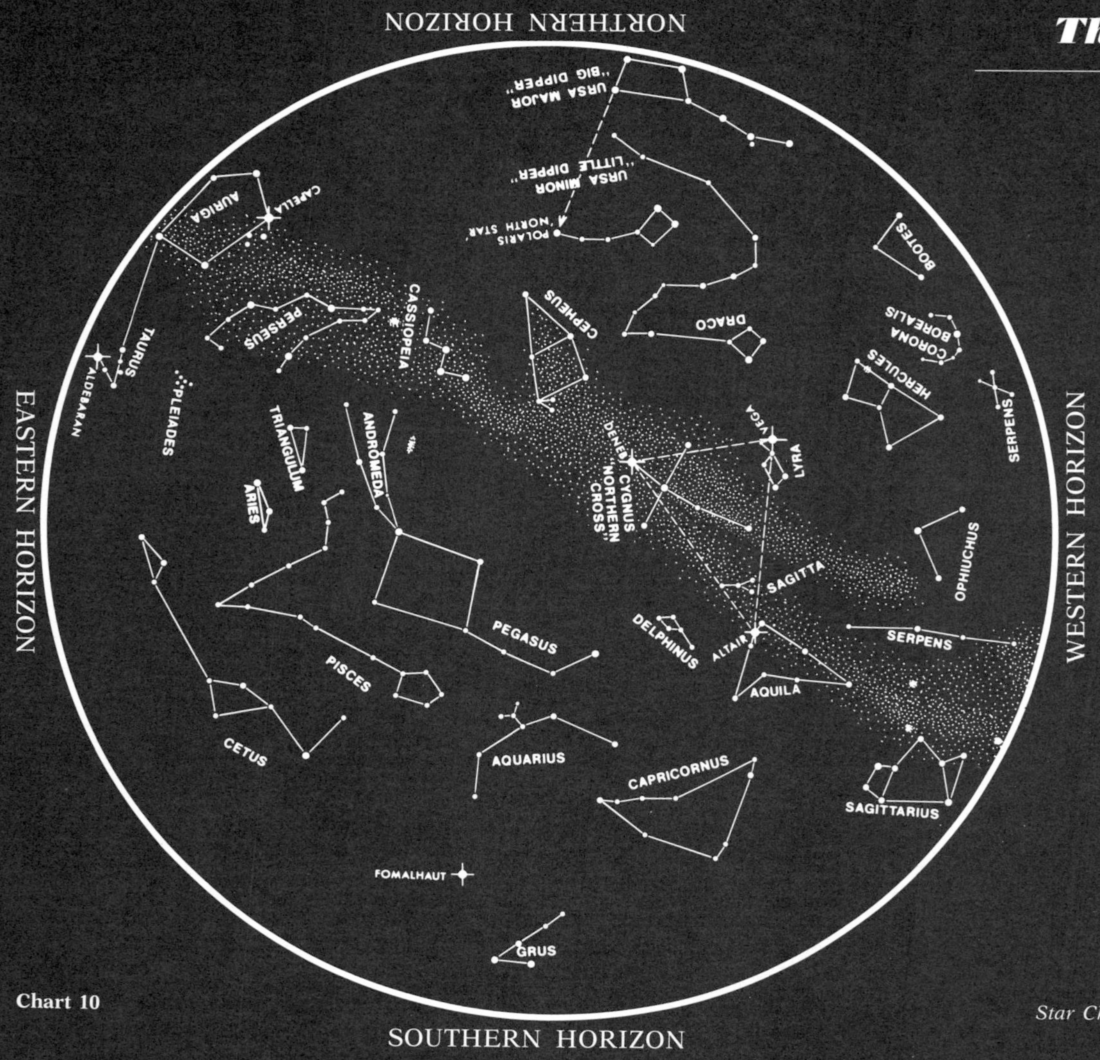

Star Chart from GRIFFITH OBSERVER monthly magazine

Chart 10

NORTHERN HORIZON
EASTERN HORIZON
SOUTHERN HORIZON
WESTERN HORIZON

Chart 11

EASTERN HORIZON

NORTHERN HORIZON

SOUTHERN HORIZON

WESTERN HORIZON

Star Chart from GRIFFITH OBSERVER monthly magazine

The Night Sky in November

Latitude of chart is 34°N, but it is practical throughout the continental United States.

To use: Hold chart vertically and turn it so the direction you are facing shows at the bottom.

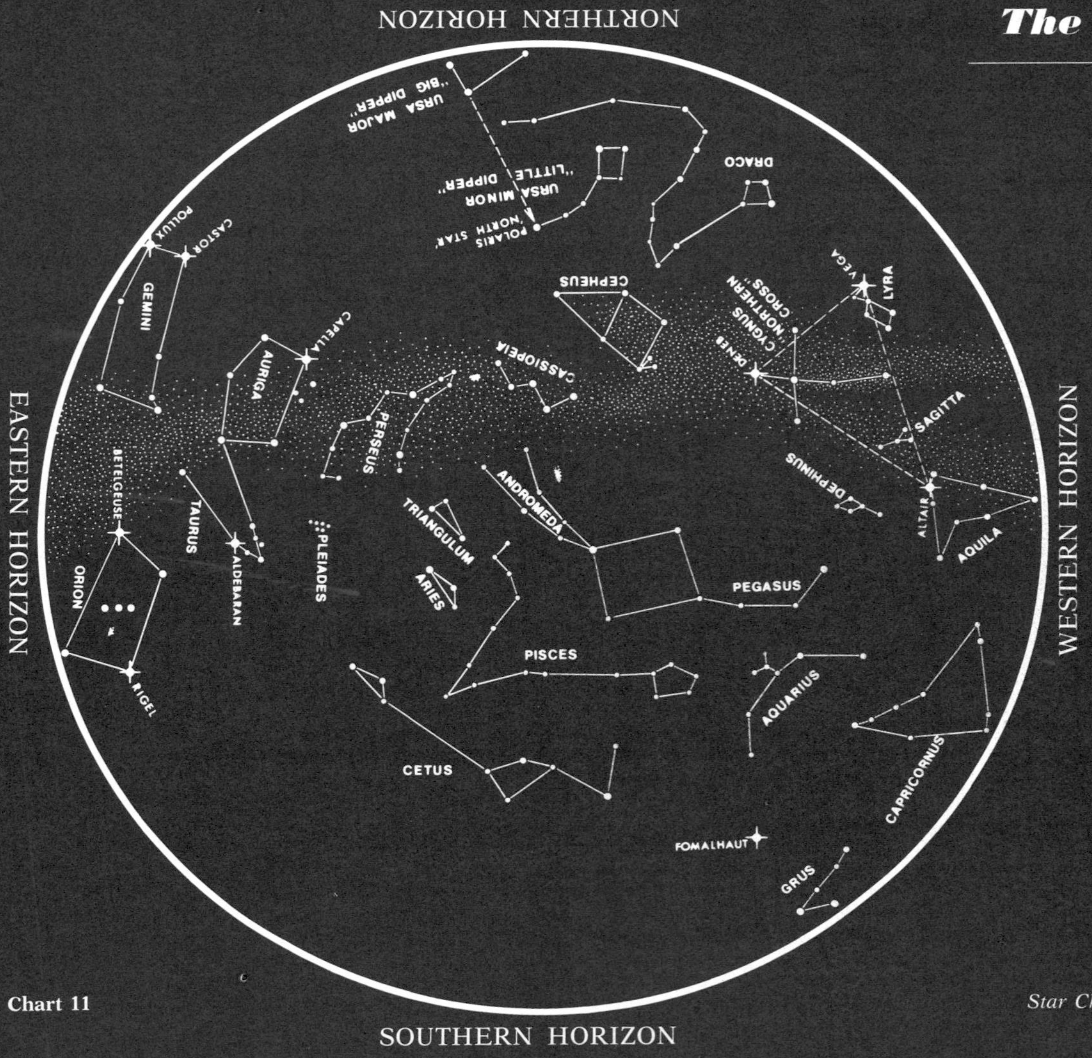

Chart time (Local Standard):
10 P.M. First of month
9 P.M. Middle of month
8 P.M. Last of month

The Night Sky in December

Latitude of chart is 34°N, but it is practical throughout the continental United States.

To use: Hold chart vertically and turn it so the direction you are facing shows at the bottom.

Chart time (Local Standard):

10 P.M. First of month
9 P.M. Middle of month
8 P.M. Last of month

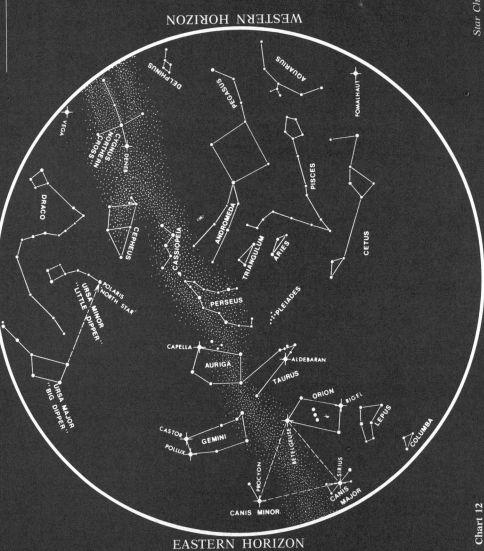

Star Chart from GRIFFITH OBSERVER monthly magazine

Chart 12